International Series in Operations Research & Management Science

Volume 355

The book series **International Series in Operations Research and Management Science** encompasses the various areas of operations research and management science. Both theoretical and applied books are included. It describes current advances anywhere in the world that are at the cutting edge of the field. The series is aimed especially at researchers, advanced graduate students, and sophisticated practitioners.

The series features three types of books:

• Advanced expository books that extend and unify our understanding of particular areas.

• Research monographs that make substantial contributions to knowledge.

• Handbooks that define the new state of the art in particular areas. Each handbook will be edited by a leading authority in the area who will organize a team of experts on various aspects of the topic to write individual chapters. A handbook may emphasize expository surveys or completely new advances (either research or applications) or a combination of both.

The series emphasizes the following four areas:

Mathematical Programming : Including linear programming, integer programming, nonlinear programming, interior point methods, game theory, network optimization models, combinatorics, equilibrium programming, complementarity theory, multiobjective optimization, dynamic programming, stochastic programming, complexity theory, etc.

Applied Probability: Including queuing theory, simulation, renewal theory, Brownian motion and diffusion processes, decision analysis, Markov decision processes, reliability theory, forecasting, other stochastic processes motivated by applications, etc.

Production and Operations Management: Including inventory theory, production scheduling, capacity planning, facility location, supply chain management, distribution systems, materials requirements planning, just-in-time systems, flexible manufacturing systems, design of production lines, logistical planning, strategic issues, etc.

Applications of Operations Research and Management Science: Including telecommunications, health care, capital budgeting and finance, economics, marketing, public policy, military operations research, humanitarian relief and disaster mitigation, service operations, transportation systems, etc.

This book series is indexed in Scopus.

Daniela Borissova

Decision-Making in Design, Maintenance, Planning, and Investment of Wind Energy

Springer

Daniela Borissova
Institute of Information and Communication Technologies
Bulgarian Academy of Sciences
Sofia, Bulgaria

ISSN 0884-8289 ISSN 2214-7934 (electronic)
International Series in Operations Research & Management Science
ISBN 978-3-031-52218-5 ISBN 978-3-031-52219-2 (eBook)
https://doi.org/10.1007/978-3-031-52219-2

This Springer imprint is published by the registered company Springer Nature Switzerland AG
The registered company address is: Gewerbestrasse 11, 6330 Cham, Switzerland

Paper in this product is recyclable.

Preface

The main objective of the book is to show how decision models can be applied to solve specific problems related to wind energy. The book begins with a description of the mathematical apparatus needed to formulate mathematical models and the problem-solving methods that support decision-making. The content of the book sequentially goes through all stages of wind farm construction, starting from site selection and wind power plant layout design, through some turbine manufacturing processes and decision-making in structural health monitoring and predictive maintenance of wind turbines, to some economic and social aspects related to the evaluation of indicators when investing in wind energy, types of risks associated with wind energy projects, and the determination of energy consumption balance index, including determination of the index of the relative share of wind energy consumption among the other renewable sources.

Chapter 1 smoothly introduces the reader to the matter of the decision-making process and types of decision support systems. It provides a classification of the multicriteria decision problems and techniques for solving them. Special attention is given to the formulation of optimization models, multi-objective optimization, and methods of their solving. This chapter also describes some specifics of combinatorial optimization and decision-making in engineering design.

Chapter 2 presents decision-making models for planning and investing in wind energy. It presents proper decision-making models that are needed to assist management in making decisions related to the assessment and selection of wind site locations considering a number of indicator groups. A ranking algorithm and model are presented to select the economy (country) to invest in and supplier selection. A group decision-making approach with three different strategies is presented too. Lastly in this chapter, the decision-making model for personnel selection or ranking taking into account differences in the expertise of group members is described. In addition, some models for the assessment of personnel skills to realize particular tasks or jobs are formulated.

Chapter 3 describes various concepts for wind farm layout design considering different specific characteristics of wind conditions and wind turbine parameters. It is shown how single-criteria and multicriteria models could be used to determine the placement of wind turbines within site areas. Also, multi-attribute decision-making techniques and multi-objective optimizations are integrated into a framework to support the design of wind farms that leads to the realization of different alternatives (different wind farm design layouts) and a selection of one of these alternatives. Two models of wind farm design with restricted areas of regular and irregular geometry for turbine placement are also presented. A model that optimizes the location of wind turbines within wind farms, taking into account the location of the common connection point, is described too. Finally, a combinatorial model for site selection for the hybrid wind-solar power plant is proposed.

Chapter 4 shows how combinatorial optimization models can help to make optimal decisions in machining parts for wind turbine manufacturing, designing new components using standard modules, and scheduling staff. The one-dimensional cutting-stock model that minimizes waste is proposed. Another optimization model that determines job shop scheduling sequence on multiple machines aiming to minimize the makespan is formulated. For the goal of designing new components, an optimization combination model based on standard modules is proposed. A model for the determination of needed personnel for details processing with different strategies to minimize staff idle, minimize processing time, or simultaneously determine minimum staff idle and minimum staff number is also described.

Chapter 5 describes various combinatorial models that could be used for monitoring and predictive maintenance of wind turbines/farms. This involves single- and multi-objective combinatorial optimization models for the determination of optimal number and sensor locations for structural health monitoring due to the vibration of the turbine. A strategy for predictive maintenance of the machine as a whole and of its individual components is presented, which is based on optimization models determining whether repair or replacement is necessary. A concept for an intelligent decision-making system for e-support that integrates different stratagems capable of expressing different uncertainties when forecasting the deterioration of the systems is proposed. Based on cost–benefit analysis, a group approach for the choice of specialized software considering different strategies of uncertainty is described too.

Chapter 6 concerns some indicators that can be used to evaluate the feasibility of investments in wind energy projects from different perspectives, such as economic aspects and the social impact. Some economic evaluation indicators of wind power projects are presented. It is shown how the jobs created and the quality of working conditions as a result of the implementation of wind energy projects can be evaluated. The major groups of risks associated with wind energy projects, namely development and construction risks, economic risks, political risks, social risks, and cyberattack risks, are discussed. Finally, the index of the balance of energy consumption formed by different combinations of conventional and renewable sources is formulated and calculated. Except for this index, another index for the relative share of wind energy consumption among other renewable sources is presented.

This book contains some results of the author's research on decision-making problems based on the formulation and solution of combinatorial optimization models. The main contribution consists of the formulated combinatorial optimization models that solve specific real-life problems, related to wind energy projects. Understanding such patterns could help in formulating models to solve other similar practical problems related to design, maintenance, planning, and investment. The discussed new perspectives in operations research and management regarding research methods and the research context of decision-making offer readers a broad range of new approaches to different real-life problems.

The results presented in the book can be used by operations research professionals, students, and a wide range of decision-makers at different levels who are dealing with practical decision-making problems.

Sofia, Bulgaria Daniela Borissova

Contents

About the Author

Daniela Borissova is a full professor at the Institute of Information and Communication Technologies. Her scientific research topics cover the areas of decision-making including mathematical modeling, integer programming, combination optimization, multicriteria optimization, multicriteria decision analysis, group decision-making, operations research, web-based applications, e-learning and distance learning, and decision support systems.

Chapter 1
General Approaches to Decision-Making

Abstract Today, in many different areas, decision-making is required at various management levels—operational, tactical, strategic, etc. The primary purpose of these solutions is to ensure the effective and efficient use of existing resources. The analogical is the situation in the wind energy sector where different decisions are to be taken starting from design, maintenance, and planning to investment. Therefore, in this chapter basic techniques and methods often used to obtain optimal and/or rational solutions are discussed. An effective tool in decision-making is scientific methods to support decision-making using various techniques. Mathematical models are one of the techniques that lead to obtaining reasoned solutions to various problems from many fields. These models are valuable in situations where the initial conditions are known, but the implementation options are sufficiently vague to make a decision. Decision support models can be considered a corrective to cognitive human deficiencies. Through them, it is possible to achieve the integration of different sources of information to determine optimal, Pareto-optimal, or rational solutions.

In this regard, this chapter provides a classification of the multicriteria decision problems and techniques for solving them. Special attention is given to the formulation of optimization models, multi-objective optimization, and methods of their solving. Some specifics of combinatorial optimization and decision-making in engineering design are described too.

Keywords Decision-making process · Multicriteria decision-making · Optimization model · Methods for multi-objective optimization

1.1 Decision-Making Process

Much of everyday activities involve problem-solving and decision-making. This process is often iterative. Many successive decisions must be made until the desired result is achieved. The goal of decision-making is to find the best option to reach a good, optimal, or Pareto-optimal solution.

D. Borissova, *Decision-Making in Design, Maintenance, Planning, and Investment of Wind Energy*, International Series in Operations Research & Management Science 355, https://doi.org/10.1007/978-3-031-52219-2_1

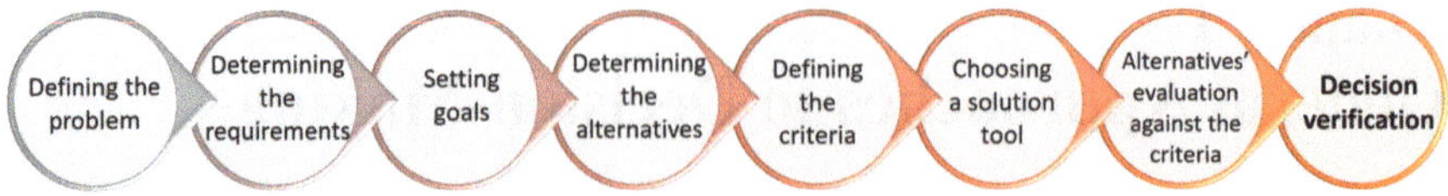

Fig. 1.1 Decision-making process

One possible approach to analyzing the decision-making process is to view it as a multi-stage process. Oftentimes, some of the stages involve the optimization of time and resources. A summary of the steps of this process can be illustrated as shown in Fig. 1.1 (Baker et al., 2002).

- *Defining the problem*—root causes, limiting assumptions, system and organizational boundaries and interfaces, and all stakeholders are to be identified.
- *Determining the requirements*—defining the conditions that any acceptable solution to the problem must satisfy.
- *Goal setting*—general statements of intent and desired values.
- *Determining the alternatives*—the alternatives offer different approaches to change the initial state to the desired state of the process. Whether they are existing or imagined, any alternative must meet certain requirements.
- *Defining the criteria*—the criteria make it possible to distinguish alternatives. It is necessary to define criteria that are objective measures of the objectives to measure how each alternative achieves the objectives.
- *Selection of a decision-making tool*—the selection of a suitable tool is not an easy task and depends on the specifics of the specific problem, as well as on the objectives of the decision-maker (DM).
- *Evaluation of alternatives against the criteria*—depending on the criterion, the evaluation can be objective (factual), by reference to some generally accepted and understandable measurements (e.g., by money), or it can be subjective, reflecting the subjective judgment of the DM.
- *Decision verification*—the selected alternative using the accepted decision-making tool must always be verified against the requirements and objectives of the problem. If the decision has not met the identified need, it is possible to gather more detailed or somewhat different information or explore additional alternatives.

1.1.1 Types of Decision Support Systems

Decision support systems (DSS) analyze large volumes of data for forming business decisions. They leverage a combination of raw data, documents, personal knowledge, and/or business models to help users make decisions. A distinguishing feature of DSS is their interactivity with decision-makers. DSS can be divided into three categories according to the relationship between users and applications: passive DSS, active DSS, and proactive DSS.

The passive models of DSS only collect data and organize it effectively and do not provide specific solutions; i.e., they only show the collected data. Active decision support systems process data and clearly show solutions based on that data. Although there are many systems that are active, in many organizations it will be difficult to believe in a computer model without human intervention. The third category is proactive DSS, which is known as ubiquitous computing technology-based DSS which contains decision-making and context-aware functionalities. They combine the human element and computer components to work together to get the best possible solution.

While the above DSS models consider the interaction with the user, another popular DSS classification is based on the way of support: model-driven DSS; communication-driven DSS; data-driven DSS; document-driven DSS; and knowledge-driven DSS.

Model-driven DSS is a complex system that helps DM to analyze or choose between different options. They use algebraic, decision analytic, simulation, and optimization models to provide decision support. Communication-driven DSS enhances decision-making by network and Internet technologies to provide an environment for group decision-making. Data-driven DSS emphasizes data collection and their manipulation to satisfy a particular decision-maker's needs. They include a database designed to store data in such a way as to allow for its querying and analysis by users. Document-driven DSS uses a variety of documents such as electronic spreadsheet documents and database records for information processing and strategy defining. Knowledge-driven DSS use specific rules coded in computerized expert systems to support the DM.

A model-driven DSS utilizes a model to solve problems or aid in decision-making. Different models can be used such as statistical, financial, mathematical, analytical, simulation, or optimization. A model-driven DSS may employ a single model or a combination of two or more models, depending on the specific needs of its users. Model-driven DSS are generally not data intensive. Rather they use parameters entered by DM and help them analyze a situation. They generate optimal solutions that are consistent with time and resource constraints. The scope of model-driven DSS is huge and can be further enhanced by integrating web-based applications.

1.2 Multicriteria Decision-Making Methods

Multicriteria decision-making (MCDM) methods can be used in the decision-making process as a tool providing flexibility to decision-makers. In decision-making for wind energy project selection, many aspects need to be analyzed using multiple qualitative and quantitative criteria and involving several stakeholders with conflicting interests. There is a need to select and classify the most suitable wind turbine type, and wind power plant layout for better use of the wind resource and reducing associated costs. Therefore, it is essential to choose a reliable approach that is able to cope with such problems considering various criteria and alternatives.

The multicriteria decision problems can be divided into four basic classifications as follows (Arslan, 2017):

- Selection between alternatives
- Rating of alternatives
- Classification of alternatives
- Identification of alternatives

The decision-making problems of selecting the best alternative, carrying out the ranking, or classification of alternatives under multiple attributes can be done by using techniques of multiple attribute decision-making (MADM). MADM expresses the process of making preference decisions via evaluating a finite number of predefined alternatives. The decision problems concerning the identification of alternatives rely on the optimization task formulation with single or multiple objectives. These types of problems are solved using multi-objective decision-making (MODM) methods.

It should be noted that MADM refers to problems with discrete decision spaces, while MODM refers to decision problems whose decision spaces are continuous. Therefore, MADM problems are suitable for solving such tasks where the alternatives are finite in number. All of these means that MADM can be applied for selection between alternatives, for rating of alternatives, or for classification of alternatives, but not for the identification of alternatives.

MCDM can also be categorized as individual or group decision-making. A simple approach is to distinguish between the MCDM methods based on whether it is individual or group decision-making by involving the points of view of multiple experts (DMs).

Other classifications can be done based on the type of data used—deterministic, stochastic, or fuzzy. There may be situations where combinations of all the above data types (such as stochastic and fuzzy data) are used.

1.2.1 Multiple Attribute Decision-Making Techniques

MADM is one of the most widely used branches of the more general class of operations research (Triantaphyllou et al., 1998). A key feature of MADM is that the solution space is discrete (Zavadskas et al., 2014; Hwang & Yoon, 1981). The set of attributes in any MADM problem represents characteristics (parameters) of the decision alternatives, and the attributes are often called "decision criteria." They act as indicators for evaluating the given set of alternatives (Zanakis et al., 1998). In such types of problems for decision-making, goals, attributes (that are criteria), and options are clear; however, the limitations are unclear and the interaction level between decision-makers is limited (Taherdoost & Madanchian, 2023). Most

Table 1.1 Decision matrix

	Weights for criteria importance		
	w_1	w_2	w_n
Alternatives	Evaluation for criteria/attributes		
A_1	a_{11}	a_{12}	a_{1n}
A_2	a_{21}	a_{22}	a_{2n}
A_3	a_{31}	a_{32}	a_{3n}
....	...	...	...
A_m	a_{m1}	a_{m2}	a_{mn}

MADM methods require attributes to be assigned corresponding weights expressing the importance of individual criteria. As a basis for posing the multi-attribute decision problem, the decision matrix A (n x m) can be used, as shown in Table 1.1 (Rao, 2007; Jahan & Edwards, 2015).

The notations used in Table 1.1 are as follows: m is the number of set alternatives, n is the number of attributes (criteria) by which each of the alternatives will be evaluated, w_n are weight coefficients expressing the relative importance between the criteria, and a_{mn} represents the evaluation of the m-th alternative to the n-th criterion.

When modeling real problems, three main formulations of MADM tasks can be distinguished (Roy, 1996):

1. A problem of choosing the best alternative in accordance with the preferences of the DM
2. A problem to rank all alternatives in descending or ascending order
3. A problem of sorting alternatives based on the classification of alternatives against predefined categories defined by DM

In decision-making, different strategies can be used to find the best choice among alternatives. These strategies are often described as compensatory or non-compensatory (Lee & Anderson, 2009). Compensatory decision strategies are based on an attribute utility function (Keeney & Raiffa, 1993; Zeleney, 1976) and are based on utility maximization. Thus, a negative value of one attribute may be offset by an equal or higher positive value of another attribute.

Unlike compensatory strategies, non-compensatory strategies apply heuristics in order quickly to evaluate alternatives with minimal effort and with acceptable losses of accuracy (Lee & Anderson, 2009).

Various approaches and models have been developed to solve MADM problems. They can be grouped into three distinct classes (Fulop, 2005; Vincke, 1992):

1. Methods based on the multi-attribute utility theory (MAUT)
2. Outranking methods
3. Interactive methods

Each of these classes has its advantages and disadvantages, which are mainly related to the ways of extracting information from the DM regarding its local and global preferences.

1.2.1.1 Methods Based on the Multi-Attribute Utility Theory

The *weighted sum model (WSM)*. In decision theory, this model is also known as *simple additive weighting* (SAW) where the alternatives' performance is expressed as a sum of the performance value of the alternative and their relative weight of importance:

$$A_{\mathrm{WSM}}^{*}=\max\left\{\sum_{j=1}^{N} w_j a_{ij}\right\}, i=1,2., \ldots, M, \tag{1.1}$$

where w_j express the weight of importance of j-th criterion and a_{ij} represents the estimation of i-th alternative with respect to j-th criterion.

SMART (simple multi-attribute rating technique). In a SMART model, criteria and alternatives are directly scored, with the scale usually ranging from 0 to 100. A value of 0 means that the alternative has no benefit, while a value of 100 means that the alternative is ideal with respect to the particular criterion. The evaluation scores for all criteria are summed into one total score by the following representation:

$$A_{\mathrm{SMART}}^{*}=\left\{\frac{\sum_{i=1}^{M} w_j\left(a_{ij}\right)_{\mathrm{norm}}}{\sum_{i=1}^{M} w_j}\right\} \tag{1.2}$$

The normalized value $(a_{ij})_{\mathrm{norm}}$ is determined depending on whether utility maximization or minimization is sought over a given attribute using normalization schemes.

The weighted product model (WPM). The main difference compared to the previous model is the fact that instead of addition there is multiplication:

$$A_{\mathrm{WPM}}^{*}=\max\left\{\prod_{j=1}^{N}\left(a_{ij}\right)^{w_j}\right\}, i=1,2., \ldots, \tag{1.3}$$

The *analytic hierarchy process (AHP)* is a theory of relative measurement with absolute scales of tangible and intangible criteria based on expert judgment, proposed by Saaty (1980). The idea of AHP is that DM determines a weighting factor for each criterion based on pairwise comparisons of the criteria, using a fundamental criteria evaluation scale proposed by Saaty (1996).

The *analytic network process* (ANP) is a more general form of AHP used in multicriteria decision-making (Saaty & Vargas, 2013). AHP structures the decision problem in a hierarchy while ANP structures the problem as a network. Both approaches then use a system of pairwise comparisons to measure the weights of the structure components and finally rank the alternatives.

The TOPSIS (technique for order of preference by similarity to ideal solution) method. TOPSIS is a compensatory approach originally proposed by Hwang and Yoon (1981). The model was further developed later by Yoon (1987) and Hwang, Lai, and Liu (Hwang et al., 1993). The compensatory TOPSIS model allows trade-offs to be made between criteria, where a poor score in one criterion can be overridden by a good score in another criterion. This provides a more realistic form of modeling than non-compensatory methods that include or exclude alternative solutions based on hard constraints. The concept of TOPSIS is based on the fact that the chosen alternative should have the shortest geometric distance from the positive ideal solution and the longest geometric distance from the negative ideal solution (Assari et al., 2012).

The VIKOR (VIseKriterijumska Optimizacija i Kompromisno Resenje) method. Multicriteria decision models, such as VIKOR and TOPSIS, are based on an aggregation function representing the proximity to the ideal (Opricovic & Tzeng, 2004). The aggregation function of the VIKOR model represents the distance from the ideal solution, taking into account the relative importance of all criteria and the balance between general and individual satisfaction. VIKOR uses linear normalization, which eliminates the dimensionality of criteria functions, and the normalized values do not depend on the evaluation unit of a given criterion. The VIKOR model determines a compromise solution providing maximum "group utility" for the "majority" and minimum utility for the "opponent."

1.2.1.2 Outranking Methods

The ELECTRE methods. The ELECTRE method was first proposed by Bernard Roy in 1965 (Roy, 1968). The ELECTRE family of methods includes ELECTRE I, ELECTRE II, ELECTRE III, ELECTRE IV, ELECTRE IS, and ELECTRE TRI (Govindan & Jepsen, 2016). There are two main phases in applying ELECTRE: (1) comprehensively comparing each pair of alternatives and (2) the recommendations obtained in the first phase are used to reject some alternatives to the problem under consideration that are unacceptable. The nature of the recommendation depends on the selection, ranking, or sorting problem at hand. Another MCDA method can then be used to select the best of the remaining alternatives.

The PROMETHEE (preference ranking organization method for enrichment evaluations) methods. The PROMETHEE family (preference ranking organization method for enrichment evaluations) includes PROMETHEE I, PROMETHEE II, GAIA, PROMETHEE V, and PROMETHEE VI and is also from the family of outranking methods (Brans & Mareschal, 2005).

There are other outranking methods besides ELECTRE and PROMETHEE such as QUALIFLEX, REGIME, ORESTE, ARGUS, EVAMIX, TACTIC, and MELCHIOR, which offer definitions and calculations of specific binary relations more or less related to the main idea of the original ELECTRE methods (Figueira et al., 2005).

The analytic hierarchy process (AHP) provides a comprehensive framework to define objectives, and their quantifying criteria, and evaluate alternative solutions. AHP performs pairwise comparisons to assess the relative importance of each criterion in a decision situation. During the comparisons, the decision-makers use their judgments to express the elements' relative meaning and importance. AHP converts the assigned evaluations to numerical values using the fundamental scale of Saaty (1990).

1.2.1.3 Interactive Methods

Interactive methods are oriented toward solving tasks of multicriteria decision analysis when a large number of alternatives and a small number of criteria are set. Interactive methods for solving this type of problem lead to obtaining the so-called Pareto-optimal solution. The solution in this case can be seen as a concept rather than a definition. This is due to the fact that there is no single global solution, and it is often necessary to identify a set of solutions that satisfy a predefined definition of optimality. The interactive MCDM methods differ by elements: weights, reference points, or constraints, which are selected for manipulations to capture the DM's preferences (Kaliszewski, 2006). Scalar problems are the basis on which various interactive methods of multicriteria optimization have been developed. The scalarization tasks determine the type of information that the DM need to specify, as well as the type and number of Pareto-optimal or weakly Pareto-optimal solutions obtained. Some of them are oriented toward solving only linear problems (Metev & Yordanova, 1993), for nonlinear problems (Metev, 1995), and others—on integer and discrete problems (Narula et al., 2003; Kirilov et al., 2015).

1.3 Multi-Objective Optimization

Multi-objective optimization (also known as multi-objective programming, vector optimization, multicriteria optimization, multi-attribute optimization, or Pareto optimization) is a field of multicriteria decision-making concerning mathematical optimization problems involving more than one objective function to be optimized at the same time (Chang, 2015). It refers to making decisions in the presence of multiple, usually conflicting, objectives. In cases of multi-objective optimization, the solution can be provided by multiple scenarios representing the trade-offs among the objective functions. MODM studies decision problems in which the decision space is continuous. To solve such type of problem, it is needed to formulate a corresponding optimization model.

1.3.1 *Optimization Model*

A basic principle when building a mathematical model is to determine some of the most important dependencies and to describe the resulting scheme of interactions using an adequate mathematical apparatus. Quantitative relationships, decision parameters, and outputs need to be defined to be used as an indicator of the effectiveness of the decision. The better the mathematical model reflects the specifics of the studied objects or processes, the more useful will be the obtained results and recommendations for the management of these objects and processes. Since in reality, the requirements for models are usually contradictory, correctly modeling problems is almost half of solving them.

In Fig. 1.2, the necessity of using abstractions in developing models is illustrated (Taha, 2010).

Assumptions about the real world are formalized by specifying basic (dominant) variables describing the behavior of the real system. Therefore, the model represents an abstraction of the real world and through the language of mathematics describes not the behavior of real systems but the assumption of their behavior. As a result, a conceptual model of the studied system is reached, which in a meaningful form describes the composition of the system, its components and the relationships between them, the main quality indicators, the variables, and the interrelationships with the quality indicators of the system, as well as a list of strategies for management (or decisions), which must be determined as a result of solving the task.

In general, a conceptual optimization model is characterized by three main components: (1) objective function, (2) variables, and (3) constraints. The objective function or performance criterion is defined as a functional dependence expressing what exactly needs to be optimized. It is usually a function of profits to be maximized or a function of costs to be minimized. Models may have more than one objective function, in which case the use of multicriteria optimization is necessary.

When formulating the mathematical model, it is necessary to determine which essential parameters of the model can be considered as variables whose values will influence the determination of the sought solution. The goal of optimization is to find the values of these variables for the defined model to generate the best value for the objective function, subject to the constraints on the input variables. Almost without exception in all models, there will be some bounded values that the variables can take, e.g., price and quantity. These constraints are expressed by mathematical

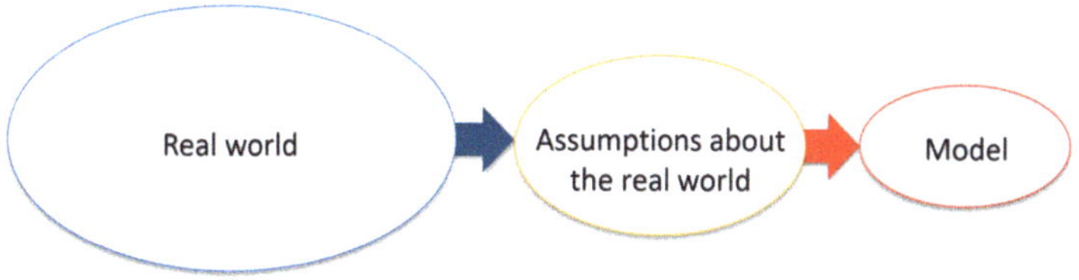

Fig. 1.2 Abstraction in model building

expressions viewed as functions of the variables in the model. Usually, these functions are defined as constraints since they actually limit the allowable values of the variables. This conceptual model could be used to further formalize the optimization problem. For this purpose, it is necessary to detail the following sets of variables:

1. A set of variables defined as *decision variables* $X = \{x_j\}$, the values of which can be selected by the DM
2. A set of variables that depend on the choice of the decision-making strategy—the so-called *output variables* of the decision-making task $Y = \{y_j\}$
3. A set of variables, the values of which depend on the system's parameters $A = \{a_k\}$—internal variables and external variables expressing external parameters for the system $\Omega = \{w_j\}$.

The objective function (the efficiency criterion) is determined depending on the adopted strategy for searching for an optimal solution, on the parameters of the modeled system, and on the parameters external to the system, such as $f: XA\Omega \rightarrow Y$. In the general case, the mathematical model of optimal decision-making tasks has the form (Taha, 2010):

$$F = f(X, Y, A, \Omega) \rightarrow Y \rightarrow \max(\min) \tag{1.4}$$

subject to

$$g_i(X, Y, A, \Omega) \leq b_i \tag{1.5}$$

$$X_{\min} \leq X \leq X_{\max} \tag{1.6}$$

$$Y_{\min} \leq Y \leq Y_{\max} \tag{1.7}$$

where the objective function (performance criterion) is given by $f(X, Y, A, \Omega) = [f_k(X, Y, A, \Omega)]$, for $k = 1, \ldots, K$; $g_i(X, A, \Omega)$ is a function of the losses of the *i-th* resource; and b_i is value of the *i-th* resource. Very often in real problems, among the constraints (1.6) and (1.7), there are also constraints of the type $X \geq 0$, $Y \geq 0$ expressing conditions for non-negativity of the variables.

In many real-world problems, the expertise of DM is used to interactively modify or fine-tune the initial optimization models. A summary representation of this process is shown in Fig. 1.3.

The optimization problem is solved by using the initial model. For this purpose, single- or multicriteria optimization is used, determining an optimal or Pareto-optimal solution. By using analysis tools, the performance indicators of the resulting solution are determined. Based on this analysis, a correction of the solution space is performed. The modification of the initial model is carried out depending on the specifics of the mathematical modeling used. Different approaches are possible based on decision analysis, for example, estimating the parameters of the initial model. The modified model replaces the initial original model and the procedure is repeated. During the search for a solution, intermediate information is stored and

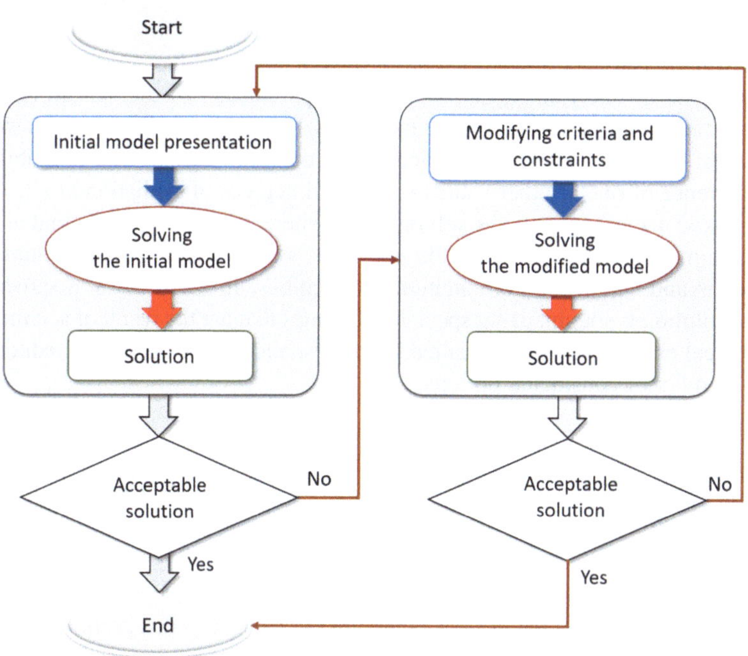

Fig. 1.3 Stages in the modeling process

used to improve the initial model. After analysis, a review of corrective functions can be performed. Both the modified model and the correction functions are suitable for searching for an optimal solution and approaching the optimal one. Such procedures reduce the computational effort for solving complex optimization problems.

To find an optimal solution to the tasks, depending on the type and structure of the objective function and the constraints, various methods of mathematical programming could be used:

1. *Linear programming*—if the functions $f(X, Y, A, \Omega)$ are linear with respect to variables $\boldsymbol{X}$
2. *Nonlinear programming*—if the functions $f(X, Y, A)$ and/or $g_i(X, Y, A, \Omega)$ are nonlinear about the variables $\boldsymbol{X}$
3. *Discrete programming*—if a discreteness condition is imposed on the control variables $\boldsymbol{X}$, e.g., integer x_j—integer ($j = 1, n$)
4. *Dynamic programming*—when for a given task it is known that the optimal solution can be constructed based on the optimal solutions of a series of subtasks, which avoids the need to repeat calculations already made
5. *Geometric programming*—if the objective function $f(X)$ and the constraints $g_k(X)$ can be represented as polynomials of the form: $g_k(X) = \sum_{i=1}^{n} c_i x_1^a x_2^a \ldots . x_m^a$

6. *Stochastic programming*—if the vector of external variables or parameters ($\Omega = \{w_s\}$) of the system $A = [a_i]$ are random
7. *Fuzzy mathematical programming*—when a decision has to be made under conditions of uncertainty, i.e., when some parameters of the system $A = [a_{ij}]$ and/or external variables $\{w_s\}$ are not precisely known and DM can either set an interval of their values or use their subjective estimates of the probability of occurrence of one or other values—so-called degrees of affiliation $\mu(a_{ij})$, $\mu(w_{isj})$
8. *Heuristic programming*—for solving tasks where it is impossible to find an exact optimum due to the nature of the problems, which leads to a huge number of options and significant computational difficulties. In these cases, a satisfactory DM solution is sought, using special techniques that are the result of accumulated practical experience—the so-called heuristics, allowing a significant reduction in the number of considered options.

1.3.2 Methods for Solving Single-Objective Optimization Problems

Methods for solving single-criteria problems can be classified into two main types—exact and approximate methods, and they are the subject of this section.

1.3.2.1 Exact Methods

The *simplex method* is proposed by Dantzig (Karp, 2008) and is one of the most widely used methods for solving linear one-dimensional problems. It has been designated as one of the top 10 algorithms of the twentieth century. The main property of the simplex method is that solving linear programming problems is done iteratively. At each iteration, the algorithm moves to a new nodal point in the convex polyhedron of the task constraints that can potentially improve the value of the objective function. This process of transition from one nodal point to another ends when further improvement of the objective function value is impossible. The implementation process of the simplex method involves a large amount of uniform and time-consuming calculations.

Interior point method developed by Narendra Karmarkar and called Karmarkar's algorithm aims to solve linear programming problems in polynomial time (Karmarkar, 1984). This algorithm is a cornerstone in linear programming and breaks the myth that there is no linear programming algorithm other than the simplex method. Due to the computational difficulties involved, it is not as widely used as the simplex method. On the other hand, the simplex method is inapplicable for solving integer or mixed-integer linear problems. This class of problems is NP hard, and their solution cannot be determined in polynomial time. This necessitates the development of special methods for their solution, the most popular of which are listed below:

Cutting plane method is the first general approach to solving the integer problem of linear optimization proposed by Gomory. The cutting plane method is an optimization method that iteratively refines the admissible set or objective function using linear inequalities that define a plane cutting the current solution in the space of feasible solutions. These procedures are used to find solutions to mixed-integer linear problems as well as to solve general, not necessarily differentiable, convex optimization problems (Balas et al., 1996).

Branch-and-bound method is developed for solving discrete and combinatorial optimization problems (Clausen, 1999). It is the most common exact algorithm for solving integer and mixed-integer programming problems. Branching divides the solution space into two parts, in each of which the optimal solution is sought. Various branching techniques are known (Wolsey, 1998), and the branch-and-bound algorithm can also be applied to nonlinear integer problems.

Branch-and-cut method is a method for solving integer combinatorial linear problems where some or all of the unknown quantities are constrained to be integers (Mitchell, 2002). Branch and cut combines the approaches of the branch-and-bound and cutting plane algorithms. In recent years, this algorithm has been particularly popular for solving combinatorial optimization problems. The main idea of the algorithm is to generate cutting planes in the entire search tree of the standard branch-and-bound algorithm. The goal is to obtain narrower limits of the domain of admissible solutions to the problem being solved at each vertex of a search tree. Different techniques can be used to implement the clipping planes in the algorithm (Atamturk & Savelsbergh, 2005).

Branch-and-price method is also a combinatorial optimization method for solving linear integer problems and mixed-integer linear problems with many variables. Its essence consists in using a column generation algorithm to solve a linear problem to the general scheme of the branch-and-bound algorithm. The next step in the development of tree-search algorithms is to combine the branch-and-cut algorithm and the branch-and-price algorithm into the so-called branch price and cut algorithm, in which branches and variables are dynamically generated (Desrosiers & Lubbecke, 2010).

1.3.2.2 Approximated Methods

Integer problems are NP-complete problems, and finding an optimal and even just an admissible solution for problems of larger dimensions is associated with large computational difficulties. Oftentimes, it is more important to find an acceptable solution quickly than to wait a long time to reach the optimal solution. In practice, in the description of the task model, some flexible constraints can be introduced that change slightly. With exact algorithms, even a small change in a constraint requires a full re-solution, which is time-consuming and can be quite expensive for real applications. Approximate algorithms are not so sensitive to small changes in some constraints. Some of them solve the problem sequentially, dividing it into parts, and then it is not necessary to completely redo the task. Approximate algorithms are used to find a suitable initial solution or to narrow the admissible area of solutions and direct the search for an optimal solution. A large number of approximate algorithms have been developed to solve large practical integer problems without guaranteeing solution optimality (Silver, 2004).

Most approximate algorithms are designed for combinatorial optimization tasks, focusing on specific applications in the fields of scheduling, logistics, routing, etc. These algorithms take advantage of the specific structure of the domain of admissible solutions of these classes of problems, achieving much better performance of the algorithm compared to algorithms that are designed to solve the general integer problem. Depending on the quality of the solutions obtained with them, approximate algorithms can be divided into the following three groups:

1. Approximate algorithms with arbitrarily set accuracy (absolute or relative)
2. Approximate algorithms with predefined accuracy, where the approximation error does not tend to zero
3. Heuristic algorithms—based on experiments or other expert evaluations, it is assumed that with a high probability, it will find a good solution to the task with acceptable computing resources, but a mathematical assessment of their accuracy cannot be guaranteed.

The main heuristic strategies that use approximate algorithms can be defined as constructive and locally improving algorithms.

Constructive algorithms generate the solution step by step using the problem data. Usually, no solution is found until the algorithm finishes its work (unlike improvement algorithms). This class of algorithms refers to the so-called greedy algorithm, where at each step the next element of the solution chosen to obtain the best local improvement is included. One of the most famous applications of the greedy algorithm is to solve the traveling salesman problem (Golden et al., 1980). These algorithms are among the fastest approximation algorithms but often lead to very poor solutions. Therefore, constructive algorithms often include forecasting procedures to assess the future consequences of the current choice (Atkinson, 1994).

The basic idea of *locally improving algorithms* is very simple—it starts with an admissible solution to the problem, obtained most often with a constructive algorithm. The feasible solutions in the neighborhood of the current solution are evaluated, and if any of them is better than the current solution, it becomes the new current solution of the task and its neighborhood is explored, etc., until no new improvement can be found and the current solution of this step is a local optimum. The importance of the way of defining a neighborhood of a point (solution) is obvious, and the question of how big this neighborhood should be is also important, i.e., at what distance from the current point to explore feasible solutions (Michalewicz & Fogel, 2000). A disadvantage of these algorithms is that only a local optimum can be guaranteed. In recent years, a new type of approximate algorithms has been developed in which combinations of different heuristic approaches are used, known as meta-heuristics in the sense of higher-level search (Glover, 1986). This class of approximate algorithms includes simulated annealing (Dowsland, 1993), tabu search (Glover & Laguna, 2002), population-based algorithms (evolutionary and genetic algorithms, and optimization) (Eiben & Smith, 2003; Dorigo & Stutzle, 2004), and others.

1.3.3 Methods of Multi-Objective Optimization

Multi-objective decision-making focuses on continuous decision spaces with an infinite number of alternatives. In wind farm design, this optimization problem does not refer to a known in advance alternative chosen as a task solution; instead, it relies on suitable restrictions and decision variables that lead to the generation of a specific new alternative. For example, the restrictions of such a problem could be the wind site area as well as the wind directions and speed while the decision variables could be associated with parameters of wind turbines' type. Here the limitations are clear and DMs have a high level of influence on the final solution.

1.3.3.1 General Formulation of Multi-Objective Optimization Problem

Real problems are usually characterized by the presence of more than one criteria, which are often contradictory and incommensurable with each other. This raises the question of how to improve the various criteria to obtain a final optimal solution. The general formulation of the multi-objective optimization problem has the form (Marler & Arora, 2004):

$$\min \mathbf{F}(\boldsymbol{x}) = [F_1(x), F_{21}(x), \ldots, F_k(x)]^T \tag{1.8}$$

subject to:

$$\mathrm{g_i}(\boldsymbol{x}) \le 0, \mathrm{j} = 1, 2, \ldots, m \tag{1.9}$$

$$\mathrm{h_l}(\boldsymbol{x}) = 0, \mathrm{l} = 1, 2, \ldots, e \tag{1.10}$$

where $\boldsymbol{F}(x)$ is a vector of criteria (objective functions), k is the number of criteria, m is the number of constraints represented as inequalities, e is the number of constraints in the form of equalities, $\boldsymbol{x} = (x_1, x_2, \ldots, x_n)^T \in E^n$ is the vector of variables, and n express the number of independent variables x_i.

In multi-objective optimization tasks, it is necessary to optimize several criteria simultaneously. In general, there is no single alternative that optimizes all criteria. Any alternative improving a certain criterion can be sought as a solution to the multicriteria problem, but for the final solution of the problem, it is necessary to select one most preferred alternative. To select the most preferred alternative, additional information is required, which is set by the DM. The information that DM sets reflects his preferences regarding the quality of the sought alternative.

1.3.3.2 Pareto Optimality

The idea of Pareto optimality has also had a significant impact on real problems. Pareto optimality is used to describe a set of choices known as a Pareto set or Pareto front. Specifying the set of these Pareto-efficient choices allows DM to express their considerations for the final solution without considering the entire set of possible solutions. Unlike single-criterion optimization, the solution of a multicriteria problem is more of a concept than a definition. There is usually no single global solution, and it is often necessary to determine a set of points that satisfy a predefined definition of optimality. The concept of determining an optimal point is Pareto optimality, which is defined as (Marler & Arora, 2004; Pareto, 1906):

Pareto-optimal solution (Marler & Arora, 2004)—a point $x^* \in X$ is Pareto optimal if there exists no other point $x \in X$ such that $F(x) \leq F(x^*)$, and $F_i(x) < F_i(x^*)$ for at least one function. In other words, a point is Pareto optimal if there is no other point that improves at least one objective function without worsening another objective function. All Pareto-optimal points are on the boundary of the admissible criterion space (Athan & Papalambros, 1996; Chen et al., 2000).

Weak Pareto-optimal solution (Marler & Arora, 2004): A point $x^* \in X$ is weakly Pareto optimal if there is no other point, $x \in X$, such that $F(x) < F(x^*)$. In other words, a point is weakly Pareto optimal if there is no other point that improves all objective functions simultaneously. Pareto-optimal points are weakly Pareto optimal, but weakly Pareto optimal points are not Pareto optimal.

The multicriteria optimization problem can be considered a problem for multicriteria choice. This problem does not have an optimal solution in the sense of the solution of a single-criterion optimization problem. Therefore, it is necessary to choose from weakly Pareto-optimal or Pareto-optimal solutions one solution that best meets the global preferences of the DM. This choice is subjective and depends entirely on the DM. The set of Pareto-optimal solutions of the multicriteria optimization problem is a subset of the set of weakly Pareto-optimal solutions of the corresponding problem.

1.3.3.3 Classification of Methods for Solving Multicriteria Tasks, Depending on the Way of Preferences Setting

The methods for solving the multicriteria optimization problem can be grouped into four classes (Fig. 1.4), depending on when the DM sets its preferences to the individual criteria—never, before, during, or after the optimization procedure (Andersson, 2000).

The information that the DM sets expresses its preferences for the sought solution (i.e., determines the importance of individual criteria). In a priori algorithms, all preferences are already known at the beginning of the decision selection process. The algorithmic techniques used aim to find the entire Pareto set or a subset of it, based on these a priori set preferences. In a posteriori algorithms, the set of all

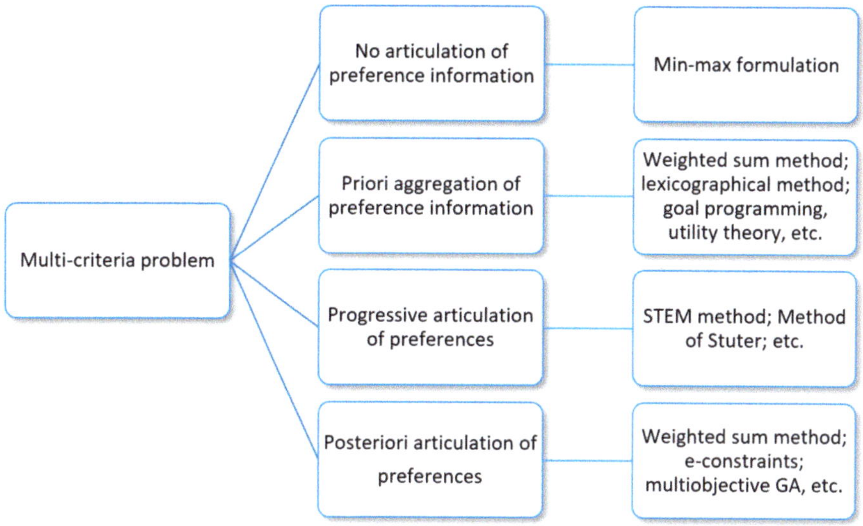

Fig. 1.4 Classification of some methods for multicriteria optimization problems

Pareto-optimal solutions for the particular task is generated, and then this set is analyzed according to the DM preferences. Many approximate algorithms have been implemented in accordance with this approach. Interactive algorithms derive Pareto-optimal solutions based on preferences progressively set by DM during the decision selection process.

The DM can express its preferences between individual Pareto-optimal solutions through the parameter values of the scalarization task. Ultimately, the DM should make a choice of the most preferred solution and is responsible for that choice.

1.3.3.4 Scalarization Approach to Solve Multi-Objective Optimization Problems

The two main approaches to solving multi-objective optimization problems are:

1. Approximation approach (Ehrgott & Wiecek, 2005)
2. Scalarization approach (Miettinen, 1999)

The first approach approximates the set of Pareto-optimal solutions (Deb, 2001; Coello et al., 2002; Ehrgott & Wiecek, 2005), and most of the methods are iterative and generate points or objects approximating this set (Vassilev et al., 2005). Some of these methods (Ehrgott & Wiecek, 2005) are precisely justified in terms of theoretical evidence and optimality, while others (Coello et al., 2002) are heuristic and often lack theoretical justification. For many multicriteria optimization problems, it is not easy to make a precise description of the set of Pareto-optimal solutions, which usually involves a very large or infinite number of solutions. For example, the solutions obtained using genetic algorithms are in general approximate Pareto-optimal solutions. The DM is isolated during the process of determining the approximating set, and finally, it is presented with too large a set of solutions for evaluation and selection, which is a relatively difficult task of multicriteria analysis. Many of these solutions are not needed by DM, and the most preferred solution may not even be found.

The second approach to solving multicriteria optimization tasks is the so-called scalarization. Scalarization is a transformation of the multicriteria optimization problem into a corresponding single-criteria problem or into a set of single-criteria optimization problems, called scalarization problems. A fundamental property of any scalarization problem is that each of its optimal solutions is a Pareto-optimal (weakly Pareto-optimal) solution of the corresponding multicriteria optimization problem. The scalarizing approach to solve multicriteria problems allows the use of the well-studied and developed theory and methods of single-criteria optimization. Different scalarization techniques are known leading to weighted sum problem, e-constraints problem, starting point problem, classification-oriented problems, etc. (Miettinen & Makela, 2002; Metev & Yordanova-Markova, 1997; Popchev et al., 1991).

A single-criteria problem can be considered as a scalarization problem of the original multicriteria problem if it meets the following two requirements (Sawaragi et al., 1985):

1. By changing the values of the parameters of the scalarization problem, it should be possible to find any Pareto-optimal solution of the multicriteria problem.
2. For each Pareto-optimal solution of the multicriteria problem, it must be possible to find suitable values of the parameters of the scalarization problem so that its optimal solution is exactly that Pareto-optimal solution of the multicriteria problem.

Multi-objective optimization inherently involves making decisions and comparing the various criteria, which are often in the form of mathematical functions. However, these functions may have different dimensions and/or significantly different values, making their comparison difficult, and sometimes even impossible. Furthermore, when the scalarization approach is used, it is possible for the same objective function with relatively large values to dominate the aggregated objective function, thus rendering the effort of using many criteria meaningless (Marler & Arora, 2005). Therefore, it is necessary to transform the objective functions so that they are commensurable, i.e., to have similar orders. This procedure is called normalization and is suitable for use on both a priori and a posteriori DM preferences about the importance of criteria.

One possible approach to criteria normalization can be presented as follows (Proos et al., 2001):

$$F_i^{\text{transf}}(x) = \frac{F_i(\text{x})}{\left|F_i^{\max}\right|} \tag{1.11}$$

This normalization results in a dimensionless objective function with an upper bound of unity (or -1) and an unbounded lower bound assuming that $F_i^{\max} \neq 0$.

A variation of the above normalization approach is the following normalization scheme (Osyczka, 1978; Salukvadze, 1979):

$$F_i^{\text{transf}}(x) = \frac{F_i(\text{x}) - F_i^o}{F_i^o}, \tag{1.12}$$

$$F_i^o(x) = \min\left\{F_i^o(x)|x \in \boldsymbol{X}\right\} \tag{1.13}$$

This approach also ensures the dimensionlessness of the objective function, with the lower bound value $F_i^{\text{transf}}(x)$ of bounded to zero and the upper bound unbounded. This transformation is also often called a relative deviation.

Another approach to normalizing the objective functions can be implemented by a transformation of the following type (Koski & Silvennoinen, 1987; Chen et al., 1999):

$$F_i^{\text{transf}}(x) = \frac{F_i(\text{x})}{F_i^o}, F_i^o > 0 \tag{1.14}$$

This approach results in obtaining dimensionless values of the objective function with a lower bound of unity.

The most commonly used approach for normalizing the objective functions, regardless of their range of variation, is obtained by the following transformation (Koski, 1984; Koski & Silvennoinen, 1987; Rao & Freiheit, 1991):

$$F_i^{\text{transf}}(x) = \frac{F_i(\text{x}) - F_i^o}{F_i^{\max} - F_i^o} \tag{1.15}$$

In this transformation, $F_i^{\text{transf}}(x)$ usually takes a value between zero and one, depending on the precision and method with which the values for $F_i^{\max}(x)$ and $F_i^o(x)$ are determined.

The following section describes the most commonly used scalarization methods for real multicriteria optimization problems.

Weighted sum method. This is the most commonly used method for solving problems of multicriteria optimization. The weighted sum method is intended for solving linear and nonlinear multicriteria problems. This method requires DM to determine weighting coefficients ω_i for the importance of each of the criteria in the objective function. The general formulation has the following form (Marler & Arora, 2010):

$$\min\left\{\sum_{i=1}^{k}\omega_i f_i(x)\right\} \tag{1.16}$$

subject to

$$\sum_{i=1}^{k}\omega_i = 1, \omega_i \geq 0 \tag{1.17}$$

In order to find multiple points of the Pareto front, several optimization problems with different weighting coefficients are solved. This method is the most intuitive and direct way to express the principle of Pareto optimality and to obtain multiple points of the Pareto-optimal front.

The weighted sum method can be considered both as a method with a priori (at the beginning of the process) set preferences of the DM and as a method with a posteriori information (at other values of the weight coefficients) about the preferences of the DM. Based on different weighting coefficients, the algorithmic techniques used aim to find the entire Pareto set or a subset of it. This method has been widely used not only to create multiple points by sequentially varying the coefficients but also to provide a single point that reflects the DM's preferences for specifically defined values of the weighting coefficients (Marler & Arora, 2010).

Compromise solution method. The compromise solution method aims to minimize the distance to an ideal (z^I) or utopian point (z^U) (Yu, 1985). The ideal point is determined according to the individual minima of all criteria such as:

$$z_j^I := \min\{\, z^j(x)\},x \in X \tag{1.18}$$

Goal programming. A special approach to multicriteria optimization is goal programming, dating from the early 1960s (Charnes & Cooper, 1957; Ignizio, 1976; Lee, 1972). More recent variants of this method are now available (Steuer, 1986; Tamiz et al., 1998). In it, for each of the criteria in the objective function, a value (goal) is set from the DM, and the general idea is to minimize the deviation from the set goals. There is usually no single point that satisfies all purposes. Different types of goal programming methods use different ways to determine which point best satisfies the given utopian values.

E-constraints method. The e-constraints method is designed for solving nonlinear multicriteria problems. In this method, one of the criteria is selected for optimization, and all the others are transformed into constraints, setting a corresponding limit for each of them. The general formulation is as follows (Mavrotas, 2009):

$$\max\{f_i(x)\} \tag{1.19}$$

subject to

$$f_i \geq e_j, \mathrm{j} = 1, 2, \ldots, k, j \neq i \tag{1.20}$$

where $\boldsymbol{x}$ is a vector of variables, $f_i(x)$ are the criteria, and S ($x \in S$) is the feasible region. Different points of the Pareto front can be obtained by changing the values of the constraints.

Lexicographic method. When the DM's preferences about the criteria are known in advance, it is called lexicographic method. According to this method, the criteria are ranked in order of importance; i.e., DM orders the criteria according to his preferences and optimizes them one by one, successively solving the corresponding single-criteria optimization problems formulated as (Marler & Arora, 2004):

$$\min\{F_j(x)\}, x \in X, j = 1, 2, \ldots, k \tag{1.21}$$

subject to

$$F_j(x) \leq F_j\left(x_j^*\right), j = \{1, 2, \ldots, i-1\}, i > 1, i = 1, 2, \ldots, k \tag{1.22}$$

In this formulation, i represents the position of the function in the preference sequence, and $F_j(x^*)$ represents the optimum of the j-th criterion found in the j-th iteration. After the first iteration ($j = 1$), $F_j(x^*)$ is not necessarily the same as the independent minimum of $F_j(x)$, as new constraints are introduced. When using the lexicographic method, the solution to each single-criterion problem yields a constraint measure for that criterion, which is used to determine an appropriate constraint in the next stage when the next most important criterion is optimized and the process is repeated. The final decision is determined at the last stage of the procedure.

1.4 Combinatorial Optimization

Many of the problems associated with practical applications are discrete in nature (Bjorndal et al., 1995). This is also the situation in the design of wind farms, where both the wind field and the number and type of turbines are to be selected. Combinatorial optimization studies problems characterized by a finite number of possible solutions. Although the optimal result for such a constrained problem can be found by simply examining all alternatives, in reality, this option is often impossible, especially for practical problems of real dimensions, where the number of possible solutions can be sufficiently large. In reality, the structural properties of problems are used to find exact or approximate algorithms to solve them. In general, combinatorial optimization problems are classified according to their computational complexity. This analysis does not always reflect the actual computational difficulty. For this reason, it is the real difficulty of the problems that lead to the development of different approaches to their solution. It is known that good solutions can be found for complex or ill-defined problems with incomplete information. This approach to finding solutions based on subjective expert judgments or based on experience is called heuristics or the heuristic method.

Heuristic methods cannot prove and guarantee the optimality of the obtained solutions. In combinatorial optimization, heuristics play an important role because they allow obtaining quasi-optimal solutions in an acceptable time by modeling the process by which one reaches a solution. When using them, it should be taken into account that there are powerful modeling tools that can be used in combinatorial optimization (mixed-integer programming, graphs, etc.) and that provide sufficient opportunities for modeling real problems with a high degree of accuracy. At the same time, the availability of software systems for solving mixed-integer problems, to which the tasks of combinatorial optimization are reduced, allows for obtaining proven optimal solutions. On the other hand, there are real problems for which the time to find this solution is unacceptably long and for which reducing this time by using heuristic techniques leading to sufficiently good solutions is advisable. In this regard, it should be noted that most heuristics are characterized by polynomial complexity.

Advances in computer computing capabilities are more important for heuristic methods than for exact methods because they allow the use of complex heuristics to solve real-world, high-dimensional problems. Since heuristic methods do not guarantee an optimal solution, another important measure is introduced—solution quality. Ideally, it can be estimated by knowing an upper bound on the solution (the optimal solution) and determining the difference between it and the resulting solution. This limit can be determined by other exact methods or by other heuristic methods. Many heuristic methods have a similar structure, leading to the development of generalized meta-heuristics such as "simulated hardening" (Aarts & Korst, 1989), "tabu search" (Glover, 1986), and "genetic algorithms" (Goldberg, 1969). These meta-heuristics provide algorithms with substantial user involvement in defining complexity. The user is the decider in determining whether solution time or solution quality is more important.

Tree-search strategies play an essential role in any exact algorithm, and different applications of them can lead to different performance. In addition, limiting procedures play a key role in any such method. The constraint procedure is based on the exact solution of a modified problem that is obtained from the original problem by removing or simplifying some of its constraints (Bjorndal et al., 1995). The solution boundary obtained by this modified problem is good but at the expense of increasing the computational effort. Classical "boundary and branching methods" are based on modified problems, which in many cases are easily solved. Some classes of problems such as "knapsack problem," "asymmetric traveling salesman problem," and "single machine schedules" provide a boundary close to the optimum. However, there are problems for which the difference between the optimum and the limit is unsatisfactorily large. The goal of polyhedral combinatorics is to provide near-optimal bounds through linear programming methods. If the combinatorial problem is defined as an integer linear programming problem, then the optimal solution corresponds to the "best" vertex of the convex polyhedron of admissible points described by a finite number of linear constraints. Using theoretical results for linear programming and available efficient algorithms, sufficiently good bounds for the solution can be determined.

There are two main problems when solving complex combinatorial optimization problems such as linear programming problems:

- It is difficult to formulate the necessary linear constraints.
- The number of these constraints is exponential, and it is not clear how to choose a polynomial number of them.

Therefore, the remainder of the monograph describes various practical combinatorial optimization models with applications in renewable wind energy.

1.5 Decision-Making in Engineering Design

The design process can be seen as the process of developing a project for a new object, which requires research, modeling, interactive setup, and re-design. The engineering design process can be represented by the following 10 steps (Gomez et al., 2004; Garrett, 1991):

1. Identification of the problem—general description.
2. Defining the criteria and objectives
3. Researching the problem—in most cases, it is necessary to obtain new knowledge and information
4. Defining the specifications—detailing the problem and clarifying what the design needs to achieve and what constraints will affect the final solution
5. Determination of possible solutions—combining the conceptual designs with the information obtained from item 3 to determine possible design solutions
6. Analysis of potential solutions—determining which option to develop—ideally the option that best meets specifications and other constraints such as time, cost, or skills
7. Preparation of working documentation and planning of all the details that are important for the realization of the chosen option
8. Making an initial model (prototype)
9. Design testing and evaluation
10. Preparation of documentation—planning and design for practical implementation

The problems of conceptual design are related to the vague nature of the initial design concepts and the availability of different options that engineers must test. Real engineering design problems are usually characterized by the presence of many often conflicting and incompatible objectives. Important tasks in engineering design are generating and refining design alternatives and then selecting a single design or set of designs to fulfill a particular need (Scott & Antonsson, 1999). During the design process, trade-offs must be made between various engineering features. The designer can use a range of modeling and simulation tools to predict the characteristics of the designed object. This process is often iterative and time-consuming and is limited by the knowledge of the particular designer. Today, customers have an increasing demand for different products. Offering a wide variety of products is important for any company to attract different consumer segments. Recent theories suggest modular product design as a key factor for product success in global competition (Antonio et al., 2007). They examine the impact of modular products on competitive manufacturing capabilities, including price, product quality, flexibility, etc.

Given the growing global competition and shortening product life cycles, there is a growing need to develop and produce affordable goods and services that can be appropriately customized (Zhang et al., 2014; Avak, 2007). This problem can be overcome by modularity in design which can provide diversity in the product range of devices (Yu et al., 2015). Modular design as a form of strategic manufacturing offers flexibility in the integration of modular components. Modularity has some significant advantages—it leads to a reduction in production costs and lead time and allows the realization of different requirements in accordance with changing markets and technologies, and in terms of disadvantages—lack of productivity optimization and excessive similarity between products (Kidd, 1994; Gershenson & Prasad, 1997; Tseng & Huang, 2008; Dorsey et al., 2006; Salhieh & Kamrani, 2008). Modular design does not involve the development of new types of components but requires knowledge of the characteristics of the modules and their compatibility with other modules. A modular system is characterized by the functional separation of modules and the use of well-defined module interfaces, as well as the use of industry standards for interfaces (AlGeddawy & ElMaraghy, 2013). This is of particular importance to apply modeling approaches in the context of using optimization models. For the formulation of a suitable optimization model, a good knowledge of the specifics of the modeled system (including the choice of control variables, constraints, and criteria) is required, as well as knowledge of the methods for solving single- and mono-criteria optimization tasks. One promising direction in design is the use of combinatorial optimization (Levin, 2009) and in particular the use of 0–1 integer variables and heuristic methods (Fujita et al., 1999).

All these premises together with operations research and decision theory can be applied to the planning, design, maintenance, and investment of wind energy.

References

Aarts, E., & Korst, J. (1989). *Simulated annealing and Boltzmann Machims*. Wiley.

AlGeddawy, T., & ElMaraghy, H. (2013). Reactive design methodology for product family platforms, modularity and parts integration. *CIRP Journal of Manufacturing Science and Technology, 6*, 34–43. https://doi.org/10.1016/j.cirpj.2012.08.001

Andersson, J. (2000). *A survey of multiobjective optimization in engineering design*. Technical Report: LiTH-IKP-R-1097.

Antonio, K. W. L., Yam, R. C. M., & Tang, E. (2007). The impacts of product modularity on competitive capabilities and performance: An empirical study. *International Journal Production Economics, 105*, 1–20. https://doi.org/10.1016/j.ijpe.2006.02.002
Arslan, H. M. (2017). Current classification of multi criteria decision analysis methods and public sector implementations. In *Current debates in public finance public administration & environmental studies* (pp. 241–261).
Assari, A., Mahesh, T., & Assari, E. (2012). Role of public participation in sustainability of historical city: Usage of TOPSIS method. *Indian Journal of Science and Technology, 5*(3), 2289–2294. https://doi.org/10.17485/ijst/2012/v5i3.2
Atamturk, A., & Savelsbergh, M. W. P. (2005). Integer-programming software systems. *Annals of Operations Research, 140*, 67–124. https://doi.org/10.1007/s10479-005-3968-2
Athan, T. W., & Papalambros, P. Y. (1996). A note on weighted criteria methods for compromise solutions in multi-objective optimization. *Optimization and Engineering, 27*, 155–176. https://doi.org/10.1080/03052159608941404
Atkinson, J. B. (1994). A greedy look-ahead heuristic for combinatorial optimization: An application to vehicle scheduling with time windows. *Journal of Operational Research Society, 45*, 673–684. https://doi.org/10.1057/jors.1994.105
Avak, B. (2007). Variant management of modular product families in the market phase. *VDI Verlag*. https://doi.org/10.3929/ethz-a-005320674
Baker, D., Bridges, D., Hunter, R., Johnson, G., Krupa, J., Murphy, J., & Sorenson, K. (2002). Guidebook to decision-making methods. In *WSRC-IM-2002-00002*. Department of Energy.
Balas, E., Ceria, S., Cornuejols, G., & Natraj, N. (1996). Gomory cuts revisited. *Operations Research Letters, 19*, 1–9. https://doi.org/10.1016/0167-6377(96)00007-7
Bjorndal, M. H., Caprara, A., Cowling, P. I., Della Croce, F., Lourenco, H., Malucelli, F., Orman, A. J., Pisinger, D., Rego, C., & Salazar, J. J. (1995). Some thoughts on combinatorial optimisation. *EJOR, 83*(2), 253–270. https://doi.org/10.1016/0377-2217(95)00005-B
Brans, J. P., & Mareschal, B. (2005). Promethee methods. In *Multiple Criteria Decision Analysis: State of the Art Surveys. International Series in Operations Research & Management Science* (Vol. 78, pp. 163–186). https://doi.org/10.1007/0-387-23081-5_5
Chang, K.-H. (2015). Chapter 19-Multiobjective optimization and advanced topics. In K.-H. Chang (Ed.), *e-Design* (pp. 1105–1173). Academic Press. https://doi.org/10.1016/B978-0-12-382038-9.00019-3
Charnes, A., & Cooper, W. W. (1957). Management models and industrial applications of linear programming. *Management Science, 4*(1), 38–91. http://www.jstor.org/stable/2627263
Chen, W., Sahai, A., Messac, A., & Sundararaj, G. (2000). Exploration of the effectiveness of physical programming in robust design. *Journal of Mechanical Design, 122*, 155–163. https://doi.org/10.1115/1.533565
Chen, W., Wiecek, M. M., & Zhang, J. (1999). Quality utility–A compromise programming approach to robust design. *Journal of Mechanical Design, 121*, 179–187. https://doi.org/10.1115/1.28294401
Clausen, J. (1999). *Branch and bound algorithms–Principles and examples*. Technical report. University of Copenhagen.
Coello, C. A. C., Veldhuizen, D. A. V., & Lamont, G. B. (2002). *Evolutionary algorithms for solving multi-objective problems*. Kluwer.
Deb, K. (2001). *Multi-objective optimization using evolutionary algorithms*. Wiley.
Desrosiers, J., & Lubbecke, M. E. (2010). *Branch-Price-and-Cut Algorithms*. Wiley Encyclopedia of Operations Research and Management Science.
Dorigo, M., & Stutzle, T. (2004). *Ant Colony optimization*. MIT Press.
Dorsey, J. T., Collins, T. J., Doggett, W. R., & Moe, R. V. (2006). Framework for defining and assessing benefits of a modular assembly design approach for exploration systems. In *Proc. Space Technology and Applications International Forum–STAIF 2006* (Vol. 813, pp. 969–981). https://doi.org/10.1063/1.2169278

Dowsland, K. (1993). Simulated Annealing. In C. R. Reeves (Ed.), *Modern Heuristic Techniques for Combinatorial Problems* (pp. 20–69). Blackwell.

Ehrgott, M., & Wiecek, M. (2005). Multiobjective programming. In J. Figueire, S. Greco, & M. Ehrgott (Eds.), *Multiple criteria decision analysis*. State of the Art Surveys.

Eiben, A., & Smith, J. (2003). *Introduction to evolutionary computing*. Springer.

Figueira, J., Greco, S., & Ehrgott, M. (2005). Multiple criteria decision analysis: State of the art surveys. *Springer Science & Business Media, 1045.*

Fujita, K., Sakaguchi, H., & Akagi, S. (1999). Product variety deployment and its optimization under modular architecture and module communalization. In *Proc. of the ASME design engineering technical conferences*. Las Vegas. DETC99/DFM-8923.

Fulop, J. (2005). *Introduction to decision making methods* (pp. 1–15) http://citeseerx.ist.psu.edu/viewdoc/download?doi=10.1.1.86.6292&rep=rep1&type=pdf

Garrett, J. (1991). *Design and technology*. Cambridge University Press.

Gershenson, J. K., & Prasad, G. J. (1997). Modularity in product design for manufacturability. *International Journal of Agile Manufacturing, 1*(1), 1–11.

Glover, F., & Laguna, M. (2002). Tabu search. In P. M. Pardalos & M. G. C. Resende (Eds.), *Handbook of applied optimization* (pp. 194–208). Oxford University Press.

Glover, F. (1986). Future paths for integer programming and artificial intelligence. *Computers and Operations Research, 13*(5), 533–549. https://doi.org/10.1016/0305-0548(86)90048-1

Goldberg, D. E. (1969). *Genetic algorithms in search. Optimization and machine learning*. Addison-Wesley.

Golden, B., Bodin, L., Doyle, T., & Stewart, W., Jr. (1980). Approximate traveling salesman algorithm. *Operations Research, 28*, 694–711.

Gomez, A., Oakes, W., & O., Les, L. (2004). *Engineering your future* (pp. 352–363). Great Lake Press.

Govindan, K., & Jepsen, M. B. (2016). ELECTRE: A comprehensive literature review on methodologies and applications. *European Journal of Operational Research, 250*(1), 1–29. https://doi.org/10.1016/j.ejor.2015.07.019

Hwang, C. L., & Yoon, K. (1981). *Multiple attribute decision making-methods and applications, A State of the Art Survey*. Springer Verlag.

Hwang, C. L., Lai, Y. J., & Liu, T. Y. (1993). A new approach for multiple objective decision making. *Computers and Operational Research, 20*, 889–899. https://doi.org/10.1016/0305-0548(93)90109-V

Ignizio, J. P. (1976). *Goal programming and extensions*. Lexington Books.

Jahan, A., & Edwards, K. L. (2015). A state-of-the-art survey on the influence of normalization techniques in ranking: Improving the materials selection process in engineering design. *Materials and Design, 65*, 335–342. https://doi.org/10.1016/j.matdes.2014.09.022

Kaliszewski, I. (2006). MCDM interactive methods-an overview. In *Soft computing for complex multiple criteria decision making* (Vol. 85, pp. 45–56). International Series in Operations Research & Management Science. https://doi.org/10.1007/0-387-30177-1_4

Karmarkar, N. A. (1984). New polynomial-time algorithm for linear programming. *Combinatorica, 4*(4), 373–395. https://doi.org/10.1007/BF02579150

Karp, R. M. (2008). George Dantzig's impact on the theory of computation. *Discrete Optimization, 5*(2), 174–185. https://doi.org/10.1016/j.disopt.2006.12.004

Keeney, R. L., & Raiffa, H. (1993). *Decisions with multiple objectives: Preferences and value tradeoffs*. Cambridge University Press.

Kidd, P. T. (1994). *Agile manufacturing: Forging new frontiers*. Addison Wesley.

Kirilov, L., Guliashki, V., Genova, K., Zhivkov, P., Staykov, B., & Vatov, D. (2015). Interactive environment WebOptim for solving multiple-objective problems using scalarising and evolutionary approaches. *International Journal of Reasoning-based Intelligent Systems, 7*(1–2), 4–15. https://doi.org/10.1504/IJRIS.2015.070907

Koski, J., & Silvennoinen, R. (1987). Norm methods and partial weighting in multicriterion optimization of structures. *International Journal for Numerical Methods in Engineering, 24*, 1101–1121. https://doi.org/10.1002/nme.1620240606

Koski, J. (1984). Multicriterion optimization in structural design. In E. Atrek, R. H. Gallagher, K. M. Ragsdell, & O. C. Zienkiewicz (Eds.), *New Directions in Optimum Structural Design* (pp. 483–503).

Lee, L., & Anderson, R. (2009). A comparison of compensatory and non-compensatory decision making strategies in IT project portfolio management. *International Research Workshop on IT Project Management, 9*. http://aisel.aisnet.org/irwitpm2009/9

Lee, S. M. (1972). *Goal programming for decision analysis*. Auerbach.

Levin, M. S. (2009). Combinatorial optimization in system configuration design. *Automation and Remote Control, 70*(3), 519–561. https://doi.org/10.1134/S0005117909030187

Marler, R. T., & Arora, J. S. (2005). Function-transformation methods for multi-objective optimization. *Engineering Optimization, 37*(6), 551–570. https://doi.org/10.1080/03052150500114289

Marler, R. T., & Arora, J. S. (2004). Survey of multi-objective optimization methods for engineering. *Structural and Multidisciplinary Optimization, 26*, 369–395. https://doi.org/10.1007/s00158-003-0368-6

Marler, R. T., & Arora, J. S. (2010). The weighted sum method for multi-objective optimization: New insights. *Structural and Multidisciplinary Optimization, 41*, 853–862. https://doi.org/10.1007/s00158-009-0460-7

Mavrotas, G. (2009). Effective implementation of the e-constraint method in multi-objective mathematical programming problems. *Applied Mathematics and Computation, 213*, 455–465. https://doi.org/10.1016/j.amc.2009.03.037

Metev, B., & Yordanova, I. (1993). Use of reference points for MOLP problems analysis. *European Journal of Operational Research, 68*(3), 374–378. https://doi.org/10.1016/0377-2217(93)90193-Q

Metev, B., & Yordanova-Markova, I. (1997). Multi-objective optimization over convex dis-junctive feasible sets using reference points. *European Journal of Operational Re-search, 98*(1), 124–137. https://doi.org/10.1016/0377-2217(95)00323-1

Metev, B. (1995). Use of reference points for solving MONLP problems. *European Journal of Operational Research, 80*(1), 193–203. https://doi.org/10.1016/0377-2217(93)E0129-L

Michalewicz, Z., & Fogel, D. B. (2000). *How to solve it: Modern heuristics*. Springer-Verlag.

Miettinen, K., & Makela, M. M. (2002). On scalarizing functions in multiobjective optimization. *OR Spectrum, 24*(2), 193–213. https://doi.org/10.1007/s00291-001-0092-9

Miettinen, K. (1999). *Nonlinear multiobjective optimization*. Kluwer Academic Publishers.

Mitchell, J. E. (2002). Branch-and-cut algorithms for combinatorial optimization problems. Handbook of applied. *Optimization*, 65–77.

Narula, S. C., Vassilev, V., Genova, K., & Vassileva, M. (2003). A partition-based interactive method to solve discrete multicriteria choice problems. *Cybernetics and Information Technologies, 3*(2), 55–66.

Opricovic, S., & Tzeng, G.-H. (2004). Compromise solution by MCDM methods: A comparative analysis of VIKOR and TOPSIS. *European Journal of Operational Research, 156*(2), 445–455. https://doi.org/10.1016/S0377-2217(03)00020-1

Osyczka, A. (1978). An approach to multicriterion optimization problems for engineering design. *Computer Methods in Applied Mechanics and Engineering, 15*(3), 309–333. https://doi.org/10.1016/0045-7825(78)90046-4

Pareto, V. (1971). *Manuale di Economia Politica. Piccola Biblioteca Scientifica, Milan, 1906. English Translation by A. Schwier, Manual of Political Economy*. Kelley Publishers.

Popchev, I., Metev, B., & Yordanova, I. (1991). A realization of reference point method using the Tchebycheff distance. In A. Lewandowski & V. Volkovich (Eds.), *Lecture notes in economics and mathematical systems* (Vol. 351, pp. 76–82). Springer-Verlag.

Proos, K. A., Steven, G. P., Querin, O. M., & Xie, Y. M. (2001). Multicriterion evolutionary structural optimization using the weighted and the global criterion methods. *AIAA Journal, 39*(10), 2006–2012.
Rao, R. V. (2007). Introduction to multiple attribute decision-making (MADM) methods. In *Decision making in the manufacturing environment* (pp. 27–41). Springer. https://doi.org/10.1007/978-1-84628-819-7_3
Rao, S. S., & Freiheit, T. I. (1991). A modified game theory approach to multiobjective optimization. *Journal of Mechanical Design, 113*, 286–291. https://doi.org/10.1115/1.2912781
Roy, B. (1996). *Multicriteria methodology for decision aiding*. Springer.
Roy, B. (1968). Classement et choix en présence de points de vue multiples (la méthode ELECTRE). *La Revue d'Informatique et de Recherche Opérationelle, 8*(5), 7–75.
Saaty, T. L. (1996). *Decision making with dependence and feedback: The analytic network process*. RWS Publications.
Saaty, T. L., & Vargas, L. G. (2013). Decision making with the analytic network process. In *International Series in Operations Research & Management Science* (Vol. 195). Springer.
Saaty, T. L. (1990). How to make a decision: The analytic hierarchy process. *European Journal of Operational Research, 48*(1), 9–26. https://doi.org/10.1016/0377-2217(90)90057-I
Saaty, T. L. (1980). *The analytic hierarchy process*. McGraw-Hill International.
Salhieh, S.'. E. M., & Kamrani, A. K. (2008). Chapter 10: Modular design. In A. K. Kamrani & E. A. Nasr (Eds.), *Collaborative engineering–Theory and practice* (pp. 207–226). Springer.
Salukvadze, M. E. (1979). *Vector-valued optimization problems in control theory*. Academic Press.
Sawaragi, Y., Nakayama, H., & Tanino, T. (1985). *Theory of multiobjective optimization*. Academic Press.
Scott, M. J., & Antonsson, E. K. (1999). Arrow's theorem and engineering design decision making. *Research in Engineering Design, 11*, 218–228. https://doi.org/10.1007/s001630050016
Silver, E. A. (2004). An overview of heuristic solution methods. *Journal of the Operational Research Society, 55*, 936–956. https://doi.org/10.1057/palgrave.jors.2601758
Steuer, R. (1986). *Multiple criteria optimization: Theory, computation and application*. John Wiley & Sons, Inc.
Taha, H. A. (2010). *Operations research: An introduction*. Prentice Hall.
Taherdoost, H., & Madanchian, M. (2023). Multi-criteria decision making (MCDM) methods and concepts. *Encyclopedia, 3*(1), 77–87. https://doi.org/10.3390/encyclopedia3010006
Tamiz, M., Jones, D., & Romero, C. (1998). Goal programming for decision making: An overview of the current state-of-the-art. *EJOR, 111*, 569–581. https://doi.org/10.1016/S0377-2217(97)00317-2
Triantaphyllou, E., Shu, B., Sanchez, S. N., & Ray, T. (1998). Multi-criteria decision making: An operations research approach. In J. G. Webster (Ed.), *Encyclopedia of electrical and electronics engineering* (Vol. 15, pp. 175–186). John Wiley & Sons.
Tseng, T. L. B., & Huang, C. C. (2008). Design support systems: A case study of modular design of the set-top box from design knowledge externalization perspective. *Decision Support Systems, 44*(4), 909–924. https://doi.org/10.1016/j.dss.2007.10.013
Vassilev, V., Genova, K., & Vassileva, M. (2005). A brief survey of multicriteria decision making methods and software systems. *Cybernetics and Information Technologies, 5*(2), 14–20.
Vincke, P. (1992). *Multi-criteria Decision-Aid*. John Wiley.
Wolsey, L. (1998). *Integer Programming*. Wiley-Interscience.
Yoon, K. (1987). A reconciliation among discrete compromise situations. *Journal of Operational Research Society, 38*, 277–286. https://doi.org/10.2307/2581948
Yu, P. L. (1985). *Multiple criteria decision making: Concepts, techniques and extensions*. Plenum Press.
Yu, S., Yang, Q., Tao, J., & Xu, X. (2015). Incorporating quality function deployment with modularity for the end-of-life of a product family. *Journal of Cleaner Production, 87*, 423–430. https://doi.org/10.1016/j.jclepro.2014.10.037

Zanakis, S. H., Solomon, A., Wishart, N., & Dublish, S. (1998). Multi-attribute decision making: A simulation comparison of select methods. *European Journal of Operational Research, 107*, 507–529. https://doi.org/10.1016/S0377-2217(97)00147-1
Zavadskas, E. K., Turskis, Z., & Kildiene, S. (2014). State of art surveys of overviews on MCDM/ MADM. *Technological and Economic Development of Economy, 20*, 165–179. https://doi.org/ 10.3846/20294913.2014.892037
Zeleney, M. (1976). The attribute dynamic attitude model. *Management Sciences, 23*, 12–25. https://www.jstor.org/stable/2629751
Zhang, M., Zhao, X., & Qi, Y. (2014). The effects of organizational flatness, coordination, and product modularity on mass customization capability. *International Journal Production Economics, 158*, 145–155. https://doi.org/10.1016/j.ijpe.2014.07.032

2.1 Wind Farm Life Cycle

Generating electricity from wind energy has the advantage of not contributing to global warming or polluting the air. Since wind energy is a renewable energy source, there is no ongoing cost of acquiring fuel. Once the wind turbine is installed, the only real cost is maintenance. Another characteristic of wind energy is the fact that it is spatially efficient and fits well with other land uses. Wind turbines can be placed in fields with enough space between them to be productive. The space below them is open for other uses, such as agriculture. The use of wind energy is economically beneficial as it relies on a local resource and therefore creates local jobs. Also worth mentioning is the fact that investment in wind power is on the rise, with wind power capacity increasing by 30 percent in recent years. When investing in wind energy, it is imperative to use not only appropriate investment evaluation methods but also knowledge and models in the field of energy, forecasting, and statistics of wind potential, including knowledge of geology and the environment. In addition, complex external factors such as emerging technologies, policy volatility, and increasing uncertainty in the international energy market make investment in wind energy challenging (Liu et al., 2021).

The life cycle of a wind farm can be represented as an iterative process passing successively through several main stages as (1) site assessment; (2) planning and design; (3) construction and commissioning; (4) operation and maintenance; and (5) decommissioning as illustrated in Fig. 2.1.

The *first stage* is related to assessing the site to decide if it has the potential to become a viable wind farm. The feasibility of a potential site must be assessed against a number of criteria, the two most important of which are the average wind speed at the site along with the area of available land and land ownership. A suitable wind farm site is identified by its geographical location and based on long-term wind records for annual wind speed variation (White et al., 1997). This can be done by using graphical charts that characterize the speed and direction of winds at a location

Fig. 2.1 Life cycle of a wind farm

known as wind roses (Arteaga-Lopez & Angeles-Camacho, 2021; Poddaeva & Fedosova, 2022). Other technological or ecological considerations are also taken into account when defining wind farm sites. The selection of appropriate sites for wind farms from the wind industry's point of view depends not only on wind characteristics but includes parameters such as landscape features, remote distance from houses, residential and industrial areas, land use of the proposed site, and adjacent land, existing and future grid infrastructure, community acceptance, construction access to the area via public roads, existing wind farms in the area and wind farms or other power generation projects under development nearby, nature protection and military areas, and other service infrastructures.

The *second stage* is related to the planning and design of wind farms including engineering consulting, wind measurement, geotechnical measurement, technical expertise, performance monitoring companies, developers, and financial institutions. Wind energy technology is evolving rapidly and in the beginning wind turbines are placed on mountains and hilltops; today, thanks to technological developments, they can be installed in more diverse settings. Wind farms can be constructed both onshore (in the terrestrial areas) and offshore (constructed on water, dominantly in seas and oceans). Higher towers and longer blades make them increasingly cost-efficient (especially offshore) and allow a broader extension. The layout design of the wind turbines plays an important role in the effectiveness of wind farms. At first, this was done heuristically on a trial-and-error basis, taking into account all social, environmental, and technical constraints (Mora et al., 2007), but today we can rely on different models for simulations. The number of wind turbines in the designed wind farm is limited by the need to be located within the selected geographical area. The separation distance between the turbines is among other things to be considered when defining the wind farm layout and depends on the needed recovery of wind energy behind the neighboring turbines (Grady et al., 2005; Sørensen, 2006). The production of industrial quantities of electricity requires the installation of as many power wind turbines as it is possible. The optimum wind farm layout solution should define the wind turbines' number and type and also their placement in the given geographical area. This is the most important stage of the life cycle of the wind power plant since the location of the turbines will depend on the generated energy, construction, maintenance, and delivery of the energy to the transmission network.

The *third stage* refers to the actual construction and inclusion in the transmission network, i.e., commissioning. The activities related to this stage concern logistics, civil engineering, foundation, turbine installation, sea-based support, marine coordination, weather forecasting and metocean data, electricity network, offshore/onshore substation, connection to the grid, etc. This includes both strategic and detailed planning for the development of the commissioning of the wind farm (a set of activities performed before an operation permit is provided to confirm that the wind turbines have been correctly installed and that they are ready for energy production). Once planning consent has been obtained construction works for the project are put out to competitive tender. It is practical to use local companies for civil works (access roads, turbine foundations, and any buildings) while the wind turbines and associated electrical equipment are supplied by one of the large turbine manufacturers.

The *fourth stage* concerns the operation and maintenance activities. The operation and maintenance phase of wind farms generally amounts to up to 25 years of a standard lifetime or 50 years of the same wind farm subjected to modernization by means of replacing the nacelle along with its accessories and the rotor after 25 years of use (Kasner, 2022). It starts after a testing period and upon approval of the purchase contract for the produced electricity by the relevant national, regional, or local authorities. This stage includes the routine servicing and repairs that are required to achieve the designed lifetime of the wind turbine and to ensure compliance with financial, safety, security, and leasing agreements, training, health, and safety inspections, turbine maintenance, and service.

The *fifth stage* refers to the end of wind power plant life. Wind power plants that have reached their service life can generally choose four treatment methods, including full decommissioning, constant repowering, repowering preceded by full decommissioning, and transition toward a tourist attraction (Szumilas-Kowalczyk et al., 2020). After a power plant's life cycle comes to an end, the wind power plant owners can decide to repower the plant by upgrading the existing infrastructure with new technology to begin operations anew, or they can decide to remove all the equipment and shut down the farm. In the second case, it is needed to decommission the turbine/s, foundation, cable, and substation. At this stage, the main objective is to pay attention to recycling and waste disposal in order to minimize their impact on the environment. In order to choose an appropriate decommissioning method, the environmental characteristics of the wind farm are to be assessed (Hall et al., 2020).

2.2 Groups of Indicators for Assessment and Selection of Wind Site Location

According to the life cycle of a wind farm, the first stage is related to the assessment and selection of a proper wind site to build a wind power plant. This process can be done considering some key indicators on the basis of which a choice of location for building a wind power plant can be made.

Choosing an optimal location for wind power plants is a difficult task that requires careful consideration of various factors (Alavi et al., 2022). It is important to note that the costs of choosing the right place to build the wind farms are a small part of the total cost of the project, but the information obtained is essential for the precise assessment of energy production. For optimal planning of wind farm projects, it is necessary to reconcile the conflicting elements related to industrial and technological aspects of wind generation with social and ecological restrictions, also considering the public opinion (Badi et al., 2023). The responsible siting aims to promote the coexistence of wind energy and local communities by minimizing or eliminating negative impacts on wildlife and people. Each geographical region has its own economic, social, and environmental characteristics and constraints; thus, selecting an appropriate location to construct a renewable energy power plant is crucial. Careful siting requires extensive research and careful public engagement. There are several groups of indicators that assist in the critical project planning step of selecting a wind site to build a wind power plant or a wind turbine location. These groups of key indicators for the selection of a wind site to build a wind power plant are as follows:

1. Technical indicators
2. Economic indicators
3. Environmental indicators

2.2.1 Technical Criteria

The first group of key indicators are the technical ones, which include the following parameters:

- Wind resources availability
- Area—onshore, offshore, floating

The design and building of a wind farm depends on many factors, including wind resources and land area (onshore or offshore). The wind resource is the most important criterion for site selection. Technical indicators for the selection of wind sites include appropriate ground conditions, access to the electricity grid, feasibility of access for abnormal loads, and suitable terrain and topography. The sea environment criteria such as sea depth, soil condition, and the transmission line length are also important factors with the wind farm to be located offshore (Fetanat & Khorasaninejad, 2015).

The availability and accuracy of meteorological data are critical to determining the wind potential of a particular location. Wind speed is an important parameter in wind speed distribution analysis. This is due to the direct relationship between the change in wind speed and the energy produced. Therefore, wind turbines are preferably located in areas where the winds are strong and constant. The probability distribution function of wind speed could be used as an essential factor in the assessment of wind energy potential at a location and could be used as an effective measure for the occurrence probability of a wind speed within a given range (Paraschiv et al., 2022). Although diurnal winds at a given location can be highly variable, monthly and especially annual averages are remarkably constant from year to year. It should be noted that it is needed to consider the socioeconomic and political environment and level of greenhouse emissions as they reflect a significant reduction in wind power density (Martinez & Iglesias, 2021).

The wind turbines are designed to generate electricity at wind speeds between 3–5 and 25 m/s. The most suitable locations for wind turbines would be in areas where the average annual wind speed is known to be moderately high or high.

Offshore wind power plants rely on stronger and steadier than onshore winds. For example, the velocity of wind at 10–15 m from shore is higher by 20–25%. Turbulence intensity is one of the main characteristics of airflow that affects the power production and lifespan of wind turbines (Asadi & Pourhossein, 2021). In addition, as a result of less turbulence in offshore wind in comparison with the onshore wind, the fatigue loads on the turbine are reduced and their service life will be increased. The consistency of electricity production from wind farms can be caused by different wind speeds and changes in wind direction on land in contrast to the higher wind speeds and constancy of wind direction offshore. It should also mention here the so-called floating farms which, unlike existing offshore wind farms, do not need to be fixed to the seabed, as they use anchors to hold them in place, similar to a boat. This means they can be located in much deeper sea areas, which means there is more room for larger turbines that generate larger amounts of power.

An advantage of onshore wind turbines is that they can be owned or operated by local cooperatives or even individuals, unlike offshore wind turbines which are usually corporate owned.

2.2.2 Economic Criteria

The second group of key indicators is the economic ones, which include the following parameters:

- Appropriate ground conditions—terrain and topography (onshore or offshore)
- Estimated capacity
- Distance to the power line
- Distance to the substation
- Distance to road and feasibility of access for abnormal loads
- Agreements with land landowners

From a building point of view, terrain slope, distance from power transmission lines, and distance from road networks are among the essential criteria in this group. Steep terrain slopes reduce access to machines such as cranes and trucks and increase the cost of power plants. Therefore, the slope of the land plays an essential role in the construction costs of wind power plants. System integration costs, which also need to be taken into account, are highly dependent on the level of penetration of intermittent renewable energy sources and the flexibility of the surrounding electricity system. In addition, the distance from power lines is one of the economic criteria, because the extended distance from the wind farms increases the electrical losses and the cabling costs. Hence, it is preferable to build wind power plants near power lines. Another essential criterion related to wind farm building is the distance of wind power plants from road networks. Given that the cost of access to wind power plants should be reasonable, wind power plants must be established near roads to reduce these costs. Moreover, it is also necessary to take into account the possibility of accessing abnormal loads like those of wind turbines. Wind farm capacity is another metric to consider when choosing a wind farm site. Larger wind farms will generate more power, and the cost of providing the necessary equipment for even a small wind farm can be comparable to the cost of a larger one. Renewable energy from onshore wind farms can be considered cheaper than offshore wind energy due to the significantly lower investment required to build infrastructure and operating costs. Cheaper infrastructure and operating costs mean that land-based farms can help the lower levelized cost of electricity. Onshore wind farms can be built in months and are relatively cheap and cost-effective to maintain compared with offshore. Last but not least, the cost of renting the land for an on-shore wind farm should be taken into account. This cost is negligible because only a small part of the land is used for installation and most of it remains usable.

2.2.3 Environmental Criteria

The third group of key indicators is the environmental ones, which include the following parameters:

- Proximity to natural and built heritage; natural protected areas; archaeological heritage
- Proximity to centers of population
- Proximity to airports/railways/rivers
- Proximity to defense interests areas and seismological recording
- Proximity to telecommunications and broadcasting installations
- Visibility in sensitive areas

During the selection of the site for the wind farm, an appropriate conservation strategy for natural protected areas and protected species such as Natura 2000 areas, if any, should be determined. That means to be identified the habitats' potential impacts on land which has high biodiversity value. One main and more distinct impact of wind turbines on the natural environment is the potential for birds to be struck by the blades. It should be noted here that the rotation speed of modern wind turbines is relatively low, which significantly reduces the risk of collisions. Wind turbine structures are complex, especially those offshore. Floating wind farms require huge mooring and anchoring systems that can affect the integrity of the seabed. Marine mammals can collide with or become entangled in the wires. In addition, wind turbines can change the dynamics of local atmospheric and ocean currents, and this can affect plankton production in the sea that is normally low in nutrients. Environmentalists have passed a law banning the construction of wind farms in the vicinity of bird and wildlife habitats, as studies have shown the harm of wind turbines to wildlife.

The location of wind farms has the potential for direct and/or indirect impacts on the historic environment due to the location of the turbines, which may affect archaeological sites. The proximity of the wind farm location to existing archaeological heritage should be assessed. Historic environment includes ancient monuments, archaeological sites and landscapes, historic buildings, gardens, and designed landscapes and their context and setting. The assessment should include direct impacts on the integrity and visual appeal of monuments where appropriate.

The social-environmental criteria refer to the distance of wind farms from residential areas such as towns and cities and the noise of the wind farm. In some countries, the installation of wind turbines close to residential areas is prohibited due to noise and visual disturbances. All projects for wind plant developments must be accompanied by a site-specific noise impact assessment undertaken. All noise impact assessments must determine the predicted noise levels at the curtilage of identified noise-sensitive receptors in the vicinity of the location of wind farm development. It should be noted that unlike onshore turbines, which are designed to operate at peak speed ratios less than optimal for reducing noise pollution, offshore systems can be designed to operate at higher speeds, sometimes 10% higher, resulting in better aerodynamic efficiency.

When evaluating the location of the wind farm, it must take into account its proximity to aviation radars and navigation systems. Wind turbines can interfere with radar signals from aviation, and this fact leads to the requirement to provide a larger buffer zone near airports. The same is the situation for proximity to the areas of defense interests, seismological recordings, telecommunications, and broadcasting installations.

It is also important to determine whether the protection of these areas is compatible with the placement of wind turbines and how the landscapes will develop in the future. Although the noise caused by the movement of the blades near a wind farm is modest, it cannot be prevented completely. Nevertheless, operating noise above regulatory levels from the most powerful wind turbines does not extend further than 700 m from the turbine even under the most conservative assumptions. For noise protection, sufficiently large buffer zones should be provided around residential areas.

The development of wind energy in countries with many decades of experience has brought the visual aspects of wind farms to the fore. People in more densely populated countries with a developed wind energy industry are more tolerant of wind turbines, but the perceptions of others are influenced by the potential economic benefits and/or environmental costs (Guo et al., 2015). The impact on the landscape is the prime concern of those living close to operating turbines, and they generate the strongest objections to wind farm developments (Botterill & Cockfield, 2016). Wind turbine arrays also affect visibility and regular arrays are considered to be better received than irregular ones. The need for visual assessment has been highlighted as a vital requirement to gain public support and reduce potential risks and impacts (Moller, 2006). In this aspect, it is shown that the visibility of the turbine varies in different viewpoints from residential buildings, protected areas, natural and archaeological sites, tourism development centers, roads, and coastal areas (Alphan, 2021). All of these consideration mean that the site selection process should include visibility as a criterion in determining a suitable wind farm location that meets both economic and environmental criteria, feasibility criteria, and visibility criteria.

In conclusion, it can be summarized that during the selection process of wind sites, some factors can be seen as restrictive and others as encouraging. Restrictive factors reduce the suitability of wind farms in the desired area, while positive factors contribute to the efficiency of the wind site. The authors found 20 restrictive factors and 30 relevant factors (Rediske et al., 2021).

2.3 Model for Selection of Wind Site Location Considering Multiple Technical, Economic, and Environmental Indicators

One of the most important decisions when starting a new investment is selecting a suitable onshore/offshore location requiring extended effort. These efforts are related to the decision-making process at the first stages of a wind farm investment. In this regard, the above-described groups of indicators could be used to assess different locations of wind power plants. To consider simultaneously all of these indicators, the following combinatorial mathematical model is proposed:

$$\max\left\{\sum_{i}^{N} x_i\left(\alpha\sum_{t}^{T} w_t^i p_t^i + \beta\sum_{c}^{C} w_c^i p_c^i + \gamma\sum_{e}^{E} w_e^i p_e^i\right)\right\} \tag{2.1}$$

subject to

$$\sum_{i}^{N} x_i = 1 \tag{2.2}$$

$$\alpha + \beta + \gamma = 1 \tag{2.3}$$

$$\sum_{t} w_i^t = 1 \tag{2.4}$$

$$\sum_{j} w_j^c = 1 \tag{2.5}$$

$$\sum_{k} w_k^e = 1 \tag{2.6}$$

In this model, x_i are binary integer variables assigned to each wind site location, coefficient α expresses the importance of criterion related to technical indicators, and coefficient β expresses the importance of criterion related to economic indicators, while coefficient γ is related to the importance of criterion about environmental indicators. The coefficients w_t^i, w_c^i, and w_e^i express the relative importance of indicators among each group (technical, economic, and environmental). The parameters p_t^i, p_c^i, and p_e^i express the performance of the corresponding indicator toward i-th alternative, i.e., corresponding location for building a wind power plant. The performance of each indicator is expressed by an evaluation of DM that shows how the particular indicator for the given alternative is performed. These evaluations are to be done using the range from 0 to 1, where the bigger value means better indicators' performance.

The objective function (2.1) seeks the maximum performance of wind site location considering three groups of criteria. The selection of the best suited location is realized by the relation (2.2) that provides a single choice. Using relation (2.3) it is possible to set the desired direction when determining the best location. The sum of these coefficients related to the technical, economic, and environmental criteria needs to be equal to 1. This is a consequence of the used ranges for the determination of the indicators' performance. Relations (2.4), (2.5), and (2.6) set a range for the weighted coefficients expressing the importance between the indicators of each group of criteria. In this way, the same range of variation is achieved for all parameters involved in the objective function.

2.3.1 Results Analysis

The proposed groups of criteria related to technical, economic, and environmental with corresponding sub-criteria expressed by the above-described indicators along with the mathematical model (2.1)–(2.6) could be used to determine the best site location for building wind plants. To verify the applicability of the modeling approach thus proposed, the following sub-criteria (indicators) for each group of criteria are used as shown in Table 2.1.

These criteria and their sub-criteria are used to evaluate the locations and potential of 3 wind power plant sites. The first of them (Site-1) is located near a populated place, the second (Site-2) is located high in the mountains, and the third (Site-3) is offshore. To assess the overall performance of these locations for wind power plants using the indicators in Table 2.1, it is necessary to determine coefficients for the

Table 2.1 Gropus of criteria and sub-criteria

No	Criteria groups	No	Sub-criteria (Indicators)
1	Technical	1.1	Wind resources
		1.2	Land area
2	Economic	2.1	Ground conditions
		2.2	Estimated capacity
		2.3	Distance to the power line
		2.4	Distance to road and feasibility of access for abnormal loads
		2.5	Agreements with land landowners
3	Environmental	3.1	Proximity to natural and built heritage
		3.2	Natural protected areas and archaeological heritage
		3.3	Proximity to centers of population
		3.4	Proximity to airports
		3.5	Proximity to defense interests areas and seismological recording
		3.6	Proximity to telecommunications and broadcasting installations

Table 2.2 Relative importance of sub-criteria and their performance

No	Sub-criteria (Indicators)	Relative importance between sub-criteria	Performance evaluations		
			Site-1	Site-2	Site-3
1.1	Wind resources	0.50	0.79	0.88	0.99
1.2	Land area	0.50	0.87	0.78	1
2.1	Ground conditions	0.20	0.92	0.68	0.6
2.2	Estimated capacity	0.20	0.7	0.6	1
2.3	Distance to the power line	0.20	0.8	0.62	0.5
2.4	Distance to road and feasibility of access for abnormal loads	0.20	0.88	0.67	0.96
2.5	Agreements with land landowners	0.20	1	1	1
3.1	Proximity to natural and built heritage	0.14	0.61	0.69	1
3.2	Natural protected areas and archaeological heritage	0.14	0.44	0.41	1
3.3	Proximity to centers of population	0.14	0.54	0.63	0.9
3.4	Proximity to airports	0.14	0.79	1	1
3.5	Proximity to defense interests areas and seismological recording	0.14	0.42	0.9	0.98
3.6	Proximity to telecommunications and broadcasting installations	0.15	0.68	0.78	0.43
3.7	Visibility in sensitive areas	0.15	0.38	0.46	0.92

relative importance between technical, economic, and environmental sub-criteria. Along with these coefficients, the evaluation performance for each sub-criteria is to be done. The summarized information about the relative importance of sub-criteria and their performance for three wind site locations is shown in Table 2.2.

The evaluations for technical, economic, and environmental performance are done using a scale from 0 to 1 where the bigger value means better performance of the evaluated indicator. It should be noted that the proposed model could integrate with other measurement units. For example, the Absolute Category Rating method could be used as it relies on the usage of a five-point overall quality rating scale (bad, poor, fair, good, excellent), which is the most widely used scale, or if required using a higher nine-point scale. Regardless of whether the five-point or the nine-point scale will be used, it is necessary to normalize the scores obtained in this way to the interval from 0 to 1. This is mandatory to provide dimensions similar to the other coefficients of the proposed model in order to obtain reliable results.

Last but not least, three additional coefficients related to the importance of technical, economic, and environmental performance need to be determined. Four different scenarios that correspond to four cases are investigated using different values for the importance of these groups of criteria. Case-1 considers technical, economic, and environmental performance with equal importance, while Case-2 gives more importance to technical performance indicators over the others. Case-3 and Case-4 emphasize on technical and environmental performance as shown in Table 2.3.

Table 2.3 Coefficients for the different scenarios

	Technical criteria	Economic criteria	Environmental criteria
Case-1	0.33	0.33	0.34
Case-2	0.60	0.10	0.30
Case-3	0.40	0.20	0.40
Case-4	0.50	0	0.50

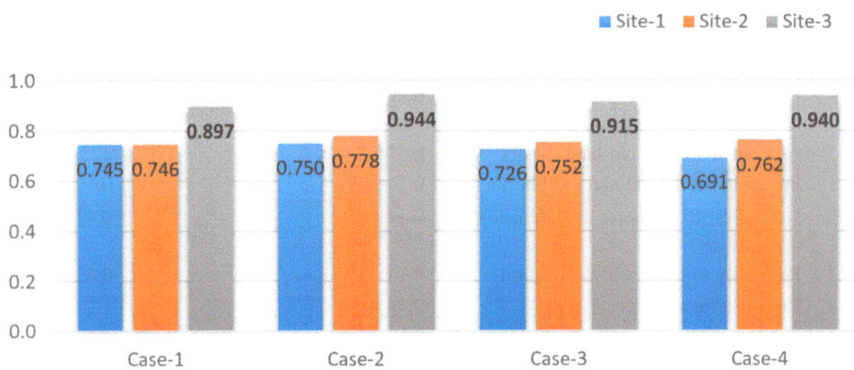

Fig. 2.2 Selection and ranking of wind sites locations

The input data from Table 3.2 and Table 2.3 are used to verify the applicability of the proposed optimization model (2.1)–(2.6) in evaluating the performance and selection of the best site location for wind power plant building according to the used technical, economic, and environmental indicators.

The obtained results of the conducted evaluation and selection of the best wind site location are visualized in Fig. 2.2.

The results of Case-1 where equal values are used for the coefficients for technical, economic, and environmental performance expressed by α, β, and γ ($\alpha = \beta = \gamma = 0.33$) along with the equal importance between indicators of these three groups of indicators show that Site-3 has a better performance than others. This is due to the obtained overall assessment being equal to 0.897 for Site-3 compared to the value of 0.746 for the assessment of Site-2 and 0.745 for Site-1.

Case-2 simulates the situation where more attention is given to the technical indicators as a whole (see Table 2.3). In this case, the best performance is also obtained for Site-3 with the overall assessment value equal to 0.944 compared to the value of 0.778 for the assessment of Site-2 and 0.75 for Site-1. A similar ranking is observed for Case-3 and Case-4 where the best performance is obtained for the Site-3 location.

It should be noted that this best performance of Site-3 is for an offshore location. Despite the large costs required to build an offshore wind farm, this location appears to be the best compared to the other two and according to the evaluation criteria and indicators used. In this regard, it is interesting to compare all three locations

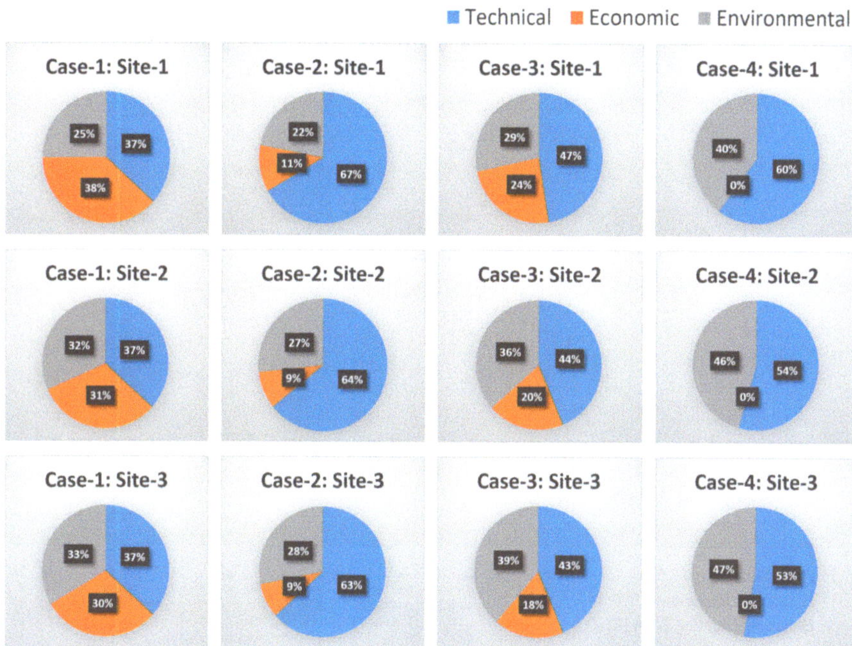

Fig. 2.3 Comparison of the locations in relation to the realized scenarios and the groups of criteria

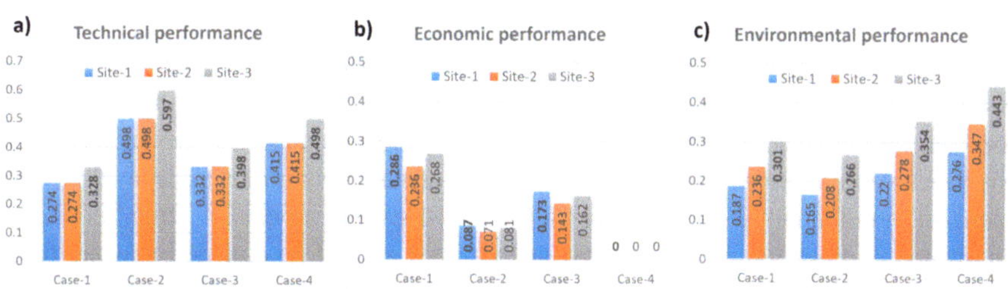

Fig. 2.4 Performance of the potential locations of wind power plant: (**a**) technical, (**b**) economic, and (**c**) environmental performance

according to the four simulated scenarios. Such a comparison of the locations in relation to the realized scenarios and the groups of criteria used is shown in Fig. 2.3.

The overall performance of a particular site's location is done by using a different number of indicators about three main groups of criteria—technical, economic, and environmental. It could be noticed from Fig. 2.3 that some locations have better performance for some criteria, and it is interesting to show the particular site's location performance according to used evaluation criteria. The corresponding performances of locations of wind power plants toward technical, economic, and environmental criteria are shown in Fig. 2.4.

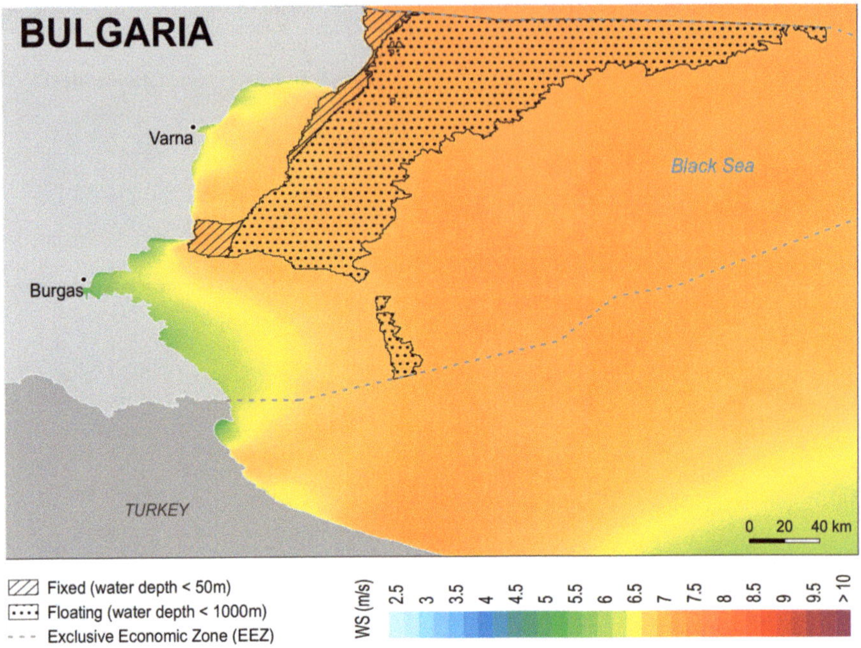

Fig. 2.5 Technical potential for fixed and floating offshore wind in Bulgaria. Source: http://documents.worldbank.org/curated/en/421291586847341111/pdf/Technical-Potential-for-Offshore-Wind-in-Bulgaria-Map.pdf

The indicators for wind resources and land area show that the best technical performance site location is Site-3 for all four cases (Fig. 2.4a). This is an expected result as offshore wind speeds are significantly higher than for the equivalent location on land due to the lack of land mass obstructions and lower water surface roughness compared to land features such as forests. This initial high-level assessment of wind conditions and available area does not consider other technical, environmental, social, or economic constraints. It should be clarified that the fixed offshore wind power plant is suitable for water depths of less than 50 m, while floating wind farms are suited for water depths between 50 and 1000 m. For offshore wind farms, only regions less than 200 km from shore are considered.

The estimated technical potential for fixed and floating offshore wind in Bulgaria in terms of installed power capacity in megawatts within 200 km of the shoreline is shown in Fig. 2.5.

Economic performance indicators show that the best location for a wind power plant building is Site-1 for all 4 cases (Fig. 2.4b). Case-2 and Case-3 also determine the location of Site-1 as the best one with the value of 0.087 for Case-2 and 0.173 for Case-3. It is important to note that behind the scenario of Case-4, the economic indicators are not taken into the overall evaluation of wind plant locations. Therefore, the results for the economic evaluation, in this case, are equal to zero (Fig. 2.4b).

Why is the location of Site-1 preferable to the others when only economic indicators are considered? The economic performance envelops several indicators such as ground conditions, estimated capacity, distance to the power line, distance to the road and feasibility of access for abnormal loads, and agreements with land landowners. The geographic location of Site-1 is located close to populated areas, unlike Site-2, which is located in the mountains, and Site-3 offshore. Building a wind power plant close to populated areas can benefit from better ground conditions, lower distance to connect to the power line, lower distance to the road, and easy access for abnormal loads. The better performance of these indicators leads to better economic performance of the location of Site-1.

Considering only environmental performance shows that the best choice is for the location of Site-3 for all four cases (Fig. 2.4c). This is an expected result as the offshore wind farm is far from natural and built heritage, centers of population, and in most cases natural protected areas and archaeological heritage, the airports and defense interests areas, and seismological recording, telecommunications, and broadcasting installations and has less impact on visibility in sensitive areas. All of these indicators have very little impact on the environmental performance of the offshore wind farm site location.

It should be noted that the results obtained are done by using the same relative importance between sub-criteria/indicators for each group (see Table 2.2). If it is necessary to emphasize some indicators or ignore others, these coefficients should be modified to reflect the specific requirement regarding wind farm site selection indicators.

The proposed optimization model could be easily modified by adding an additional group of criteria for the selection of wind power plant location. Also, in each group of criteria, it is possible to change the number of indicators to more precisely selection for a suitable location for a wind farm. For example, some additional indicators could be related to economic policies, local laws and regulations, or possibilities for public–private partnership participation.

In conclusion, the first step in determining the location for the construction of a wind farm is to assess and analyze the wind resource. It is important to find a site with good wind conditions, but a good wind site is not sufficient. On the other hand, it is needed to consider the distance to and capacity of the nearby transmission grid, environmental aspects, and access to land to identify the most suitable location to start an onshore/offshore wind energy project. Transport traffic is another key constraint both onshore and offshore. The transportation and access refer to both the construction and operation phases. Another essential factor that has to be considered is which turbine technology is best suited for the given site and wind conditions. Last but not least is social acceptance and public–private partnership participation in wind energy projects. Green energy public–private partnership projects play an important role not only in the development of green energy but also in achieving environmental benefits by contributing to the construction of the necessary infrastructure (Liu et al., 2021). It is found that public–private partnerships that are smaller and younger, where the private partner takes more responsibilities, and the prime revenue source comes from the electricity consumers' payments, show a greater degree of private investors' participation (Fleta-Asin & Munoz, 2021). Thus, these three factors prove to be of essential importance in implementing projects for constructing wind power plants.

2.4 Pros and Cons in Location Selection for Onshore and Offshore Wind Farms Building

Wind energy as part of the European Green Pact plays an essential role, and therefore, it is important to outline some main advantages and disadvantages related to the construction of onshore and offshore wind farms.

Apart from the obvious sustainability advantages, the benefits of onshore wind power plants can be summarized as follows:

- *Less expensive*—The infrastructure required to build an onshore wind farm is significantly cheaper than that required for an offshore wind farm. Due to the good profitability, onshore wind farms are usually larger, they produce more energy per site, and the investment can pay for itself in as little as 2 years.
- *Quicker installation*—Not only is the installation of onshore wind farms cheaper, but it is quicker too. Onshore wind farms can be constructed and ready for commercial operations in a matter of months. They require shorter cables due to the smaller distance between the turbines and the consumer; there is less voltage drop in the wiring. During their operation, onshore wind turbines have low maintenance costs compared to offshore wind turbines.
- *Stimulates the local economy*—Onshore wind power plant companies looking for the best location for wind turbines will consider the benefits their farms can bring to the local economy. Economic value can be expressed by the income generated for onshore landowners and jobs created for people in the community.

Along with the advantages, some disadvantages of land-based wind power plants should be noted that are related to:

- *Wind speed variability*—Wind speed and direction on land vary and achieving consistent power generation can be challenging. Therefore, the wind speed and direction are to be carefully monitored in order to estimate the energy generation well enough. Exactly how large the variation depends both on local surface conditions and obstacles and on the weather, including the change of day and night and the change of seasons.
- *Potential physical wind blockages*—Physical blockages are due to the presence of buildings and surrounding landscapes such as hills or mountains, which can cause inconsistencies in energy production. For this reason, onshore wind cannot produce power all year round and can only achieve about 2.5 MW, compared to offshore wind's approximately 3.6 MW.
- *Visual and sound factors*—Landscape perception and visual impact are key indicators in the selection of a location for a wind power plant building. In

order to generate more energy, it is necessary for the wind turbines to be positioned high, which in turn leads to an imposing visual effect on the surrounding residential areas. It should also be mentioned that in operation wind turbines cause noise pollution if they are located near a residential area. On average, large land-based wind turbines produce sounds that fall in the 35–45 dB range when heard from 300 m away. That means they create far less noise pollution than average city car traffic. In addition, multiple laboratories are coordinating activities to address wind turbine radar interference.

In summary, the advantages of onshore wind and the sustainable energy it can create outweigh the listed potential disadvantages.

The pros of offshore wind power plants can be summarized as follows:

- *More energy generated*—Offshore wind farms have the potential to produce on average over 1 MW more than onshore per year. Unlike onshore, where turbines can be limited to a certain height, offshore turbines can be scaled much larger, allowing more energy to be captured. Offshore wind speeds are generally higher than onshore, and even small increases in speed can lead to large increases in wind power generation.
- *More efficient*—Offshore wind farms are considered more efficient than onshore in terms of the energy they can produce. This is due to the fact that offshore farms can generate the same amount of energy as onshore but with fewer turbines. Wind speeds offshore do not vary as much and the wind direction does not change as often, so offshore turbines are more consistent, i.e., which means more reliable energy can be generated. Offshore turbines can also be built taller than those onshore, which means they can harness more wind energy and produce more electricity.
- *Less visual impact*—Offshore turbines do not have as much visual impact as onshore turbines. Offshore wind farms do not interfere with neighboring countries, or land uses, or have buildings or obstructions in their path to interrupt wind flow. For these reasons, offshore wind farms can be larger and generate more energy than onshore ones, with less physical impact.

Along with the advantages, some disadvantages of land-based wind power plants should be noted that are related to:

- *Higher cost*—Setting up infrastructure for offshore wind farms can be expensive and complex, especially in deeper waters. When developing wind farm projects, the overall project costs are mostly accumulated during the construction phase due to the very expensive turbines, foundations, and transmission assets. Offshore projects tend to be generally more complex and around 2–3 times more expensive than onshore with higher percentages of other key infrastructure costs other than the turbine (Deloitte, 2014). Turbines and wind farms are growing in size and many are being built further offshore. Locations are often remote and conditions are harsh, especially given the increasing severity of weather events.
- *Maintenance and repairs*—Offshore constructions are prone to high technical risks through mechanical force, corrosion, and biofouling, so the expected

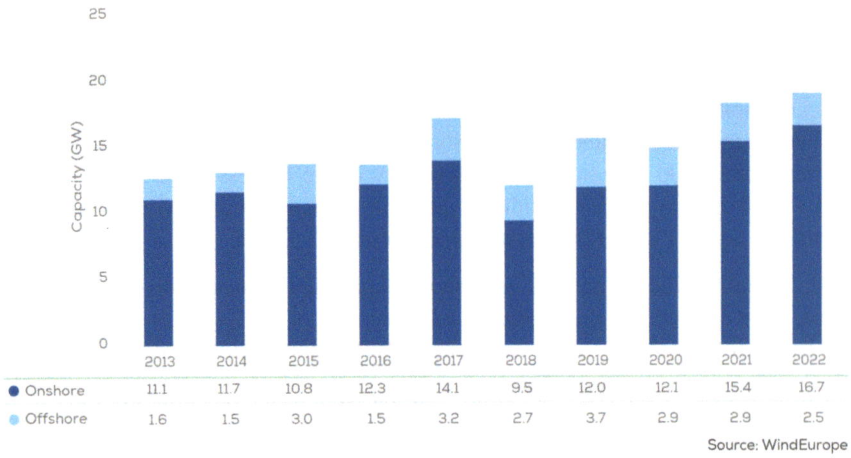

Fig. 2.6 Wind installations in Europe

lifetime of an offshore structure is to a great extent determined by the risk of failures (Rockmann et al., 2017). Sea waves and very strong winds can damage turbines, so they need more maintenance than their counterparts on land. Offshore wind farms are also difficult to access, meaning longer waiting times for repairs.

- *Noise and visibility*—The increase in low-frequency noise from offshore wind turbines raises concerns about the cumulative contribution of wind farms to the underwater soundscape and its possible impact on marine ecosystems (Tougaard et al., 2020). Also, not all offshore wind farms are built out of public view, so they can still be an eyesore for locals.
- *Fewer local jobs*—Unlike onshore wind farms, offshore wind farms have limited capacity to benefit local economies. As the manufacturer's offices are located inland and often far from the offshore site, no jobs are created in the local community and no other investment is made.

Europe now has 255 GW of installed wind power, of which 225 GW is onshore and 30 GW offshore. On the other hand, the EU-27 has 204 GW installed, of which 188 GW are onshore and 16 GW offshore. It should be mentioned that 87% of wind power plant installations in 2022 came from the onshore wind as shown in Fig. 2.6 (Costanzo et al., 2023).

The contribution of each EU-27 country in both onshore and offshore wind installations is shown in Fig. 2.7 (Costanzo et al., 2023).

It can be seen that the main contribution in wind energy is due to Germany with the most newly installed wind power capacity in 2022 with 2.7 GW, of which 2.4 GW onshore and 0.3 GW offshore. Sweden and Finland have the same installed wind capacity of 2.4 GW for 2022, both onshore. The next country with significant advances in wind energy is France which installed 2.1 GW of which 1.6 GW was onshore and 480 MW offshore. The investment of the UK during 2022 is mainly offshore wind power farms with an installed capacity of 1.7 GW and 0.5 GW

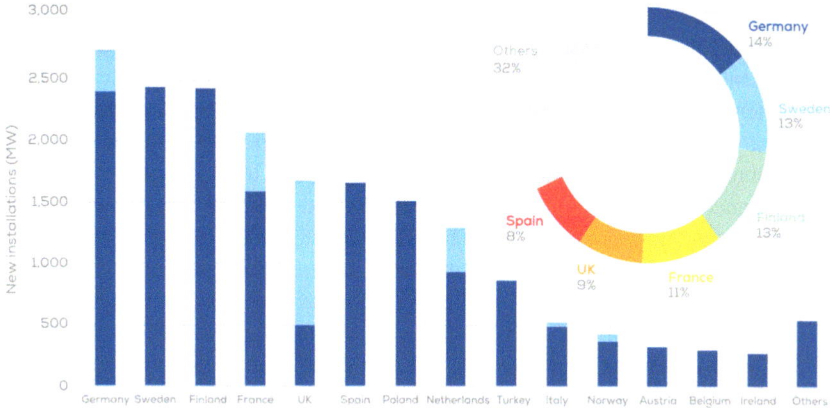

	Germany	Sweden	Finland	France	UK	Spain	Poland	Netherlands	Turkey	Italy	Norway	Austria	Belgium	Ireland	Others
Onshore	2,403	2,441	2,430	1,590	502	1,659	1,517	933	867	496	372	328	303	280	546
Offshore	342	0	0	480	1,179	0	0	369	0	30	60	0	0	0	0

Source: WindEurope

Fig. 2.7 Wind installations in Europe per country in 2022

installed capacity onshore. Spain installed 1.7 GW of onshore capacity, and Poland has installed an onshore wind capacity of 1.5 GW. The Netherlands installed a total of 1.3 GW, of which 0.9 GW onshore and 369 MW offshore. In 2022, Turkey has installed 867 MW only onshore.

Such an analysis is not only useful but also mandatory to determine in which country wind energy projects have or have not yet reached the required capacity. On the basis of this analysis, the potential countries and the potential wind sites on land or in the sea, where the construction of a new wind power plant begins, can now be considered. In the presence of more than one potential country, it is necessary to study the specific regulatory documents and indicators for doing business. These indicators of doing business can be used as criteria for ranking countries (economies) to choose the most suitable one to invest in wind energy.

2.5 Ranking of Economies for Investment

Many countries liberalize and promote foreign investment in various industries and introduce proper regulatory and restrictive measures. They adjust entry policies for local and foreign investors. Mobilizing investments and ensuring sustainable development is a priority for all countries. More and more governments pursue a broader and more intricate development policy agenda while building or maintaining a generally favorable investment climate.[1]

[1] Investment Policy Framework for Sustainable Development, United Nations Conference on Trade and Development—UNCTAD, http://unctad.org/en/PublicationsLibrary/diaepcb2012d5_en.pdf

With globalization, the competitiveness of a territory (or country) is acquiring a very important role. When a particular territory is competitive it is also attractive for investments. In particular, the investor will try to benefit from those assets and competencies that make the territory competitive. Such a territory attracts capital and knowledge, and this implies a comparison of local enterprises with other external realities. As a result, this comparison stimulates local enterprises to improve and enhance their competitiveness and to take international investment decisions (Baltzer et al., 2013). Financial managers need to make decisions under many restrictions and often have to deal with vague and imprecise information. Therefore in such situations, fuzzy logic could provide an adequate solution to a multicriteria investment problem (Popchev & Georgieva, 2008; Georgieva & Popchev, 2013).

The decision-making for investing in a particular country depends largely on the current and anticipated economic situation but also depends on some political and social factors. An investment climate also includes economics location-specific factors (incentives and opportunities) that determine if investors are willing to risk their capital in financing a particular project.[2]

2.5.1 *Multicriteria Combinatorial Optimization Model for Ranking of Economies with Investment Purpose*

A well-known approach for defining investment attractiveness is ranking of the economics (countries) in accordance with different economic, political, social, etc., indicators. Most of the published results of ranking approaches rely on statistical averages, surveys of executives, and experts' estimations used as input variables key indicators[3,4] (Mia et al., 2007; Groh et al., 2013). In contrast, a generalized multicriteria combinatorial optimization model for ranking economies could be used (Mustakerov & Borissova, 2013):

$$\begin{cases} \max\{a_1, a_2, \ldots, a_j\} \\ \min\{b_1, b_2, \ldots, b_k\} \end{cases} \tag{2.7}$$

subject to

[2]Investment Climate—data and indicators, World Bank Group (US), International Finance Corporation, https://www.wbginvestmentclimate.org/research-and-diagnostics/

[3]Back to Business: Optimism Amid Uncertainty. The 2013 A.T. Kearney Foreign Direct Investment Confidence Index®, http://www.atkearney.com/research-studies/foreign-direct-investment-confidence-index

[4]Global cities investment monitor 2012 new rankings, trends and criteria. Greater Paris Investment Agency, http://www.kpmg.com/FR/fr/IssuesAndInsights/News/Documents/GPIA-KPMG-CIM-2012.pdf

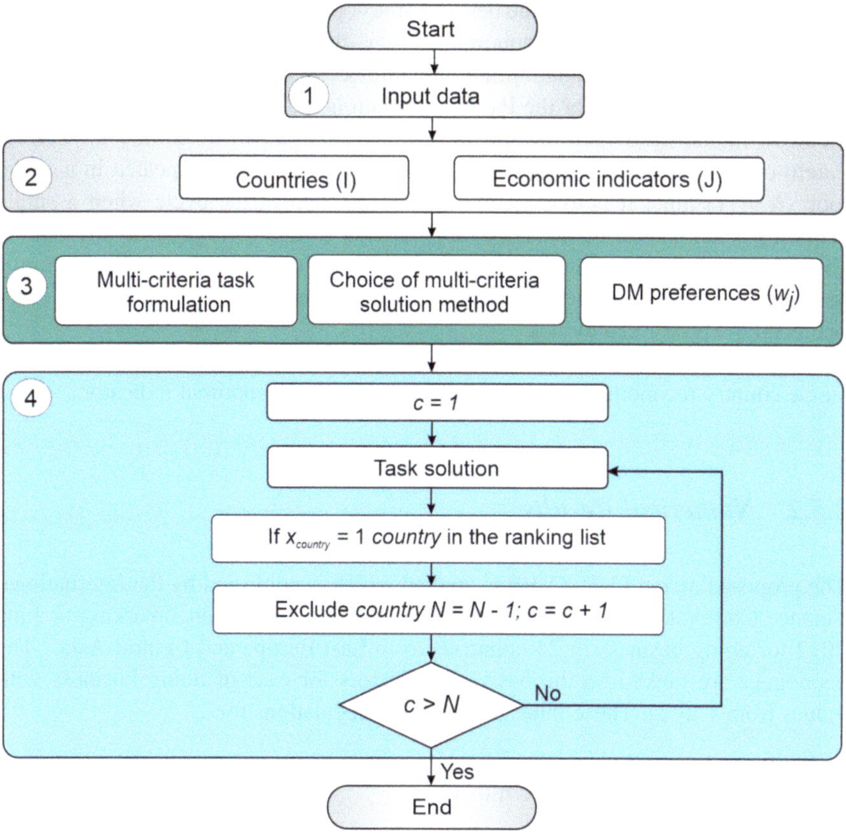

Fig. 2.8 Algorithm for countries ranking

$$\forall j = 1, 2, \ldots, J : a_j = \sum_i^I a_j^i x_i \tag{2.8}$$

$$\forall k = 1, 2, \ldots, K : b_k = \sum_i^I b_k^i x_i \tag{2.9}$$

$$\sum_i^N x_i = 1, x_i \in \{0, 1\} \tag{2.10}$$

where J is the number of the objective functions (economic indicators) a that have to be minimized, K is the number of objective functions (economic indicators) b that have to be maximized, x_i are binary integer decision variables, and N is the number of economics (countries) that are to be ranked.

The basic idea of the proposed algorithm for ranking is to solve sequentially (N – 1) multicriteria optimization tasks as shown in Fig. 2.8.

The first task solution gives the best in Pareto-optimal selection of country and the ranking list starts with this economics. Next, this economics is excluded from the optimization task formulation; i.e., the number of countries (economics) where the new search will start for the Pareto optimum is reduced from N to $(N - 1)$. The modified multicriteria optimization task is solved again to determine the second Pareto-optimal economics in the ranked list. This procedure is repeated in a cyclic loop $(N - 1)$ time. It is evident that in the last step of the cycle when a single economy is left and the choice is obvious the task solution will provide information about the objective function value that could be useful in some cases. For example, the objective function values can be used as an estimation of how far from the best one is the particular economics.

The solution of the multicriteria optimization problem (2.7)–(2.10) determines a single country (economics) with a combination of Pareto-optimal indicators.

2.5.2 Numerical Results

The proposed model (2.7)–(2.10) is applied which is published by the International Finance Corporation, a member of the World Bank Group, benchmarking to June 2012 for doing business in 24 countries from East Europe and Central Asia.[5] The economies are ranked on the basis of indicators for ease of doing business with values from 1 to 24. These indicators estimate regulations for:

- Starting a business (1)
- Dealing with construction permits (2)
- Getting electricity (3)
- Registering property (4)
- Getting credit (5)
- Protecting investors (6)
- Paying taxes (7)
- Trading across borders (8)
- Enforcing contracts (9)
- Resolving insolvency (10)

The low value of the indicator for the ease of doing business means the regulatory environment is more conducive to the starting and operation of an enterprise. The numerical data of indicators for countries of Eastern Europe and Central Asia in alphabetical order are shown in Table 2.4.

The optimization model (2.7)–(2.10) is used to formulate corresponding tasks for ranking the economies toward the indicators from Table 2.4. A priory preference information of the DM point of view about the relative indicators' importance according to the weighted sum method is shown in Table 2.5.

[5] Economy rankings, The World Bank, 2013, http://www.doingbusiness.org/rankings

Table 2.4 Economic indicators for Eastern Europe and Central Asia countries

#	Country/economy	Indicators									
		(1)	(2)	(3)	(4)	(5)	(6)	(7)	(8)	(9)	(10)
1	Albania	14	24	14	22	6	3	20	9	17	10
2	Armenia	4	3	10	3	10	6	14	14	19	9
3	Azerbaijan	6	19	21	5	15	6	10	20	5	18
4	Belarus	3	2	20	2	21	16	17	18	2	7
5	Bosnia and Herzegovina	24	16	15	18	18	19	16	12	22	14
6	Bulgaria	11	9	13	14	10	11	13	10	18	17
7	Croatia	18	12	2	20	10	23	5	13	14	19
8	Cyprus	8	7	9	19	15	9	3	2	21	1
9	Georgia	2	1	1	1	1	4	4	4	8	13
10	Kazakhstan	7	15	7	8	19	1	1	22	7	6
11	Kosovo	23	13	12	16	6	19	6	15	24	15
12	Kyrgyz Republic	5	6	22	6	4	2	23	21	13	23
13	Latvia	13	8	8	9	1	13	7	1	4	2
14	Lithuania	22	4	5	4	15	13	8	3	3	3
15	Macedonia	1	5	10	13	6	4	2	7	15	8
16	Moldova	20	17	16	7	10	16	15	16	6	16
17	Montenegro	12	18	4	21	1	9	12	5	23	4
18	Romania	15	10	19	15	4	11	18	6	16	20
19	Russian Federation	21	20	24	12	21	21	9	19	1	5
20	Serbia	9	21	6	10	10	16	19	11	20	21
21	Tajikistan	17	22	23	17	24	6	24	23	11	12
22	Turkey	16	11	3	11	19	13	11	8	9	22
23	Ukraine	10	23	17	24	6	21	22	17	10	24
24	Uzbekistan	19	14	18	23	23	23	21	24	12	11

Table 2.5 DMs' preference toward the criteria/indicators

DMs	w1	w2	w3	w4	w5	w6	w7	w8	w9	w10
DM-1	0.10	0.10	0.10	0.10	0.10	0.10	0.10	0.10	0.10	0.10
DM-2	0.15	0.05	0.05	0.05	0.05	0.10	0.15	0.10	0.05	0.25
DM-3	0.05	0.25	0.05	0.25	0.05	0.05	0.10	0.05	0.10	0.05

The preferences of DM-1 correspond to the equivalent importance of all economic indicators. The preferences of DM-2 illustrate the dominant importance of the resolving insolvency indicator. The preferences of DM-3 express strong preferences about the indicators dealing with construction permits and registering property.

As a result of the algorithm performance, three lists with ranking according to different DMs' preferences are obtained as shown in Table 2.6.

It is interesting to compare the results obtained by using the proposed modeling approach with already published data. For the goal, the ranking of the countries as a

Table 2.6 Countries rankings according to DMs's preferences

Ranking in accordance with DM-1		Ranking in accordance with DM-2		Ranking in accordance with DM-3	
1	Georgia	1	Georgia	1	Georgia
2	Latvia	2	Macedonia	2	Lithuania
3	Macedonia	3	Latvia	3	Belarus
4	Lithuania	4	Cyprus	4	Latvia
5	Kazakhstan	5	Kazakhstan	5	Macedonia
6	Cyprus	6	Lithuania	6	Armenia
7	Armenia	7	Montenegro	7	Kazakhstan
8	Montenegro	8	Armenia	8	Kyrgyz Republic
9	Belarus	9	Belarus	9	Cyprus
10	Turkey	10	Azerbaijan	10	Turkey
11	Azerbaijan	11	Albania	11	Azerbaijan
12	Kyrgyz Republic	12	Bulgaria	12	Bulgaria
13	Bulgaria	13	Russian Federation	13	Moldova
14	Romania	14	Turkey	14	Romania
15	Croatia	15	Croatia	15	Croatia
16	Albania	16	Kyrgyz Republic	16	Russian Federation
17	Moldova	17	Romania	17	Kosovo
18	Serbia	18	Kosovo	18	Belarus
19	Kosovo	19	Moldova	19	Serbia
20	Russian Federation	20	Serbia	20	Bosnia and Herzegovina
21	Bosnia and Herzegovina	21	Tajikistan	21	Albania
22	Ukraine	22	Bosnia and Herzegovina	22	Uzbekistan
23	Tajikistan	23	Uzbekistan	23	Tajikistan
24	Uzbekistan	24	Ukraine	24	Ukraine

result of the algorithm execution in accordance with the preferences of DM-1 and the ranking according to the World Bank[6] are graphically illustrated in Fig. 2.9.

The explanation for the differences in Fig. 2.9 is that ranking averages the country's percentile rankings on 10 indicators while giving equal weight to each topic. In general, the average results of each country are fractions and they are rounded to get the index (position in the list) of that country. For example, the averages for Macedonia and Latvia are 7.1 and 6.6. When these averages are rounded, both countries have an index equal to 7 and Macedonia is positioned above Latvia or vice versa on an equal basis. The solution of the formulated multicriteria optimization task defines values of aggregated objective function equal to 0.2434783 for the case of Latvia and 0.2652174 for Macedonia. This means Latvia has a better objective function value and should be ranked above Macedonia.

[6] Economy rankings, The World Bank, 2013, http://www.doingbusiness.org/rankings

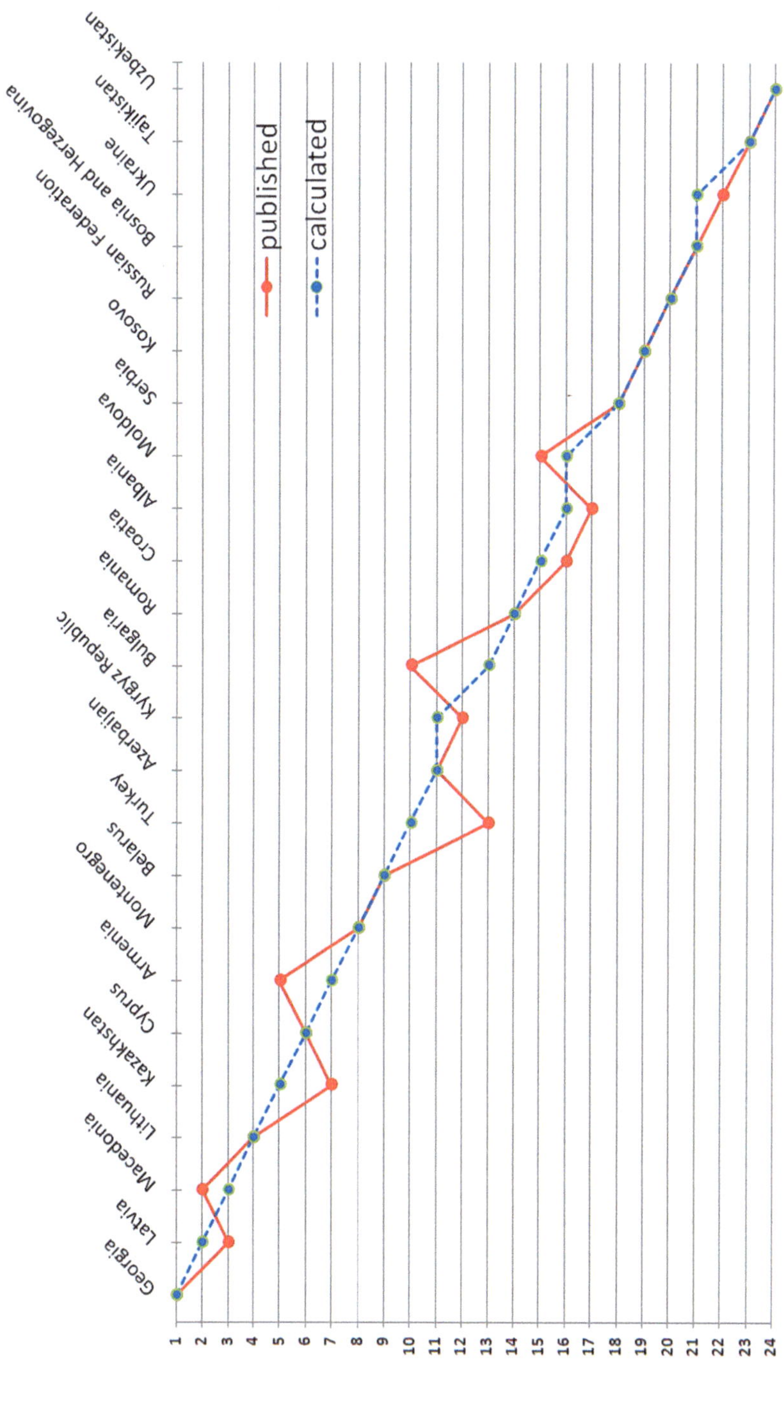

Fig. 2.9 Comparison of results for equal weights of all indicators

It is possible to add or exclude some indicators when looking for an economy to invest in, also taking into account some specific provisions to promote investments in renewable or wind energy or the possibility of public–private investments. In the presence of such circumstances, the proposed model can be easily modified by including additional indicators. This way we can make an informed decision about the particular investment or not. For example, to realize a project such as a wind energy project, the abovementioned 10 indicators can be reduced to 5 related to starting a business, dealing with construction permits, registering property, paying taxes, and enforcing contracts. Only these indicators could provide a sufficient preliminary assessment for starting or not a similar investment project. That is why these indicators were chosen as criteria in the numerical testing aiming to rank some countries of the Organisation for Economic Co-operation and Development (OECD). The scores according to the World Bank[7] data for the considered indicators are shown in Table 2.7.

The preferences of the DM toward the importance of these indicators are shown in Table 2.8.

Taking into account the optimization model (2.7)–(2.10) and DM's preferences regarding the evaluated indicators, the following ranking is obtained, as shown in Fig. 2.10.

The first four countries in the ranked list are Estonia with a value of 0.899, followed by Lithuania (0.852), Denmark (0.816), and Norway (0.801). This scenario illustrates how, by using the proposed multicriteria model (2.7)–(2.10), it is possible, as a second step, to determine a country where a wind farm can be built. The first step remains to determine a sufficient wind resource, on the basis of which the relevant wind power plant can be designed.

Having a wind energy project that takes into account the available resources and the chosen country for investment can move to the next stage, which refers to the decision to determine the appropriate seller (supplier) of the selected wind turbines.

2.6 Group Decision-Making for Supplier Selection

In contemporary economics, proper vendor (supplier) determination becomes a key strategic decision for the success of different business activities. This is true for the selection of the right vendor for wind turbines too. The supplier selection problem is recognized as a complex problem consisting of both quantitative and qualitative criteria (Sarkar et al., 2018; Danielson et al., 2016). In most cases, these evaluation criteria are in conflict. The availability of qualitative and quantitative criteria requires them to be evaluated simultaneously in the decision-making process (Tavana et al., 2016). To be more transparent, the selection process should involve a group of

[7] https://archive.doingbusiness.org/en/rankings?region=oecd-high-income

Table 2.7 Restricted set of OECD high-income countries

Economy	Starting a business	Dealing with construction permits	Registering property	Paying taxes	Enforcing contracts
Denmark	18	1	6	2	8
UK	6	10	21	15	17
Norway	9	9	7	18	2
Sweden	16	13	5	17	19
Lithuania	14	3	2	8	4
Estonia	5	8	3	5	5
Latvia	10	22	10	7	9
Finland	12	18	16	4	20
Germany	31	12	29	24	7
Ireland	7	15	26	1	29
Iceland	23	26	8	20	16
Austria	32	20	14	22	6
Spain	27	27	25	19	15
France	15	21	32	29	10
Switzerland	25	25	9	10	26
Slovenia	17	32	24	23	32
Portugal	22	23	17	21	18
Poland	33	16	30	32	24
Czech Republic	34	34	15	26	31
Netherlands	8	29	13	12	27
Slovak Republic	30	33	4	27	21
Belgium	19	19	33	30	25
Hungary	26	31	12	28	14
Italy	28	30	11	34	33
Luxembourg	24	6	31	13	12
Greece	4	28	34	31	34

Table 2.8 Restricted set of OECD high-income countries

#	Indicators	DM preferences
1	Starting a business	0.25
2	Dealing with construction permits	0.21
3	Registering property	0.12
4	Paying taxes	0.17
5	Enforcing contracts	0.25

experts with different skills, experience, and knowledge capable of evaluating all of the criteria (Borissova et al., 2016; Mustakerov & Borissova, 2014).

Supplier selection for the goal of public procurement is an important case of group decision-making where the spending of funds has to be public. An increased interest in public procurement has been observed over the last few decades, as the

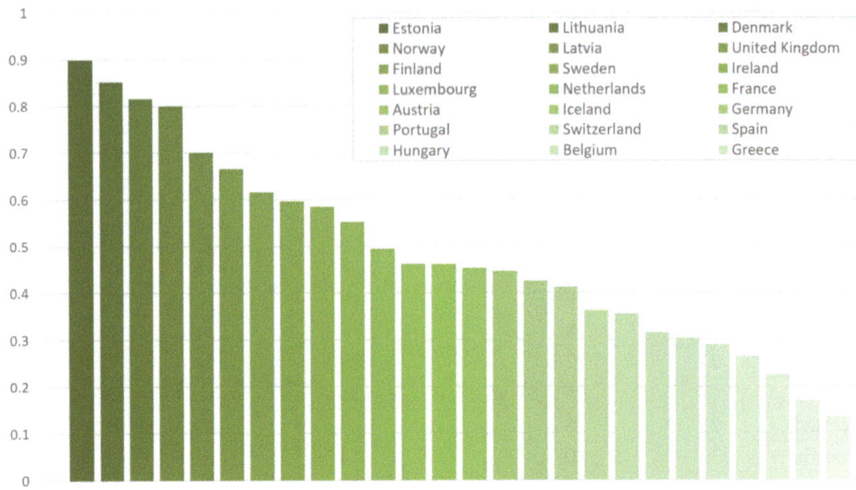

Fig. 2.10 Ranking of countries using different relative importance between indicators

purchase of goods by the public sector is increased. The public procurements are to be transparent and in accordance with legislation and administrative regulations and should follow particular public procurement procedures.

To tackle group decision-making, the following generalized algorithm that considers the differences in experts' knowledge and expertise could be used (Korsemov et al., 2018) as shown in Fig. 2.11.

The first stage concerns the description of the existing group decision-making problem. The second stage considers the determination of a set of acceptable alternatives appropriate to cope with the goal of the problem. The next stage is focused on the determination of important criteria for the evaluation of the alternatives. Stage 4 deals with the determination of a relevant group of competent experts in the area of the problem. In the 5th stage, the corresponding weighted coefficients for the experts' knowledge and experience are to be set. The next stage requires the determination of relative importance between criteria by assigning corresponding weighted coefficients in accordance with the point of view of each expert.

In the 7th stage, the evaluation of alternatives toward criteria from all experts is to be done. At this stage, normalization may be needed depending on the units of dimension for evaluation criteria. The common approach is to apply normalization to transform experts' evaluations in the range of 0 and 1 in dimensionless units.

Stage 8 concerns the choice of proper utility function and formulation and solving of the corresponding optimization task. This task should incorporate the information from the above stages including (1) weights for experts from the group, (2) performance of alternatives in accordance with the evaluation criteria, and (3) weights for relative importance of the evaluation criteria. In the last stage, the solution of the optimization task will give information for the best alternative performance considering the different points of view of the experts.

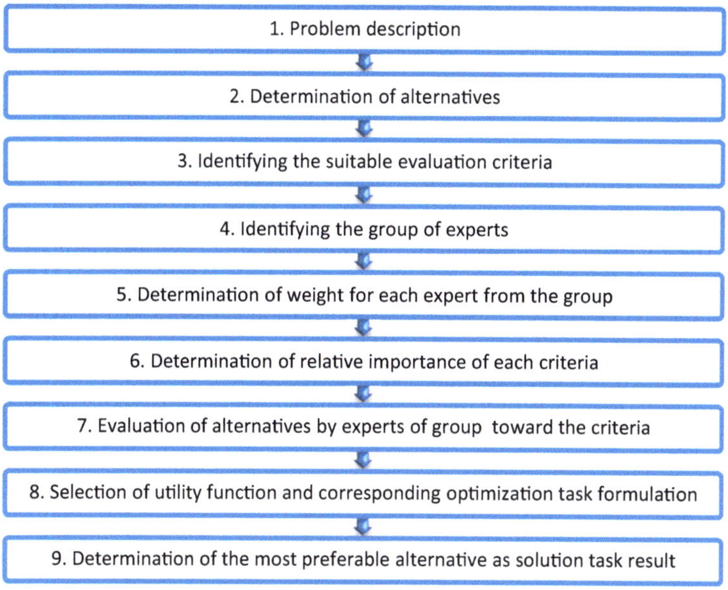

Fig. 2.11 Generalized algorithm for group decision-making

2.6.1 Combinatorial Optimization Model for Group Decision-Making in Selection of Supplier

Two types of utility functions based on SAW and WPM are used to define corresponding combinatorial optimization models for group decision-making. The proposed mathematical pure integer linear model based on a modification of SAW is as follows (Korsemov et al., 2018):

$$\max\left\{\sum_{i}^{M}\sum_{k}^{K}\sum_{j}^{N} w_j^k a_{i,j}^k \lambda^k x_i\right\} \tag{2.11}$$

subject to

$$\sum_{j}^{N} w_j^k = 1, \forall k = 1, 2, \ldots, K \tag{2.12}$$

$$\sum_{k}^{K} \lambda^k = 1 \tag{2.13}$$

$$\sum_{i}^{M} x_i = 1 \tag{2.14}$$

$$x_i \in \{0, 1\} \tag{2.15}$$

where w_j^k are coefficients for relative importance between criteria for k-th expert point of view, $a_{i,j}^k$ represents the i-th alternative evaluation to j-th criterion from k-th expert. The following sets used in this formulation are (1) set of alternatives to get the final selection $\{i = 1, 2, \ldots, M\}$; (2) set of evaluation criteria $\{j = 1, 2, \ldots, N\}$; (3) set of experts $\{k = 1, 2, \ldots, K\}$; (4) set of decision variables $\{x_i\}$ assigned to each alternative; and (5) set of weighted coefficients for experts $\{\lambda^q, q = 1, 2, 3, \ldots, K\}$ representing their level of expertise and knowledge.

The modification of WPM for group decision-making is as follows (Korsemov et al., 2018):

$$\max\left\{\sum_{i}^{M}\sum_{k}^{K} x_i \lambda^k \prod_{j}^{N}\left(a_{i,j}^k\right)^{w_j^k}\right\} \tag{2.16}$$

subject to the same restrictions (2.12)–(2.15).

Here multiplication $\prod_{j=1}^{N}\left(a_{i,j}^k\right)^{w_j^k}$ expresses the performance of i-th alternative toward j-th evaluation criteria according to k-th expert's opinion.

2.6.2 Numerical Results

Three offers from different vendors (suppliers) for the delivery of wind turbine components are considered as three possible alternatives (A-1, A-2, A-3). Each supplier is evaluated by 12 quality and numeral criteria: (1) technical performance; (2) bid price; (3) price breaks and quantity discounts; (4) payment terms—the possibility of deferred payment; (5) warranty; (6) out-of-warranty service; (7) number of available repair shops; (8) availability of experienced staff; (9) certifications; (10) previous experience; (11) lead time; and (12) customer recommendations.

A predefined group of five authorized experts relevant to the problem (excluding the leader of the group) is involved to evaluate the alternatives: E-1, E-2, E-3, E-4, and E-5. Each expert in the group determines the importance of all criteria by assigning corresponding weighted coefficients in accordance with his point of view.

All of the described data together with normalized evaluation scores from experts' points of view toward the performance of alternatives in accordance with the given criteria are shown in Table 2.9.

The obtained results by using the data from Table 2.9 along with three cases about the importance of experts' opinions in the group are shown in Table 2.10.

Table 2.9 Modified weighted decision matrix

Experts	Alternatives	Evaluation criteria											
		C1	C2	C3	C4	C5	C6	C7	C8	C9	C10	C11	C12
Criteria weights		0.08	0.10	0.09	0.09	0.08	0.07	0.08	0.08	0.09	0.08	0.07	0.10
E-1	A-1	0.76	1.00	0.72	0.82	0.76	0.56	0.69	1.00	0.68	0.66	0.66	0.56
	A-2	0.83	0.88	0.78	0.93	0.77	0.58	0.65	0.93	0.79	0.76	0.68	0.55
	A-3	0.81	0.90	0.76	1.00	0.72	0.66	0.72	0.92	0.74	0.75	0.70	0.63
Criteria weights		0.11	0.07	0.07	0.06	0.09	0.09	0.09	0.09	0.08	0.08	0.08	0.09
E-2	A-1	0.81	0.85	0.73	0.67	0.67	0.89	0.73	0.67	0.74	0.85	0.80	0.93
	A-2	0.84	0.78	0.76	0.65	0.77	0.91	0.75	0.69	0.78	0.82	0.81	0.85
	A-3	0.82	0.74	0.73	0.72	0.72	0.82	0.78	0.65	0.79	0.78	0.73	0.91
Criteria weights		0.15	0.06	0.06	0.06	0.11	0.10	0.09	0.09	0.06	0.06	0.07	0.09
E-3	A-1	0.86	0.81	0.78	0.77	0.95	0.82	0.84	0.82	0.81	0.76	0.96	0.79
	A-2	0.72	0.79	0.76	0.81	0.83	0.76	0.70	0.81	0.78	0.79	0.89	0.82
	A-3	0.81	0.78	0.79	0.69	1.00	0.80	0.79	0.88	0.83	0.72	1.00	0.76
Criteria weights		0.13	0.06	0.06	0.06	0.12	0.11	0.09	0.08	0.06	0.07	0.08	0.08
E-4	A-1	1.00	0.85	0.61	0.62	0.95	0.93	0.73	0.73	0.94	0.85	0.85	0.93
	A-2	0.88	0.74	0.66	0.66	0.96	1.00	0.75	0.79	0.88	1.00	0.92	0.88
	A-3	0.92	0.76	0.73	0.75	0.83	0.91	0.78	0.76	1.00	0.88	0.78	0.90
Criteria weights		0.10	0.10	0.10	0.10	0.10	0.06	0.06	0.07	0.06	0.09	0.10	0.06
E-5	A-1	0.92	0.79	0.88	0.73	0.83	0.78	1.00	0.73	0.72	0.76	0.63	0.84
	A-2	0.78	0.88	1.00	0.72	0.84	0.71	0.88	0.74	0.75	0.78	0.68	1.00
	A-3	0.84	0.76	0.82	0.68	0.88	0.73	0.91	0.79	0.81	0.82	0.69	0.86

Table 2.10 Solution results—selected alternative

Case	Coefficients for expertise of the group members					Selected alternative by modified SAW	Selected alternative by modified WPM
	E-1	E-2	E-3	E-4	E-5		
Case-1	0.20	0.20	0.20	0.20	0.20	A1	A3
Case-2	0.25	0.10	0.15	0.25	0.25	A3	A3
Case-3	0.27	0.12	0.13	0.18	0.30	A2	A3

Using different weighted coefficients for the expertise of group members' influence essentially the final selection of the most preferable alternative. The consideration of equal expertise for all group members (Case-1) leads to the choice of alternative A-1 for modified SAW and A-3 for modified WPM. The weighting coefficients from Case-2 put more weight on the opinions of experts E-1, E-4, and E-5 followed by the importance of experts E-2 and E-3. In this case, the defined preferable alternative is A-3 for both models. When the major attention is paid to the opinions of experts E-5 and E-1 and less on experts E-2, E-2 and E-4, the most preferable alternative is A-2 for modified SAW and A-3 for modified WPM.

From Table 2.10, it is seen that the modified SAW is more sensitive than the modified WPM in the determination of different choices of alternatives when different weighted coefficients for the expertise of the experts are defined. Additional experiments are to be done to prove this fact. Nevertheless, both proposed modifications of SAW and WPM could be used for group decision-making to determine the most preferable alternative. In this particular case, the use of these models contributes to the clear determination of the supplier who will supply the wind turbines.

When making such important decisions, it is good for executive managers to be offered more than one vendor/supplier to choose from. For such situations, it is good to apply another approach that is based on different strategies with group decision-making for z-best selections or ranking.

2.7 Group Decision-Making by Using of Different Strategies for Selection or Ranking

Due to globalization and a growing economy, today it is difficult to make successful business decisions without the assistance of experts at different levels (strategic, tactical, and operational). That is why different strategies should be available to help in the determination of (1) the most preferable alternative, (2) several good alternatives simultaneously, or (3) to rank all given set of alternatives. Similar activities are present in the selection of equipment for monitoring or management software suites in the maintenance of wind farms. Such decisions often involve business intelligence to gain maximum information from available data in order to make effective business decisions (Borissova et al., 2016). Business intelligence relies on analytics experts with capabilities to provide believable information to help in effective and high-quality business decision-making. The core of any decision support system is to help the higher management in different decision problems. Increasing competition and developing ICT make the problem of business decisions more significant and more complex. The complexity is related to the essence of the selection process which involves various quantitative and qualitative criteria. For such problems, multicriteria decision-making methods can be used to tackle different and conflicting criteria. The performance evaluation and selection of the most appropriate alternative have multi-level and multi-factor features. This defines essential difficulties in making trade-offs between evaluation criteria to get the best compromise solution.

When modeling real-world decision problems, three basic problem formulations can be distinguished: choice, ranking, and sorting problems. With respect to this classification, the chapter deals with the problems of choice and ranking. An appropriate quantitative tool for group decision-making that can be applied to such problems is the multi-attribute utility theory (MAUT) (Rao, 2007). MAUT seeks to overcome different dimensions via one-dimension values followed by aggregation through a weighted linear average. The simple additive weighting (SAW) method is a common aggregation approach in MAUT. A simple multi-attribute rating technique (SMART) is another form of MAUT with the possibility to be applied to any type of weight assignment technique like absolute, relative, etc. (Edwards, 1977; Barfod & Leleur, 2014). It should be noted that different methods of MAUT could lead to different solutions (Korsemov et al., 2018).

The selection problem could be viewed as a multi-attribute group decision-making (MAGDM) problem as often the executives have no confidence in an untested supplier and to overcome this different indicators are needed to guarantee objectives performance. There are different approaches proposed to tackle the problem of selection (Wetzstein et al., 2016). For example, an integrated approach is proposed to select using the AHP-ARAS-MCGP methodology (Fu, 2019). Other authors emphasize the usage of the multicriteria intuitionistic fuzzy TOPSIS method (Memari et al., 2019). A group decision-making approach by extended SMART is used in the evaluation and ranking of students (Borissova & Keremedchiev, 2019). To cope with the different competencies of the group members, modifications of simple additive weighting and weighted product models are proposed (Korsemov & Borissova, 2018). These modifications could be easily implemented in the spreadsheet as a tool to support group decisions (Borissova et al., 2018). The comparison between approaches of individual and group decision-making toward multicriteria decision problems shows that the individual point of view of each expert significantly influences the final decision (Borissova et al., 2019). In this respect, to express more transparently the expertise of each group member a two-component function of objective and subjective parts seems to be useful (Borissova, 2018).

In processes of selection and ranking, the group decision-making could be applied to find a solution—a single alternative or set of alternatives according to the preferences provided by the authorized group of experts. While most publications considered the problem for handling different expressions of experts' preferences by utility-based individual preferences (Huang et al., 2013), fuzzy relations (Peneva & Popchev, 2007; Peneva & Popchev, 2009), decision Markov logic (Sgurev, 2014), fuzzy logic based on multicriteria evaluation (Buyukozkan, 2004), multidimensional preference (Zhang et al., 2016), linguistic information (Cabrerizo et al., 2013), etc., the proposed approach emphasizes on the evaluation and aggregation phase of the group decision-making process.

2.7.1 Generalized Approach to Support Business Group Decision-Making by Three Different Strategies

The current section describes a generalized approach to support business decisions incorporating different strategies to make better decisions by analyzing the existing alternatives. The different strategies aim to determine the most preferable alternative, to determine several good alternatives simultaneously, or to rank all given alternatives. Furthermore, the described approach takes into account the experts' knowledge and experience when forming the final group decision. Each strategy is based on the formulated optimization model.

The possibility to incorporate a wide variety of quantitative and qualitative criteria makes SMART a popular and widely used method. The original SMART describes the whole process as rating the alternatives and weighting of attributes (Edwards, 1977). The proposed approach to support the business by group decision-making with three different strategies is based on SMART. The basic stages of the proposed modification of SMART are graphically illustrated and presented in Fig. 2.12.

Preparation stage 1 of the proposed approach starts with the identification of the main goal of group decision-making that includes (1.1) identification of evaluation criteria and alternatives, (1.2) identification of needed experts (DMs) for group decision-making, and (1.3) determination of corresponding weights for each DM in accordance with its knowledge and experience. This stage starts with the determination of alternatives and suitable evaluation criteria. Next, the group of experts responsible for evaluating different aspects of the decision goal is to be selected. According to Yetton and Botter, groups of five experts are the most effective in group decision-making closely followed by a group of seven experts. Weights of DMs play a very important role in multiple-attribute group decision-making, and in the proposed algorithm, the corresponding weights for each expert reflect their expertise.

Evaluation stage 2 is composed of two steps: (2.1) determination of weighted coefficients for the evaluation criteria by DMs and (2.2) evaluation of the given alternatives against the criteria by each DM. The weighted coefficients for criteria importance and alternatives evaluations are expressed by scores derived by the experts where the higher value means better performance.

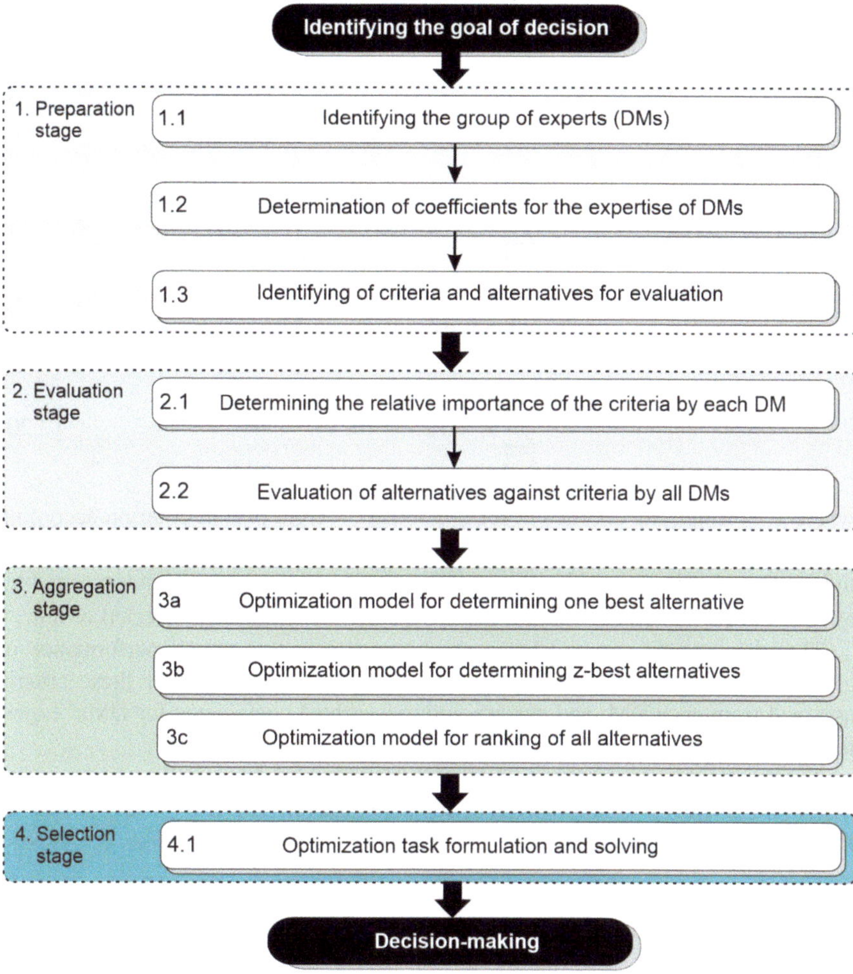

Fig. 2.12 Generalized approach to support group decision-making by three different strategies

The next and the most essential stage is *aggregation stage* 3. A key issue in MAGDM is the aggregation of individual preferences. In this chapter, the aggregation stage can be realized by three different optimization models depending on the particular goal of decision-making strategy for (3a) selection of a single alternative, (3b) defining of z best alternatives, and (3c) ranking of all alternatives.

The first decision-making strategy of the aggregation stage (Fig. 2.12) illustrates the situation where the selection of single alternative is to be done considering the alternatives performance and the corresponding formulation of the combinatorial optimization problem is as follows:

$$\max\left\{\sum_{q=1}^{K}\lambda^q\left(\sum_{j=1}^{N}\sum_{i=1}^{M}w_j^q p_{ij}^q x_i\right)\right\} \tag{2.17}$$

subject to

$$\sum_{q=1}^{K}\lambda^q=1, \lambda^q\in(0,1) \tag{2.18}$$

$$\sum_{j=1}^{N}w_j^q=1, q=\{1,2,\ \ldots, K\} \tag{2.19}$$

$$\sum_{i=1}^{M}x_i=1, x_i\in\{0,1\} \tag{2.20}$$

where λ^q is weighting coefficient for q-th expert; weight of n-th criterion according to the q-th expert is denoted by w_n^k; the performance of i-th alternative in respect of j-th by q-th expert is denoted by p_{ij}^q where $0\leq p_{ij}^q\leq 1$ and x_i are binary integer variables assigned to each alternative that indicate whether the item i-th is selected or not.

The objective function (2.17) seeks to maximize the overall performance of alternatives in accordance with the evaluation criteria, weights of these criteria assigned from each DM, and corresponding weighted coefficients for DMs' expertise. Constraint (2.20) provides the selection of only one alternative.

The second decision-making strategy of the aggregation stage (Fig. 2.12) aims to determine several good alternatives by replacing the relation (2.20) with the following statement (Borissova, 2016):

$$\sum_{i=1}^{M}x_i=z, x_i\in\{0,1\}, 1<z<M \tag{2.21}$$

where z is integer number bigger than 1 and less than number of alternatives (M).

The third decision-making strategy of the aggregation stage of the proposed group decision-making approach makes it possible to rank all alternatives using the following algorithm for ranking as shown in Fig. 2.13.

This algorithm for ranking of all alternatives is based on the sequential solution of multiple tasks of the type (2.17)–(2.20). The solution on the first iteration provides the best alternative which goes on the top of the ranked list. On the second iteration, this alternative is excluded from the set of alternatives and the modified optimization task is solved again to get the next alternative in the ranked list. This process is repeated until only one alternative remains.

During selection stage 4, the optimization tasks corresponding to the simulation of three different decision-making scenarios (strategies) are to be formulated and solved. Depending on the obtained solution for one best alternative, several good

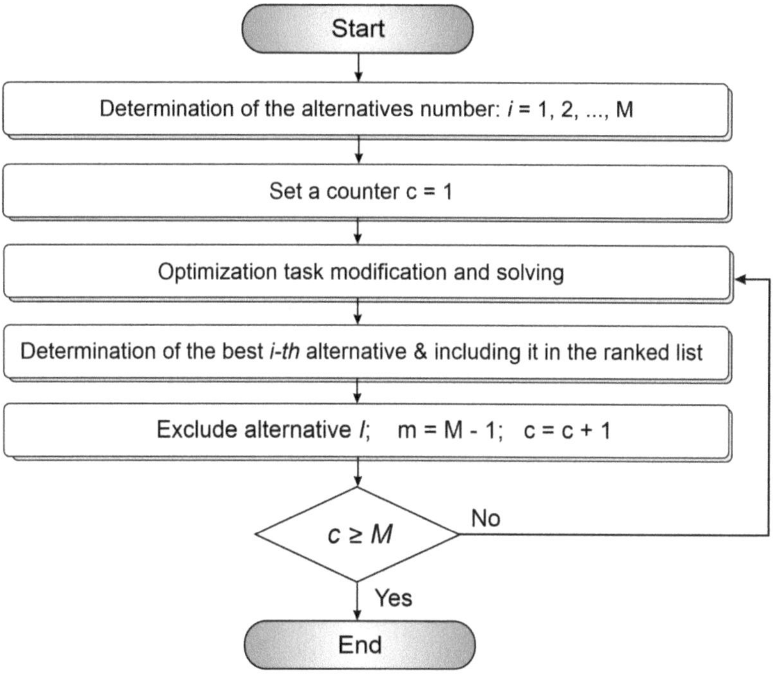

Fig. 2.13 Flowchart of the algorithm for ranking of alternatives

alternatives, or ranking of all alternatives, the higher management could make the most appropriate choice from the three scenarios. The higher management could make a decision based on the obtained solution from stage 3 taking into account also the defined goal and policy of the company.

2.7.2 Numerical Application

The numerical example refers to five predefined vendors (alternatives) of wind turbines, among which one must be selected. The decision should be taken based on the evaluations of a group of five experts: E-1, E-2, E-3, E-4, and E-5 taking into account 10 evaluation criteria concerning the parameters denoted by C-1, C-2; C-3, C-4, C-5, C-6, C-7, C-8, C-9, and C-10 for simplicity. The evaluations of these 5 potential alternatives in terms of 10 criteria from 5 experts are shown in Table 2.11.

In accordance with stage 2 of the proposed generalized group decision-making algorithm (Fig. 2.12), all potential alternatives are evaluated with respect to the given criteria (attributes) by a group of experts. Furthermore, the relative importance between criteria from the particular expert's point of view has to be done by corresponding weights as shown in Table 2.11.

Table 2.11 Evaluations of alternatives from the experts

Experts	Alter-natives	Criteria/weights/evaluations									
		C-1	C-2	C-3	C-4	C-5	C-6	C-7	C-8	C-9	C-10
E-1		0.04	0.13	0.14	0.08	0.15	0.15	0.08	0.1	0.07	0.06
	A1	0.2	0.9	0.9	1	0.3	0.4	0.8	0.6	0.5	0.4
	A2	0.8	0.9	1	0.9	0.7	0.8	0.9	0.8	0.6	0.7
	A3	0.8	0.3	0.6	0.8	0.3	0.5	0.7	0.5	0.6	0.8
	A4	0.2	0.6	0.2	0.4	0.3	0.5	0.7	1	0.4	0.4
	A5	0.4	0.8	0.7	0.8	0.6	0.8	0.6	0.3	0.2	0.8
E-2		0.05	0.14	0.16	0.05	0.09	0.06	0.08	0.15	0.16	0.06
	A1	0.4	0.8	0.2	0.6	0.1	0.5	0.9	0.4	0.6	1
	A2	0.5	0.8	0.5	0.6	1	0.5	0.7	0.9	1	0.8
	A3	0.4	0.7	0.5	0.8	0.7	0.9	0.4	1	0.5	1
	A4	1	0.5	0.8	0.1	0.6	0.4	0.8	0.9	0.4	0.2
	A5	0.8	0.9	0.1	0.2	0.5	1	0.9	0.8	0.6	0.8
E-3		0.1	0.1	0.1	0.1	0.11	0.1	0.08	0.1	0.12	0.09
	A1	0.6	0.7	0.8	0.4	0.9	0.3	0.4	0.6	0.8	0.8
	A1	0.8	0.7	0.7	0.8	0.7	0.7	1	0.8	0.6	0.8
	A3	0.6	0.5	0.7	0.8	0.2	0.2	0.6	0.6	0.6	0.6
	A4	0.8	0.6	0.2	0.2	0.2	0.2	0.1	1	0.1	0.7
	A5	0.7	0.8	0.7	0.8	0.8	0.6	0.5	0.4	0.3	0.6
E-4		0.11	0.11	0.06	0.08	0.14	0.06	0.11	0.11	0.14	0.08
	A1	0.1	0.7	0.7	0.4	0.1	0.5	0.2	0.3	0.4	0.1
	A2	0.9	0.5	1	0.8	0.7	0.8	0.7	0.9	0.8	0.7
	A3	0.8	0.1	0.8	0.6	0.6	0.7	0.7	0.7	0.5	0.6
	A4	0.5	0.7	0.6	0.3	0.5	0.6	0.4	0.7	0.4	0.3
	A5	0.7	0.8	0.6	0.7	0.7	0.6	0.8	0.5	0.1	0.5
E-5		0.07	0.1	0.1	0.08	0.15	0.08	0.1	0.1	0.15	0.07
	A1	0.1	0.6	0.6	0.4	0.2	0.2	0.2	0.2	0.2	0.1
	A2	0.9	0.6	0.9	0.8	0.7	0.7	0.5	0.6	0.7	0.6
	A3	0.8	0.2	0.6	0.7	0.9	0.7	0.6	0.8	0.7	0.9
	A4	0.5	0.8	0.5	0.7	0.6	0.6	0.1	0.5	0.9	0.2
	A5	0.6	0.7	0.6	0.6	0.8	0.8	0.7	0.4	0.3	0.6

Three different cases, representing three scenarios about the experts' knowledge and experience, are used to demonstrate the importance of experts in group decision-making. Case-1 expresses the situation where the knowledge and experience of all experts in the group are equal, while Case-2 and Case-3 make a difference between involved experts. Case-2 considers expert E-5 as the most trusted expert, while the dominant role in Case-3 is assigned to expert E-4. The results of the solution of optimization tasks, formulated in step 3.1a, are shown in Table 2.12.

The proposed approach to support the business by group decision-making with different strategies is applied for (1) the determination of the best alternative, (2) the determination of two good alternatives, and (3) the ranking of all alternatives.

Table 2.12 Solution results of tasks on step 3

Cases	Weighting coefficients of experts					The best alternative	Two good alternatives	Ranking of alternatives
	E-1	E-2	E-3	E-4	E-5			
Case-1	0.20	0.20	0.20	0.20	0.20	A-2	A-2 & A-5	A-2, A-5, A-3, A-1, A-4
Case-2	0.10	0.15	0.18	0.22	0.35	A-3	A-3 & A-5	A-3, A-5, A-2, A-1, A-4
Case-3	0.16	0.13	0.11	0.44	0.17	A-5	A-3 & A-5	A-5, A-3, A-2, A-4, A-1

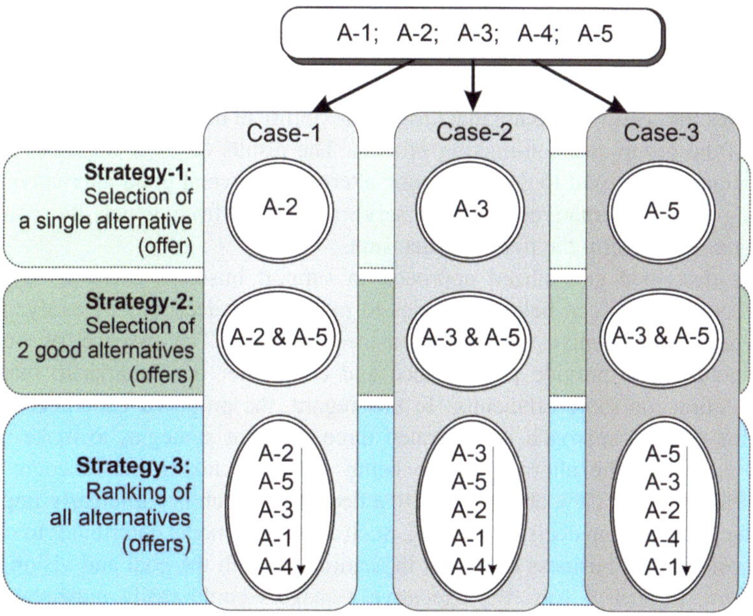

Fig. 2.14 Selection and ranking of alternatives (offers)

According to the input data from Table 3.8 and three different cases for the weights of experts from Table 2.12, the obtained solutions for selection and ranking are illustrated in Fig. 2.14.

The main idea of the proposed approach for group decision-making is to aggregate different experts' evaluations of the alternatives into a final group decision. The flexibility of the modification of SMART is realized by formulation and solving combinatorial optimization tasks for the determination of a single alternative, for several good alternatives, or for ranking all alternatives in accordance with the selected strategy. Regardless of the strategy used, all of the optimization models (respectively tasks) take into account the knowledge of the experts by corresponding weighting coefficients.

The numerical application when using a strategy for selection of one alternative determines the best option alternative A-2 when all experts are considered with equal level of competency and knowledge. For the same strategy, but considering the differences in experts' knowledge and experience, the solution of the tasks determines alternative A-3 (in Case-2) and alternative A-5 (in Case-3) as the best decision.

In the second strategy, there is no information on which of the selected alternatives A-2 and A-5 (in Case-1) or A-3 & A-5 (in Case-2 and Case-3) is better than the other, but both alternatives are certainly better than the others. It should be noted that several good alternatives are obtained simultaneously as a result of the solution via a single run of the task. The usage of the third strategy allows the ranking of all alternatives according to the point of view of all experts and takes into account the level of the expertise of group members.

The implementation of the described approach allows the simulation of different strategies for decision-making that provide flexibility in the achievement of the main goal of the group decision-making process. The results of these strategies' implementation can be used to determine one alternative, several good alternatives, or a ranking of all alternatives that can serve as a base from which the executive managers could form the final best decision.

The described generalized approach to support business decisions by using different strategies can help companies to make better decisions by analyzing the existence of alternatives within a business context. The essence of business intelligence is to provide performance and competitor benchmarks to make the organization run more efficiently. In this regard, the proposed generalized group decision-making approach incorporated three different strategies to make a clear basis on which the alternatives are being evaluated taking into account many different points of view about alternative decisions which is particularly important in situations of group decision-making. Such a business model contributes to making the most suitable business decisions in accordance with the goal and vision of the company. In such a way, the executive managers could easily understand data without the technical know-how to dig into the data themselves.

Similar models are also applicable in the field of wind energy, where they can be used not only in the selection of a specific supplier of turbines, but the same model can also be applied in the selection of different subcontractor companies in the construction of wind farms, for the selection of equipment for monitoring, for software products for managing the maintenance of wind farms, etc.

2.8 Decision-Making in Personnel Selection

Last but not least in the implementation of a wind power plant project are the activities related to hiring personnel. During all stages of the wind farm life cycle, it is necessary to have the right professionals from scientists and engineers to project managers, truck drivers, crane operators, and many more. To realize such a complex task, different selection methods can be used from curriculum vitae and interviews to knowledge and sociotechnical tests. Personnel selection techniques are increasingly focused on the competencies and skills of candidates. In this regard, this section describes a group decision-making model considering the competence of personnel selection experts.

In modern society, qualified personnel is of great importance due to rapidly developing technologies. That makes the problem of personnel evaluation and selection a very important activity for every enterprise including sub-contracting companies to consult, construct, or maintain wind farms. Thus, personnel selection can be considered a complex multiple-attribute group decision-making problem due to the requirement for different abilities for a particular position. Different authors have proposed various methods for MAGDM. An intuitionistic fuzzy multicriteria group decision-making method with gray relational analysis can be applied to the ranking and selection. An intuitionistic fuzzy weighted averaging operator is utilized to aggregate individual opinions of decision-makers into a group opinion (Zhang & Liu, 2011). A decision support tool is developed using an integrated analytic network process and fuzzy data envelopment analysis approach to effectively deal with the personnel selection problem (Lin, 2010). Based on the Karnik-Mendel algorithm, an analytical solution to the fuzzy TOPSIS method is proposed (Sang et al., 2015). To make the recruitment process more reasonable, analyzing the analytic hierarchy process based on the fuzzy multiple criteria decision-making model is used to achieve the goal of personnel selection (Afshari et al., 2017). The preferences of more than one DM are internally aggregated into the TOPSIS procedure (Shih et al., 2007).

It should be noted that experts in the group may have different backgrounds and knowledge of the investigated problem. This requires the use of some metrics to express the different qualifications of group experts. Despite a variety of proposed MAGDM methods, there is no unique direction to make differences between experts in group decision-making. Most of the authors emphasize the aggregation of utility functions to express the alternatives' performance (Korsemov & Borissova, 2018) or deal with the linguistic or fuzzy expression of preferences (Tao et al., 2015; Peneva & Popchev, 2009). Therefore, the aim is to propose adequate decision-making models for personnel selection or ranking.

2.8.1 Group Decision-Making Model Taking into Account Differences in the Expertise of Group Members

The basic formulation of the MCDM problem is usually expressed in the following matrix format:

$$D^k = \begin{matrix} A_1 \\ \cdots \\ A_m \end{matrix} \begin{bmatrix} r_{11}^k & \cdots & r_{1n}^k \\ \cdots & \cdots & \cdots \\ r_{m1}^k & \cdots & r_{mn}^k \end{bmatrix} \begin{bmatrix} w_1^k \\ \cdots \\ w_n^k \end{bmatrix} \tag{2.22}$$

where index i expresses the number of alternatives ($i = 1, 2, \ldots, m$), index j expresses the number of evaluation criteria ($j = 1, 2, \ldots, n$), the performance rating of i-th alternative toward j-th criterion from k-th expert is expressed by r_{ij}^k, and w_j^k represents the weight for j-th criterion from k-th expert. Each expert determines weights for criteria w_j^k, such that $\sum_{j=1}^{n} w_j^k = 1$.

To make a more effective selection in the determination of the most preferable alternative for the group, the rating of alternatives and criteria weights of each group member should be considered with different importance. This could be expressed by different levels of competency including previous experience, level of position and responsibilities, availability of certificates for qualifications, etc. For this purpose, the corresponding relevant competency of each group member is proposed to be expressed by weighted coefficients. These considerations are taken into account by formulation of the following optimization and group decision-making problem:

$$\max \begin{bmatrix} x_1 \\ \cdots \\ x_m \end{bmatrix} \left(\alpha^1 \begin{bmatrix} r_{11}^1 & \cdots & r_{1n}^1 \\ \cdots & \cdots & \cdots \\ r_{m1}^1 & \cdots & r_{mn}^1 \end{bmatrix} \begin{bmatrix} w_1^1 \\ \cdots \\ w_n^1 \end{bmatrix} + \ldots + \alpha^k \begin{bmatrix} r_{11}^k & \cdots & r_{1n}^k \\ \cdots & \cdots & \cdots \\ r_{m1}^k & \cdots & r_{mn}^k \end{bmatrix} \begin{bmatrix} w_1^k \\ \cdots \\ w_n^k \end{bmatrix} \right) \tag{2.23}$$

subject to

$$\sum_{i=1}^{m} x_i = 1, \quad x_i \in \{0, 1\} \tag{2.24}$$

$$\sum_{k=1}^{q} \alpha^k = 1 \tag{2.25}$$

$$\sum_{j=1}^{n} w_j = 1 \tag{2.26}$$

where r_{ij}^k is the performance rating of i-th alternative toward j-th criterion according k-th expert (group member), w_j^k is coefficient for relative importance between criteria for k-th expert, α^k is weighted coefficient that expresses the level of expertise of k-th expert, and x_i are binary decision variables assigned to each alternative to perform a single selection.

The weighted coefficients for the relevant competent expertise of group members consist of objective and subjective parts. The objective part concerns the experience of experts in a particular field. This includes previous experience involving acquired knowledge, responsibilities related to different projects, availability of publications, etc. The subjective part is formed by the advisability of relevance area to the particular decision-making problem. All of these considerations are represented as aggregation function of experience in the particular field and advisability of relevant areas to the particular problem as follows:

$$\alpha^k = \beta^k + \gamma^k \tag{2.27}$$

where β is a coefficient that expresses the experience of expert in years and γ is coefficient for advisability of expertise field competency to the particular problem.

The coefficient for experience is an objective coefficient, while the coefficient for advisability is subjective and is determined by stakeholders or by higher managers authorized to manage the decision process. The values for the coefficient of advisability are within the range of 0 to 1.

Obviously, there are no restrictions to compose other functions to express the expertise of the group's members. Regardless of the used function for the determination of weighted coefficients for relevant competent expertise of group members, some normalization is to be done to obtain a dimensionless coefficient.

The criteria could also require some normalization when having different scales to obtain a common scale to enable aggregation and to make comparable criteria to get the overall score for each alternative.

2.8.1.1 Numerical Results

The applicability of the proposed decision-making model considering the experience and knowledge of group members is numerically tested in the case of personnel selection. The input data for numerical testing are adopted from human resources selection examples where some relevant tests with benefit attributes are to be evaluated (Shih et al., 2007). Four experts have been appointed to evaluate 17 qualified candidates in regard to results from knowledge tests, skill tests, and interviews, and the results are shown in Table 2.13.

The performance rating score of candidates and weights for criteria importance are represented in normalized values. The criteria for knowledge tests and skill tests are considered objective, while the interviews are considered subjective. The scores for objective criteria are obtained as a sum of corresponding scores for tests. The score for the performance of candidates toward subjective criteria from interviews is determined by each particular expert in the same range as the test results. In this evaluation process, each expert determines his own coefficient for relative importance between evaluation criteria.

To test the proposed model (2.23)–(2.26), three different cases of weighted coefficients for the expertise of the group members are used as shown in Table 2.14.

The input data from Tables 2.13 and 2.14 are used to formulate the corresponding optimization task, the solution of which will determine the most preferable alternative. The proposed group decision-making model (2.23)–(2.26) takes into account

Table 2.13 Weighted decision matrix

	Candidates																	E-1	E-2	E-3	E-4
Criteria	1	2	3	4	5	6	7	8	9	10	11	12	13	14	15	16	17	Weights for criteria			
Language test	80	85	78	75	84	85	77	78	85	89	65	70	95	70	60	92	86	0.066	0.042	0.060	0.047
Professional test	70	65	90	84	67	78	83	82	90	75	55	64	80	80	78	85	87	0.196	0.112	0.134	0.109
Safety rule test	87	76	72	69	60	82	74	72	80	79	68	65	70	79	87	88	80	0.066	0.082	0.051	0.037
Professional skills	77	80	80	85	75	81	70	80	88	67	62	65	75	80	70	90	70	0.130	0.176	0.167	0.133
Computer skills	76	75	85	65	85	79	71	78	90	77	70	60	70	85	66	85	72	0.130	0.118	0.100	0.081
Panel interview	80	65	90	65	75	80	65	70	80	70	50	60	75	80	70	90	80	0.216	–	–	–
1-on-1 interview	75	75	85	70	80	80	70	60	85	75	60	65	75	70	65	95	85	0.196	–	–	–
Panel interview	85	60	80	55	75	75	70	75	95	75	62	65	80	75	75	92	70	–	0.215	–	–
1-on-1 interview	80	70	85	60	80	85	60	65	85	80	65	75	80	72	70	90	75	–	0.255	–	–
Panel interview	75	70	80	68	50	77	65	75	90	68	60	50	65	80	65	85	75	–	–	0.203	–
1-on-1 interview	70	77	90	72	55	82	72	67	85	78	65	60	75	70	70	80	80	–	–	0.285	–
Panel interview	90	60	90	62	70	75	67	82	90	65	65	45	70	75	60	88	70	–	–	–	0.267
1-on-1 interview	85	70	95	72	75	75	75	85	92	70	70	50	75	75	65	90	75	–	–	–	0.326

Table 2.14 Weighted coefficients for expertise of the group members

Case	Weighted coefficients for expertise			
	E-1	E-2	E-3	E-4
Case-1	0.25	0.25	0.25	0.25
Case-2	0.12	0.15	0.35	0.38
Case-3	0.18	0.27	0.33	0.22

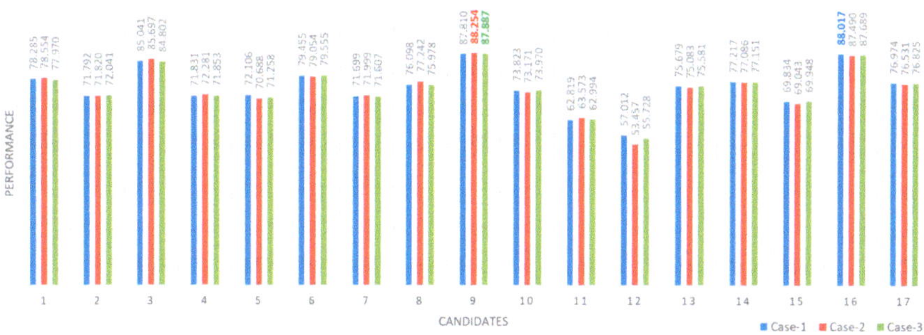

Fig. 2.15 Candidates' performance for three cases for expertise of group members

not only alternative evaluations and weights of criteria importance but also considers also the relevant competent expertise of members within the group. The used binary variables allow for determining the most preferable alternative by a single run of the task. Using data from Table 2.13 and three different cases for the expertise of group members (Table 2.14) the following results, as shown in Fig. 2.15, are determined.

Case-1 illustrates the situation with an equal level of expertise for all group members. This is the case equivalent to other group decision-making approaches that do not make a difference between group members. The solution, in this case, is determined as the most preferable alternative corresponding to candidate #16 with a value for the objective function of 88.017 (Fig. 2.13). The same candidate is determined in Shih et al. (2007) where an extension of the TOPSIS model is used and the same input data. Case-2 and Case-3 refer to situations where each group member has a different level of relevant competency and expertise expressed by weighted coefficients given in Table 2.14. When the level of competency for experts E-4 and E-3 is dominant for the experts E-1 and E-2, the solution is determined as the most preferable alternative corresponding to candidate #9 with a value for the objective function of 88.254 (Fig. 2.13). In Case-3, experts E-2 and E-3 are with a higher level of competency than experts E-1 and E-4 and the results show that candidate #9 is the preferable solution with an objective function value of 87.887 (Fig. 2.15).

The wind energy sector needs a skilled and qualified wind energy workforce to produce clean power. Wind turbine service technicians will be the fastest-growing job in the wind energy sector. Vital positions in the wind industry include researchers, scientists, engineers, trade workers, educators, transportation workers, business and sales forces, and many others. Creating a skilled workforce, such as project managers, scientists, engineers, technicians, electricians, welders, pipefitters, truck drivers, crane operators, and many others, is also essential (IRENA, 2021). The need for labor inputs and skills changes over time as technologies mature, the scale and complexity of operations grow, learning takes effect, and automation progresses, raising demand for people with backgrounds in fields such as artificial intelligence, computer science and engineering, and telecommunications radio telemetry, among others. Balancing skills demand and supply requires close coordination among industry, government, and educational and training institutions to attract a broader and more diverse set of candidates for the future workforce.

Therefore, there are two ways to recruit personnel for the construction of wind energy projects: (1) by the selection through the proposed combinatorial model (2.23)–(2.26) of already skilled personnel or (2) by conducting specialized training.

2.8.2 *Mathematical Model for Personnel Ranking Taking into Account the Presence of Hard and Soft Skills*

Labor market skills requirements are a relatively new area of research and skills requirements vary from company to company and industry to industry. Before describing the most preferable pretender for a particular job position, it is necessary to understand the difference between hard skills and soft skills. In the general case, hard skills refer to these abilities that can be measured, while soft skills are difficult to measure and are considered more universal and could refer not only to one specific job. In the field of software project management, authors discuss soft skills that HR should possess to improve the co-sourcing process (Schlichter & Buchynska, 2021). Other authors describe how IT companies look for candidates for their open positions (Montandon et al., 2021). They concluded that programming languages are the most demanded hard skills, while communication, collaboration, and problem-solving are the most demanded among soft skills. There are many studies focused on the evaluation of acquired hard skills while soft skills are neglected (Borissova & Keremedchiev, 2019). It is found that high-level hard skills requirements are positively correlated to differences in productivity in firms (Lyu & Liu, 2021). Recently, a growing interest in soft skills has been focused on exploring the main factors to understanding how they are influenced by socioeconomic and family background (Marcenaro-Gutierrez et al., 2021). The authors combine econometric with multi-objective optimization to provide an overarching framework for assessing the trade-offs between the different dimensions of non-cognitive skills.

In order to be able to assess each candidate's knowledge and ability to adapt to teamwork, it is necessary to consider two separate parts related to hard and soft skills. This can be realized through the proposed mathematical model for the assessment of personal performance ($P_i^{\text{performance}}$), formulated as follows (Borissova et al., 2022b):

$$P_i^{\text{performance}} = \max\left\{\alpha \sum_{h=1}^{H} w_h e_{ih} + \beta \sum_{s=1}^{S} w_s e_{is}\right\}, \forall i = 1, \ldots, N \tag{2.28}$$

subject to

$$\alpha + \beta = 1 \tag{2.29}$$

$$\sum_{h=1}^{H} w_h = 1 \tag{2.30}$$

$$\sum_{s=1}^{S} w_s = 1 \tag{2.31}$$

where $i = \{1, 2, \ldots, N\}$ represents the set of candidates, coefficient α expresses the importance of hard skills while coefficient β expresses the ability for teamwork (soft skills), the coefficients w_h and w_s express the relative importance between criteria related to the hard and soft skills, and e_{ih} and e_{is} represent evaluation scores of i-th candidate about h-th criterion related to the hard skills and corresponding i-th candidate about s-th criterion related to the soft skills. It should be mentioned that the range for evaluation scores e_{ih} and e_{is} has to correspond to the range of other variables of the proposed model (2.28)–(2.31). Therefore, the acceptable range for these scores is to be between 0 and 1 to have a comparable scale. The relation (2.29) makes it possible to aggregate the separated two parts of the evaluation regarding hard skills (acquired knowledge) and soft skills (teamwork ability) in the final generalized assessment. As could be seen from relation (2.28), the ranking of the candidates is realized considering hard and soft skills and these two types of components could be taken into account with different importance. This allows the model to be flexible enough to consider hard and soft skills with different proportions in determining the final complex ranking.

The proposed mathematical model (2.28)–(2.31) could be simplified by using a value equal to zero for the coefficient β ($\beta = 0$). In this case, the model (2.28)–(2.31) will rely on only the hard skills (acquired knowledge) for the ranking.

2.8.2.1 Numerical Results

In order to verify the applicability of the proposed mathematical model (2.28)–(2.31) for ranking an experiment was conducted with a group of 25 students. The ranking of these students is done by using the 5 hard skills, namely (1) knowledge in sensors and signal processing (h-1); (2) knowledge in programming language/s (h-2); (3) knowledge in communications protocols (h-3); (4) knowledge in data mining (h-4); (5) knowledge in decision-making models and algorithms (h-5); and 3 soft skills concerning (1) teamwork (s-1), (2) motivation (s-2), (3) time management (s-3). These several skills are used as evaluation criteria for the ranking. The

Table 2.15 Evaluations of students' skills

	Hard skills					Soft skills		
Student #	h-1	h-2	h-3	h-4	h-5	s-1	s-2	s-3
1	0.94	0.78	0.81	0.94	0.86	0.78	0.98	0.91
2	0.95	0.91	0.79	0.87	0.88	0.92	0.75	0.81
3	0.88	0.96	0.79	0.83	0.89	0.95	0.77	0.92
4	0.87	0.93	0.8	0.82	0.89	0.9	0.85	0.81
5	0.91	0.87	0.79	0.86	0.81	0.82	0.92	0.79
6	0.89	0.93	0.78	0.86	0.79	0.89	0.85	0.77
7	0.90	0.93	0.72	0.81	0.79	0.88	0.94	0.86
8	0.80	0.97	0.82	0.78	0.9	0.82	0.9	0.79
9	0.86	0.94	0.86	0.86	0.8	0.79	0.85	0.82
10	0.88	0.87	0.86	0.88	0.76	0.88	0.83	0.8
11	0.89	0.85	0.81	0.89	0.72	0.95	0.86	0.85
12	0.90	0.82	0.78	0.81	0.87	0.88	0.92	0.9
13	0.81	0.86	0.8	0.79	0.88	0.85	0.81	0.92
14	0.79	0.86	0.77	0.83	0.84	0.84	0.88	0.91
15	0.79	0.91	0.87	0.85	0.86	0.93	0.79	0.8
16	0.80	0.78	0.91	0.77	0.86	0.9	0.84	0.86
17	0.87	0.85	0.87	0.84	0.81	0.88	0.91	0.77
18	0.78	0.77	0.95	0.79	0.88	0.81	0.88	0.92
19	0.76	0.80	0.89	0.83	0.86	0.93	0.98	0.86
20	0.79	0.86	0.9	0.8	0.86	0.87	0.91	0.79
21	0.86	0.97	0.8	0.86	0.81	0.88	0.89	0.93
22	0.88	0.92	0.86	0.86	0.78	0.95	0.9	0.82
23	0.89	0.96	0.88	0.81	0.80	0.92	0.81	0.9
24	0.89	0.92	0.89	0.78	0.80	0.91	0.72	0.87
25	0.90	0.87	0.86	0.83	0.77	0.81	0.92	0.88

corresponding evaluations about the hard skills (e_{ih}) and soft skills (e_{is}) expressed by scores regarding these skills are shown in Table 2.15.

Along with the evaluation score about the hard (e_{ih}) and soft (e_{is}) skills according to the proposed model (2.28)–(2.31), there is also a need to determine the ratio between hard and soft skills by the coefficients (α) and (β) and also the value for the coefficients w_h and w_s. Three different cases are identified and shown in Table 2.16.

Case-1 considers hard and soft skills with equal importance ($\alpha = \beta = 0.5$) and the distribution between criteria concerning both hard and soft skills is also taken with equal importance. Case-2 expresses the scenario where the hard skills are predominant ($\alpha = 0.65$) in respect of soft skills ($\beta = 0.35$), but distribution between particular criteria remains the same. Case-3 illustrates the situation, where hard skills are more preferred than soft skills; each criterion from hard and soft skills is considered with different importance.

The obtained results for the ranked students based on the data of Case-1 are visualized in Fig. 2.16.

Table 2.16 Coefficients for hard and soft skills and their distribution among criteria

	Case-1	Case-2	Case-3
Hard skills	$\alpha = 0.5$	$\alpha = 0.65$	$\alpha = 0.65$
h-1	0.2	0.2	0.22
h-2	0.2	0.2	0.2
h-3	0.2	0.2	0.22
h-4	0.2	0.2	0.18
h-5	0.2	0.2	0.18
Soft skills	$\beta = 0.5$	$\beta = 0.35$	$\beta = 0.35$
s-1	0.33	0.33	0.38
s-2	0.33	0.33	0.20
s-3	0.34	0.34	0.42

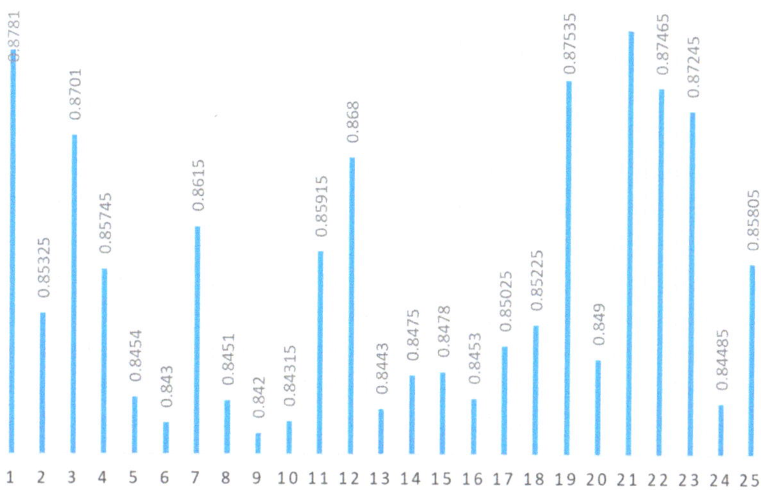

Fig. 2.16 Students' ranking under different importance of evaluation criteria

From this ranking, it can be seen that the first three places are occupied by student #21 with overall performance score of 0.88015, followed by student #1 with score of 0.8781 and student # 19 with score 0.87535.

If the strategy is changed and the preferences presented in Case-2 are used, then the ranking of students acquires a different appearance, as shown in Fig. 2.17.

In this ranking, the first three places are occupied by student #1 with overall performance score of 0.87447, followed by student #21 with score of 0.874015 and student # 3 with score 0.87364.

When the preferences expressed through the scenarios of Case-3 are used, the following ranking of the students is obtained, as shown in Fig. 2.18. In this ranking, the first best three students in the ranking are as follows: student # 3 with a score of 0.88034, followed by student #23 with a score of 0.87764, and student #21 with a score of 0.87492.

The comparison between these students' rankings is shown in Fig. 2.19.

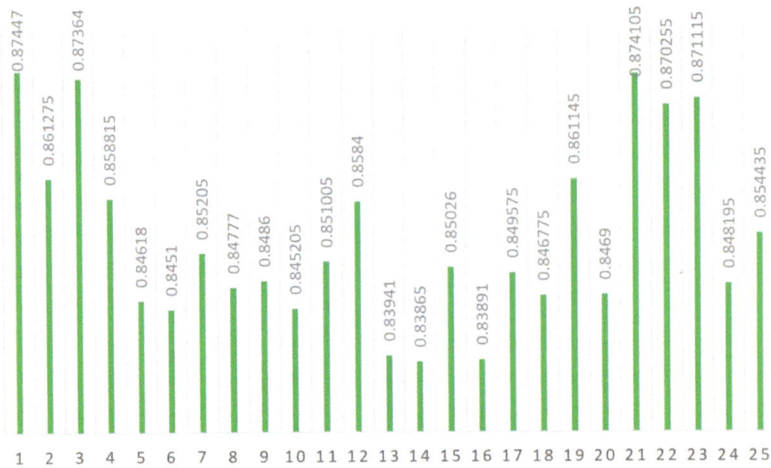

Fig. 2.17 Students' ranking under different importance of evaluation criteria

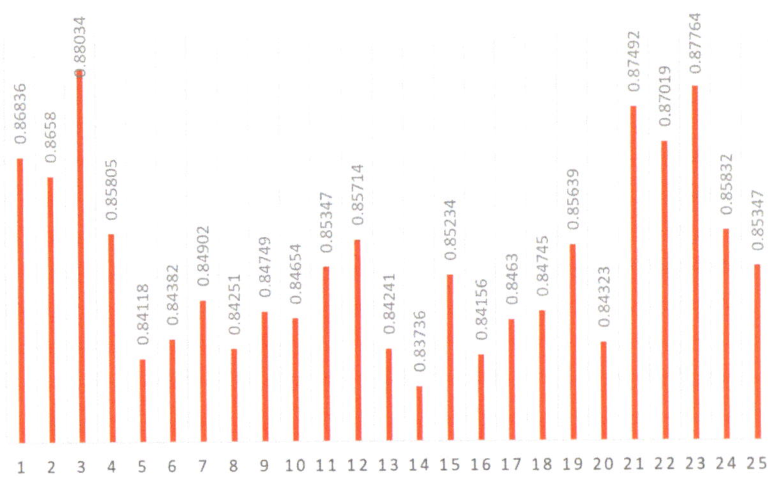

Fig. 2.18 Students' ranking under different importance of evaluation criteria

Depending on the purpose of the ranking, it is possible to make a proper list for students with more ability of hard skills, a suitable list for the students with soft skills, or some combination of hard and soft skills. These lists can be used to recommend specific students to find a suitable job, according to their level of knowledge and adaptation to working in teams. It is important to prepare personnel capable of implementing wind energy projects from school, as it is more difficult to retrain them through appropriate courses. On the other hand, maintaining a certain level of knowledge and skill must be periodically tested to prevent unwanted incidents.

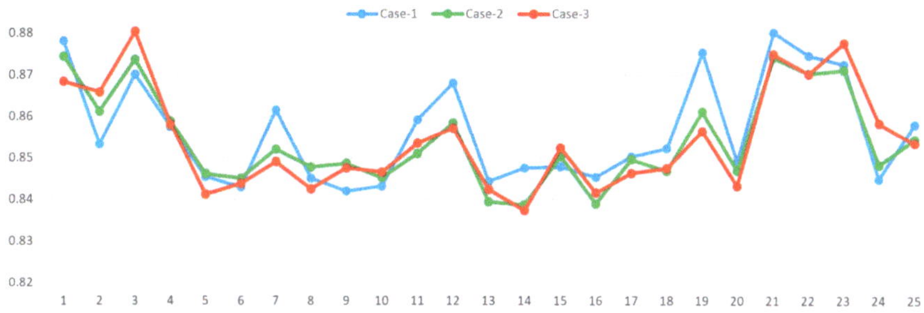

Fig. 2.19 Comparison of students' ranking

In cases where the job description allows working alone or working in small teams, soft skills can be neglected compared with hard skills. The vice versa is also true; the larger team always will require abilities related to the soft skills, while the missed hard skills could be overcome by proper seminars or workshops. Considering the problems in wind energy, various job positions can be found without teamwork requirements such as configuring embedded software and SCADA monitoring.

2.9 Assessment of Personnel Skills Using Model-Driven Framework for Generating a Variety of Online and Print Tests

The personnel selection can be realized in assessment centers where the assessment is carried out through the application of various types of situational tests or simulations, management exercises, role-playing games, group dynamics, business games, etc. New business challenges require the incorporation of state-of-the-art technology into business intelligence tools to improve the quality of service/product. Therefore, the current trends are focused not only on the provision of content but as an integral part of the training are the tools for verification and assessment level of knowledge (Saeed, 2021; Petrova et al., 2022). The process of assessment could be executed differently including proper tests, writing code, or constructing algorithms. It is worth mentioning that some courses in the area of mathematics and programming require special kinds of tools for algorithms visualization (Kostadinova et al., 2017) or automated design algorithms (Borissova & Mustakerov, 2015). Another suitable tool useful for testing in programming is the existence of a proper environment where some software code can be tested (Meng & Qu, 2021).

Tests can be designed and developed with automatic classification options. Although setting up a question bank is a time-consuming activity, the test elements can be reused and students' exams are automatically assessed. Moreover, the immediate online assessment contributes to receiving quick feedback on achieved results. It is shown that students who use the tests during training have better results than other students (Hennig et al., 2019). Different types of questions could be used like questions of multiple-choice type, true-false selection, short-answer, and essay all of them provided by test banks. It should be noted that true-false questions have a 50% chance of guessing the correct answer, so multiple-choice questions are often preferred. The use of multiple-choice questions provides an opportunity to learn new information from the test: for example, the learner may not know the answer to the question, but may be able to justify the answer, based on the knowledge gained (Marsh & Cantor, 2014). Another promising direction in question construction is the usage of some kind of gamification (Borissova et al., 2020). An analysis of algorithms used to generate test questions in e-testing systems can be found in Kostadinova et al. (2019).

To guarantee needed flexibility a promising approach is the usage of different mathematical models. In this regard, the current section describes an integrated model-driven framework for generating and evaluating online tests with different levels of complexity is proposed.

2.9.1 Integrated Models-Driven Framework for Tests Generation

The proposed integrated framework for test generation is based on different scenarios, each of them realized by a specific mathematical model. Each of these models could be tuned in a such way to generate tests with different levels of complexity. The basic workflow algorithm of the proposed framework for generating various tests with different levels of complexity is illustrated in Fig. 2.20 (Borissova et al., 2022a).

This approach for the generation of tests is applicable in cases (1) of using multiple-choice, short-answer, or true-false questions; (2) when the total number of questions; and (3) their corresponding difficulty level score, available through databases in different subject areas, are known. In the presence of already prepared databases with questions, the user can choose the appropriate subject area for which it will be tested to establish and assess the level of acquired knowledge.

The next step is to select the level of complexity of the generated test. Tests with different levels of complexity can be generated to match different grades. Several levels of complexity of the generated tests have been proposed in accordance with the rating scale, namely an average test (3), a good test (4), a very good grade test (5), and an excellent test (6). It should be noted that there are no obstacles to defining more levels of complexity of the generated tests, if necessary. The choice of the level of complexity of the test is not related to the condition that the user has successfully passed the lower test level; i.e., there is no limit to the sequence when choosing a test, according to its complexity.

The next step is to choose a model to generate a test. Three main models are proposed: (1) model for optimization of the number of scores in the generated tests, (2) model for optimization of the number of questions in the generated tests, and

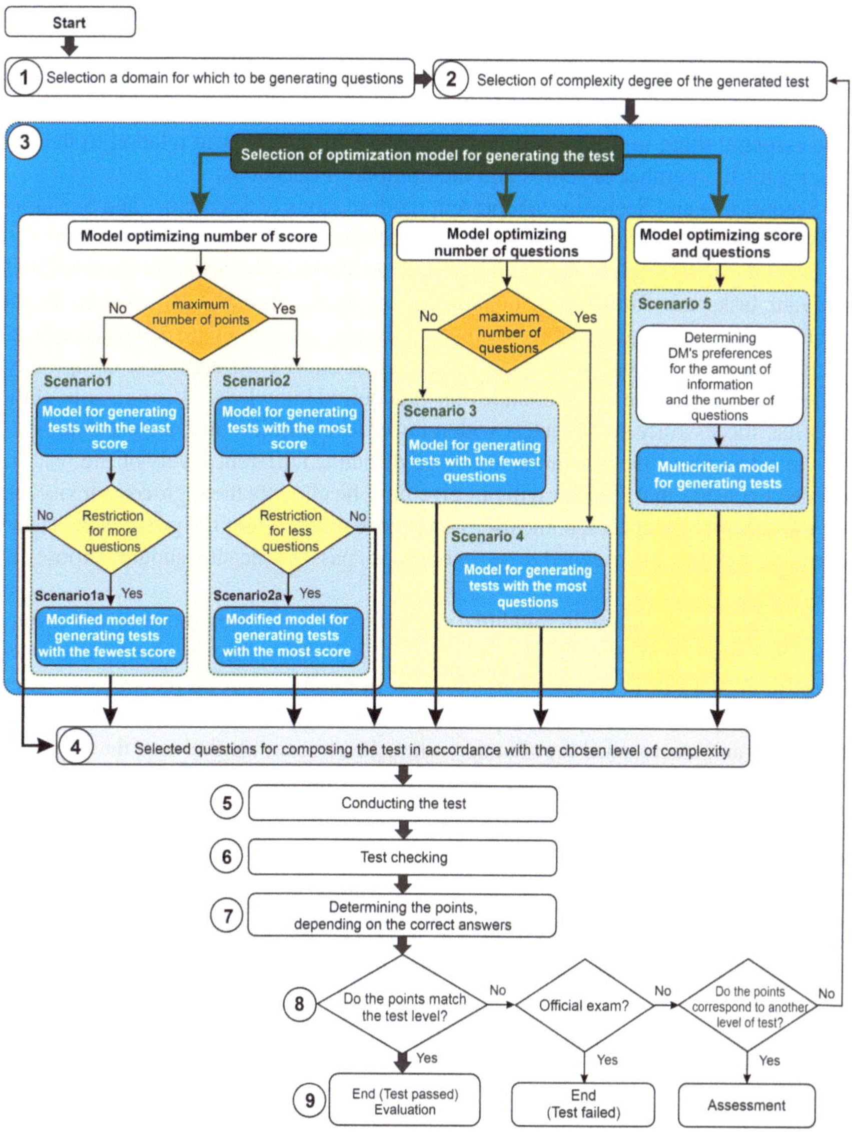

Fig. 2.20 Basic workflow algorithm of a framework for tests generation

(3) model for simultaneous optimization of the number of scores and the number of questions in the generated tests. The first model can be implemented through two separate scenarios. In Scenario-1, the situation in which questions with the lowest score can be selected is realized, while in Scenario-2 the goal is to choose questions with the highest score to satisfy the selected level for testing/evaluation. The model that optimizes the number of questions in the generated tests can also be

implemented through two scenarios. In Scenario-3, the case of generating a test with the fewest questions is realized, and in Scenario-4 a test with the most questions is generated. The third model for simultaneous optimization of the number of scores and the number of questions in the generated tests is implemented by Scenario-5. The essential thing in it is to determine the user's preferences in relation to the two criteria for the number of scores and the number of questions.

Scenario-1a and Scenario-2a represent modifications of Scenario-1 and Scenario-2 by introducing restrictions on the number of questions to compose the tests.

Regardless of which of the models and the respective scenarios are implemented, relevant tasks are formulated and solved, the solutions of which determine the quantity and type of questions for compiling the tests, according to the selected level.

For the purposes of self-testing, the user can choose different test complexity. This allows the testing of different levels of acquired knowledge. For example, if the user has successfully passed a test with a lower level of difficulty, the user may take the next test when the questions are more difficult. Different levels of the test are needed to ensure that the user will not give up if he chooses the high-complexity test for the first time. In this case, the user can choose the test from the previous level and check if the user's knowledge is enough to pass or he/she should choose an easier test.

It should also be borne in mind that regardless of the chosen mode of self-testing or official assessment, the time for each test is fixed. This means that the time to solve a test with more difficult questions, but less in number, will be the same as when solving a test with easier questions, but more in number. When the user is in official test mode and has reached the situation where he chooses another test, the total time of the two tests will not be the sum of the times of the two tests but will have the set time provided for the test.

2.9.2 Mathematical Models for Generating Tests with Different Levels of Complexity

The needed input data for the realization of the different levels of complexity of tests are shown in Table 2.17.

The set of questions is denoted by $Q = \{q_1, q_2, \ldots, q_M\}$ and is composed of M number of questions with different difficulty levels. Each question is assigned a corresponding level of difficulty, and which occupies values in the interval $1 \leq d_i \leq N$, where the lowest level of difficulty of the questions is denoted by 1, and the value corresponding to the highest level of difficulty is expressed by N. It should be noted that there is no relationship between the total number of questions (Q) and the levels of difficulty.

The difficulty index (p) of the questions can be determined using the proposed formula (Bachman, 2004):

Table 2.17 Questions, complexity of questions, levels of complexity of tests

Total number of questions	Degree of questions' difficulty	Binary integer variables	Level of complexity of the test (L^k)		
			L^1 Excellent	L^2 Very good	L^3 Good
q_1	d_1	x_1	$L^1_{\min} < L^1$	$L^2_{\min} < L^2$	$L^3_{\min} < L^3$
q_2	d_2	x_2	$L^1 < L^1_{\max}$	$L^2 < L^2_{\max}$	$L^3 < L^3_{\max}$
...	...	...			
q_i	d_i	x_i			
....	....	....			
q_M	d_M	x_M			

$$p = \left(U_p + L_p\right)/2n \tag{2.32}$$

where U_p and L_p are the number of learners who answered the question correctly in the upper and lower group, and n is the value of the total number of learners in the groups.

Three levels of difficulty are often used in the implementation of such tests, and the corresponding allowable intervals for these difficulties are as follows: easy (1.00 ÷ 0.65), medium (0.65 ÷ 0.35), and difficult (0.35 ÷ 0.0) (Tasci et al., 2014).

An essential feature in determining the questions that make up the respective test is that the sum of the score for the difficulty of the questions must be greater than the required value, corresponding to an excellent grade, i.e., $L^k < \sum d_M$. Each question is assigned control binary variables x_i, through which the choice of questions is realized, according to the required level of complexity of the test. In the general case, as many levels of complexity can be used, but to demonstrate the applicability of described approach, three types of levels of complexity are defined, namely level for excellent assessment, level for very good, and level for good. For each level, it is necessary to determine the permissible range of score obtained from the tests performed by setting the respective lower ($L_{\min}$) and upper limits ($L_{\max}$) for the total number of score. There are no restrictions on the levels of tests that can be generated, but for each level it is necessary to determine the relevant limits.

The proposed mathematical model that realizes Scenario-1 which optimizes the amount of information and determines the number of questions for tests with different levels of complexity is as follows (Borissova et al., 2022a):

$$\max\left(L^k = \sum_{i=1}^{M} x_i d_i\right) \tag{2.33}$$

subject to

$$\sum_{i}^{M} d_i = D \tag{2.34}$$

$$L_{\max}^{k} < D \tag{2.35}$$

$$L^{k} \geq L_{\min}^{k} \tag{2.36}$$

$$L^{k} \leq L_{\max}^{k} \tag{2.37}$$

$$x_i \in \{0, 1\} \tag{2.38}$$

where L^k is the level of complexity of the selected test, x_i is the binary integer variables that can take a value of 0 or 1, d_i is the level of difficulty of the questions, and D is the total number of score for all questions, of which the test will be generated.

The use of binary integer variables x_i guarantees the single choice of questions in the generation of the specific test. The objective function (2.33) seeks to maximize the total score number of questions for the test taking into account the predetermined degree of difficulty for each of the questions. The total amount of possible score for the respective difficulties of all questions (D) is represented by the constraint (2.34). It should be noted that this total sum of the difficulty score of all questions (D) is not the upper limit ($L_{\max}^{k}$) for the highest complexity of tests (test for excellent grade). Therefore, this upper limit $L_{\max}^{k}$ for the test with the selected level of difficulty must always be less than the total number of points for all questions.

Constraints (2.36) and (2.37) determine the lower and upper limit for the number of score that form the chosen level of complexity of the test for which the learner is tested. The constraint (2.38) was used to determine the questions for the specific test using the binary integer variables x_i.

The formulated model (2.33)–(2.38) can be modified by changing the direction of the objective function (2.33) and instead of looking for a maximum of the number of score for the generated test, we can look for its minimum (Scenario-2).

In order to implement Scenario-1a and Scenario-2a, it is necessary to add additional restrictions on the number of selected questions that make up the specific test. An upper limit on the number of questions in the generated tests can be added to the first formulated model (2.33)–(2.38). In this case (Scenario-1a), the model will use the objective function (2.33) subject to constraints (2.34)–(2.38) and an additional constraint:

$$\sum_{i=1}^{M} x_i \leq K_{\max} \tag{2.39}$$

To realize Scenario-2a where objective function (2.33) is minimizing, the additional restriction is about the lower limit for the number of questions for the generated tests as follows:

$$\sum_{i=1}^{M} x_i \geq K_{\min} \tag{2.40}$$

Constraints (2.39) and (2.40) provide flexibility by providing options for choosing a small number of questions, but with high difficulty and vice versa—choosing more questions and not so difficult.

For the implementation of Scenario-3, another mathematical model is used, which minimizes the number of questions that make up the respective test and has the following form:

$$\min\left(K^k = \sum_{i=1}^{M} x_i\right) \tag{2.41}$$

subject to

$$\sum_{i}^{M} x_i d_i \leq L^k \tag{2.42}$$

$$L^k \leq L_{\max}^k \tag{2.43}$$

$$L^k \geq L_{\min}^k \tag{2.44}$$

$$x_i \in \{0, 1\} \tag{2.45}$$

$$L_{\max}^k < D \tag{2.46}$$

$$\sum_{i}^{M} d_i = D \tag{2.47}$$

It is also possible to introduce an additional restriction on this model regarding the minimum number of generated questions ($K > K^*$).

To implement Scenario-4, where as many test questions as possible are identified, the target function is changed to maximize (2.41) under the same constraints. In this realization, it is also possible to add additional restriction that could be introduced regarding the admissible maximum number of questions that will constitute the generated test ($K < K^*$).

Scenario-5 requires using multicriteria optimization as two parameters need to be optimized simultaneously. The corresponding model that realizes this scenario is as follows:

$$\begin{cases} \max\left(L^k = \sum_{i=1}^{M} x_i d_i\right) \\ \min\left(K^k = \sum_{i=1}^{M} x_i\right) \end{cases} \tag{2.48}$$

subject to

$$L^k \le L^k_{\max} \tag{2.49}$$

$$L^k \ge L^k_{\min} \tag{2.50}$$

$$L^k_{\max} < D \tag{2.51}$$

$$\sum_{i}^{M} d_i = D \tag{2.52}$$

$$x_i \in \{0, 1\} \tag{2.53}$$

The essence of this formulation is that the objective function simultaneously seeks to maximize the score (L^k) and minimize the number of questions (K^k) of the questions that will make up the test. The two criteria formulated are contradictory, as the first criterion for score will seek to rely on as many questions as possible, and the second will seek to minimize the total number of questions for the test. The limitations of this model are similar to the previous models and are related to the satisfaction of the required number of score for the respective level of complexity of the test.

In multicriteria optimization problems, several criteria are optimized simultaneously, and in the general case, there is no single alternative to optimize all criteria. Unlike single-criteria optimization, the solution of a multicriteria problem can be considered as a concept rather than a definition. In this type of optimization, there is usually no optimal solution and it is necessary to determine a set of ones that satisfy the user's preferences toward the optimality expressed as Pareto optimality (Marler & Arora, 2004).

Among the variety of methods able to cope with such problems, the weighted sum approach is the easiest to understand. For simplicity, it is assumed that the used two criteria are of equal importance. The usage of the weighted sum approach requires transforming multiple objective functions (2.48) into single aggregated objective function as follows:

$$\max\left(w_1\left(\frac{L^{\max} - L}{L^{\max} - L^{\min}}\right)_k + w_2\left(\frac{K - K^{\min}}{K^{\max} - K^{\min}}\right)_k\right) \tag{2.54}$$

An additional restriction is also added:

$$w_1 + w_2 = 1 \tag{2.55}$$

where w_1 and w_2 are weighted coefficients that express the criteria importance.

2.9.3 Numerical Application

All described models that express specific scenarios for generating different tests are numerically tested. The applicability of the proposed integrated model-driven framework is demonstrated by using a setup test bank with 40 questions from computer science area. The corresponding difficulties of the questions are shown in Table 2.18.

A specific feature of the proposed approach to generating different tests is the fact that each time the questions from the specific domains are updated, it is necessary to check the total amount of points that form the difficulty of the individual questions. This is necessary to ensure that the set upper and lower limits of the respective test can be realized by the questions used and their difficulties. This is due to the core of the combinatorial nature of all models included in the proposed integrated models-driven framework.

A distinguishing feature of all models is the requirement for setting up upper and lower limits for the different tests, according to their degree of difficulty. For the formation of the different tests it is necessary to achieve a certain number of score. Considering the given input data about the questions and their difficulties the used admissible limits about three levels of complexity for generated e-tests are shown in Table 2.19.

The number of selected questions for the different levels of complexity of generated e-tests using all the proposed single-criteria and multicriteria optimization models is shown in Fig. 2.21.

From the obtained results, it can be established that the number of questions constituting the generated tests when using the multicriteria model differs from the

Table 2.18 Set of question and their difficulty

#	Question difficulty	#	Question difficulty	#	Question difficulty	#	Question difficulty
1	5	11	6	21	6	31	8
2	7	12	7	22	3	32	4
3	10	13	8	23	5	33	8
4	9	14	3	24	4	34	10
5	7	15	10	25	7	35	6
6	6	16	5	26	5	36	5
7	3	17	7	27	10	37	10
8	5	18	10	28	6	38	6
9	8	19	9	29	8	39	7
10	9	20	8	30	3	40	7

Table 2.19 Upper and lower limits for the different tests

Test for excellent		Test for very good		Test for good	
L_{max}^{E}	L_{min}^{E}	L_{max}^{VG}	L_{min}^{VG}	L_{max}^{G}	L_{min}^{G}
200	190	189	176	175	160

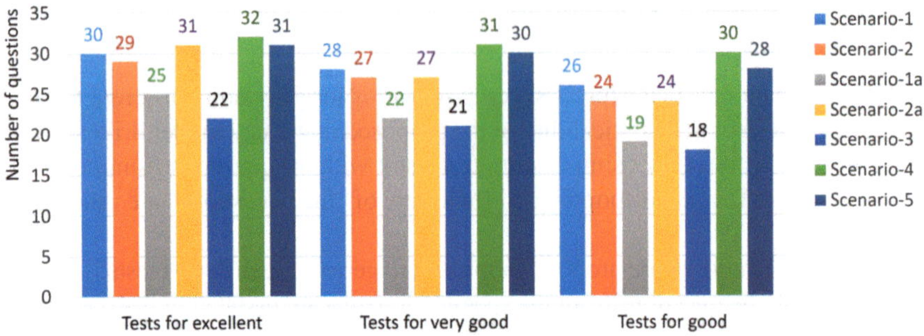

Fig. 2.21 Selected number of questions for the generated tests by using different models

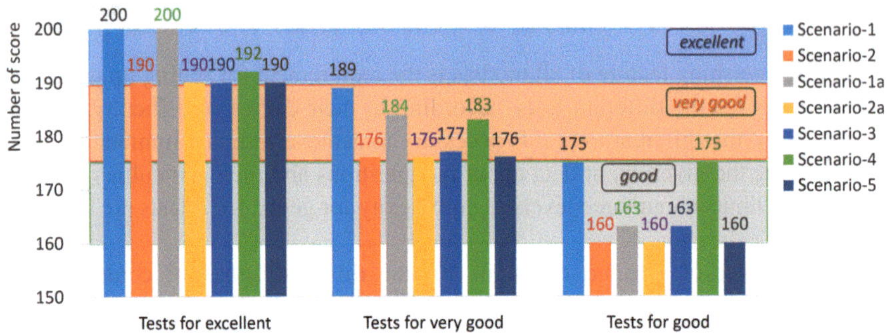

Fig. 2.22 Permissible point limits for the different levels of complexity of the tests and the results obtained using the proposed models

number of questions determined by the single-criteria models. This proves that the use of multicriteria optimization can be used as well to generate tests as single-criteria optimization, as it allows finding other combinations of questions that satisfy the corresponding level of complexity of the tests. For example, in a test for good, the number of questions when using the multicriteria model is set at 28, in contrast to the results in single-criteria models—respectively 26, 24, 19, 24, 18, and 30 questions. In the tests for very good and good, the result is similar (see Fig. 2.21).

A similar comparison of the score for the generated tests using the formulated single-criterion and multicriteria optimization models is shown in Fig. 2.22.

When using the model implementing Scenario-1, the results show that the generated tests at all three levels of complexity tend to reach the upper limit for the respective level. This is an expected result, as the objective function seeks to satisfy the maximum level of complexity. In the model implementing Scenario-2, the result for the value of the objective function is not unambiguous. For the model implementing Scenario-1a and the model implementing Scenario-2a they use additional limits for the number of questions included in the generated tests.

When using the model implementing Scenario-3, the aim is to determine the smallest possible number of questions satisfying the required number of points for the respective level, which is evident from the results of Fig. 2.19. In this model, the main idea is to choose issues that are more difficult. The model implementing Scenario-4 seeks to identify as many questions as possible that will have less difficulty, but will still be able to satisfy the required number of points for the respective level. In the model that implements Scenario-5, the objective function seeks to satisfy simultaneously both criteria expressed with equal importance. That means a compromise between number of questions and number of score is needed that should conform to the required test complexity level.

For two of the formulated models (implementing Scenario-1a and Scenario-2a), using limitations for the number of questions, it may not be possible to find a solution if the set lower/upper limits for the questions cannot be satisfied. If such a situation arises, the relevant restriction needs to be changed. The use of other models allows only by changing the direction (min/max) and the expression for the objective function (number of score/number of questions) to generate the relevant tests.

From the obtained results in Fig. 2.22, it is easy to establish that the number of score for the generated tests when using the multicriteria model does not differ from the others. This is a confirmation that multicriteria optimization can be used just as well to generate tests, as the score of the defined combinations of questions satisfies the set levels of test complexity (Fig. 2.22).

A distinctive feature of the proposed approach is the ability to choose different mathematical models, leading to the determination of different combinations of questions that make up the generating tests. It should be noted that in self-test mode, when another test level or a new test of the same level is selected, it is possible to use appropriate flags for the already selected questions, so that the next generated test consists of questions other than the previous test. This would guarantee the uniqueness of the tests and crawl the entire range of content if the number of questions for the specific area is large enough.

All formulated models are of the type of mixed-integer linear programming. Determining whether a question will be part of the test or not is realized through the use of binary integer variables that uniquely define the question. From the conducted numerical testing, it can be established that tests with the same level of complexity can be realized by using a different number of questions. This is due to the fact that the questions in the database have different degrees of difficulty and it is possible to implement different combinations of questions that meet the set limits for the number of scores for different levels of complexity of the tests.

The advantage of the proposed approach, and of the formulated mathematical models, is the flexibility to generate tests of different degrees of complexity using binary integer variables.

Electronic testing is the most widely used tool not only in universities and schools but also as a means of establishing the level of competence of employees from various fields. Similar tests of varying degrees of complexity can be used when hiring personnel with varying degrees of responsibility in the implementation of a wind farm project. Thus, depending on the obtained results, it will be possible to have confidence that the right person has been selected for the relevant position.

References

Afshari, A. R., Nikoloc, M., & Akbari, Z. (2017). Personnel selection using group fuzzy AHP and SAW methods. *Journal of Engineering Management and Competitiveness, 7*(1), 3–10.

Alavi, S. M. S., Maleki, A., & Khaleghi, A. (2022). Optimal site selection for wind power plant using multi-criteria decision-making methods: A case study in eastern Iran. *International Journal of Low-Carbon Technologies, 17*, 1319–1337. https://doi.org/10.1093/ijlct/ctac009

Alphan, H. (2021). Modelling potential visibility of wind turbines: A geospatial approach for planning and impact mitigation. *Renewable and Sustainable Energy Reviews, 152*, 111675. https://doi.org/10.1016/j.rser.2021.111675

Arteaga-Lopez, E., & Angeles-Camacho, C. (2021). Innovative virtual computational domain based on wind rose diagrams for micrositing small wind turbines. *Energy, 220*, 119701. https://doi.org/10.1016/j.energy.2020.119701

Asadi, M., & Pourhossein, K. (2021). Wind farm site selection considering turbulence intensity. *Energy, 236*, 121480. https://doi.org/10.1016/j.energy.2021.121480

Bachman, L. F. (2004). *Statistical analyses for language assessment*. Cambridge University Press.

Badi, I., Pamučar, D., Stevic, Z., & Muhammad, L. J. (2023). Wind farm site selection using BWM-AHP-MARCOS method: A case study of Libya. *Scientific African, 19*, e01511. https://doi.org/10.1016/j.sciaf.2022.e01511

Baltzer, M., Stolper, O., & Walter, A. (2013). Is local bias a cross-border phenomenon? Evidence from individual investors' international asset allocation. *Journal of Banking & Finance, 37*(8), 2823–2835. https://doi.org/10.1016/j.jbankfin.2013.04.009

Barfod, M. B., & Leleur, S. (Eds.). (2014). *Multi-criteria decision analysis for use in transport decision making*. DTU Lyngby: Technical University of Denmark, Transport.

Borissova, D., Buhtiyarov, N., Yoshinov, R., Garvanova, M., & Garvanov, I. (2022a). Integrated models-driven framework to generate various online and print tests. In K. Saeed & J. Dvorský (Eds.), *Computer information systems and industrial management. CISIM 2022. Lecture Notes in Computer Science* (Vol. 13293, pp. 316–329). https://doi.org/10.1007/978-3-031-10539-5_23

Borissova, D., Danev, V., Garvanova, M., Garvanov, I., & Yoshinov, R. (2022b). Key indicators to measure student performance in IoT and their teamwork ability. In M. E. Auer & T. Tsiatsos (Eds.), *New realities, mobile systems and applications. IMCL 2021. Lecture Notes in Networks and Systems* (Vol. 411, pp. 711–720). https://doi.org/10.1007/978-3-030-96296-8_64

Borissova, D., Keremedchiev, D., & Tuparov, G. (2020). Multi-criteria model for questions selection in generating e-education tests involving gamification. *TEM Journal – Technology, Education, Management, Informatics, 9*(2), 779–785. https://doi.org/10.18421/TEM92-47

Borissova, D., & Keremedchiev, D. (2019). Group decision making in evaluation and ranking of students by extended simple multi-attribute rating technique. *Cybernetics and Information Technologies, 18*(3), 45–56. https://doi.org/10.2478/cait-2019-0025

Borissova, D., Korsemov, D., & Mustakerov, I. (2018). Multi-attribute group decision making considering difference in experts knowledge: An Excel application. In *Proceedings of 12th International Management Conference – Management Perspectives in the Digital Era, Bucharest, Romania* (pp. 387–395).

Borissova, D., Korsemov, D., & Mustakerov, I. (2019). Multi-criteria decision making problem for doing business: Comparison between approaches of individual and group decision making. In K. Saeed, R. Chaki, & V. Janev (Eds.), *Computer information systems and industrial management. CISIM 2019. Lecture Notes in Computer Science* (Vol. 11703, pp. 385–396). https://doi.org/10.1007/978-3-030-28957-7_32.

Borissova, D., Mustakerov, I., & Korsemov, D. (2016). Business intelligence system via group decision making. *Cybernetics and Information Technologies, 16*(3), 219–229. https://doi.org/10.1515/cait-2016-0045

Borissova, D., & Mustakerov, I. (2015). E-learning tool for visualization of shortest paths algorithms. *Trends Journal of Sciences Research, 2*(3), 84–89.

Borissova, D. (2018). A group decision making model considering experts competency: An Application in personnel selections. *Comptes Rendus de l'Académie Bulgarie des Science, 71*(11), 1520–1527.
Borissova, D. (2016). Group decision making for selection of k-best alternatives. *Comptes Rendus de l'Académie Bulgarie des Science, 69*(2), 183–190.
Botterill, L. C., & Cockfield, G. (2016). The relative importance of landscape amenity and health impacts in the wind farm debate in Australia. *Journal of Environmental Policy and Planning, 18*, 447–462. https://doi.org/10.1080/1523908X.2016.1138400
Buyukozkan, G. (2004). Multi-criteria decision making for e-marketplace selection. *Internet Research, 14*(2), 139–154. https://doi.org/10.1108/10662240410530853
Cabrerizo, F. J., Herrera-Viedma, C. E., & Pedrycz, W. (2013). A method based on PSO and granular computing of linguistic information to solve group decision making problems defined in heterogeneous contexts. *European Journal of Operational Research, 230*(3), 624–633. https://doi.org/10.1016/j.ejor.2013.04.046
Costanzo, G., Brindley, G., & Cole, P. (2023). *Wind energy in Europe: 2022 Statistics and the outlook for 2023–2027* (Ed. R. O'Sullivan). https://www.connaissancedesenergies.org/sites/default/files/pdf-actualites/WindEurope-report-wind-energy-in-europe-2022.pdf
Danielson, M., Ekenberg, L., Gothe, M., & Larsson, A. (2016). A decision analytical perspective on public procurement processes. In *Real-World Decision Support Systems. Integrated Series in Information Systems, 37.*
Deloitte. (2014). *Establishing the investment case Wind power*. Deloitte.
Edwards, W. (1977). How to use multiattribute utility measurement for social decision making. *IEEE Transactions on Systems, Man, and Cybernetics, 7*(5), 326–340. https://doi.org/10.1109/TSMC.1977.4309720
Fetanat, A., & Khorasaninejad, E. (2015). A novel hybrid MCDM approach for offshore wind farm site selection: A case study of Iran. *Ocean & Coastal Management, 109*, 17–28. https://doi.org/10.1016/j.ocecoaman.2015.02.005
Fleta-Asin, J., & Munoz, F. (2021). Renewable energy public–private partnerships in developing countries: Determinants of private investment. *Sustainable Development, 29*(4), 653–670. https://doi.org/10.1002/sd.2165
Fu, Y.-K. (2019). An integrated approach to catering supplier selection using AHP-ARAS-MCGP methodology. *Journal of Air Transport Management, 75*, 164–169. https://doi.org/10.1016/j.jairtraman.2019.01.011
Georgieva, P., & Popchev, I. (2013). Fuzzy logic q–measure model for managing financial investments. *Comptes Rendus de l'Académie Bulgarie des Science, 66*(5), 651–658.
Grady, S. A., Hussainia, M. Y., & Abdullah, M. M. (2005). Placement of wind turbines using genetic algorithms. *Renewable Energy, 30*, 259–270. https://doi.org/10.1016/j.renene.2004.05.007
Groh, A., Liechtenstein, H., & Lieser, K. (2013). *The venture capital and private equity country attractiveness index 2013 annual*. http://blog.iese.edu/vcpeindex/
Guo, Y., Ru, P., Su, J., & Anadon, L. D. (2015). Not in my backyard, but not far away from me: Local acceptance of wind power in China. *Energy, 82*, 722–733. https://doi.org/10.1016/j.energy.2015.01.082
Hall, R., Joao, E., & Knapp, C. W. (2020). Environmental impacts of decommissioning: On-shore versus offshore wind farms. *Environmental Impact Assessment Review, 83*, 106404. https://doi.org/10.1016/j.eiar.2020.106404
Hennig, S., Staatz, C. E., Bond, J. A., Leung, D., & Singleton, J. (2019). Quizzing for success: Evaluation of the impact of feedback quizzes on the experiences and academic performance of undergraduate students in two clinical pharmacokinetics courses. *Currents in Pharmacy Teaching and Learning, 11*, 742–749. https://doi.org/10.1016/j.cptl.2019.03.014
Huang, Y. S., Chang, W. C., Li, W. H., & Lin, Z. L. (2013). Aggregation of utility-based individual preferences for group decision-making. *European Journal of Operational Research, 229*, 462–469. https://doi.org/10.1016/j.ejor.2013.02.043

IRENA, Renewable Energy and Jobs Annual Review. (2021). SPECIAL EDITION Labour and Policy Perspectives. https://www.ilo.org/wcmsp5/groups/public/%2D%2D-dgreports/%2D%2D-dcomm/%2D%2D-publ/documents/publication/wcms_823807.pdf

Karakislak, I., & Schneider, N. (2023). The mayor said so? The impact of local political figures and social norms on local responses to wind energy projects. *Energy Policy, 176*, 113509. https://doi.org/10.1016/j.enpol.2023.113509

Kasner, R. (2022). The environmental efficiency of materials used in the lifecycle of a wind farm. *Sustainable Materials and Technologies, 34*, e00512. https://doi.org/10.1016/j.susmat.2022.e00512

Korsemov, D., Borissova, D., & Mustakerov, I. (2018). Group decision making for selection of supplier under public procurement. In S. Kalajdziski & N. Ackovska (Eds.), *ICT Innovations 2018. Communications in Computer and Information Science* (Vol. 940, pp. 51–58). https://doi.org/10.1007/978-3-030-00825-3_5

Korsemov, D., & Borissova, D. (2018). Modifications of simple additive weighting and weighted product models for group decision making. *Advanced Modeling and Optimization, 20*(1), 101–112.

Kostadinova, I., Rasheva-Yordanova, K., Garvanova, M. (2019). Analysis of algorithms for generating test questions in e-testing systems. In *11th International Conference on Education and New Learning Technologies, EDULEARN'19* (pp. 1714–1719). https://doi.org/10.21125/edulearn.2019.0498

Kostadinova, I., Rasheva-Yordanova, K., Ivanov, I., & Petrova, P. (2017). Automated system for generating and validation a learning tests. In *ICERI2017 Proceedings* (pp. 414–424). https://doi.org/10.21125/iceri.2017.0160

Lin, H.-T. (2010). Personnel selection using analytic network process and fuzzy data envelopment analysis approaches. *Computers & Industrial Engineering, 59*, 937–944. https://doi.org/10.1016/j.cie.2010.09.004

Liu, J., Yu, J., Yin, Y., & Wei, Q. (2021). An evolutionary game approach for private sectors' behavioral strategies in China's green energy public–private partnership projects. *Energy Reports, 7*(7), 696–715. https://doi.org/10.1016/j.egyr.2021.09.201

Lyu, W., & Liu, J. (2021). Soft skills, hard skills: What matters most? Evidence from job postings. *Applied Energy, 300*, 117307. https://doi.org/10.1016/j.apenergy.2021.117307

Marcenaro-Gutierrez, O. D., Lopez-Agudo, L. A., & Henriques, C. O. (2021). Are soft skills conditioned by conflicting factors? A multiobjective programming approach to explore the trade-offs. *Economic Analysis and Policy, 72*, 18–40. https://doi.org/10.1016/j.eap.2021.07.008

Marler, R. T., & Arora, J. S. (2004). Survey of multi-objective optimization methods for engineering. *Structural and Multidisciplinary Optimization, 26*, 369–395. https://doi.org/10.1007/s00158-003-0368-6

Marsh, E. J., & Cantor, A. D. (2014). Learning from the test: Dos and don'ts for using multiple-choice tests. In M. McDaniel, R. Frey, S. Fitzpatrick, & H.L. Roediger (Eds.), *Integrating cognitive science with innovative teaching in STEM disciplines.* https://doi.org/10.7936/K7Z60KZK

Martinez, A., & Iglesias, G. (2021). Wind resource evolution in Europe under different scenarios of climate change characterised by the novel Shared Socioeconomic Pathways. *Energy Conversion and Management, 234*, 113961. https://doi.org/10.1016/j.enconman.2021.113961

Memari, A., Dargi, A., Jokar, M. R. A., Ahmad, R., & Rahim, A. R. A. (2019). Sustainable supplier selection: A multi-criteria intuitionistic fuzzy TOPSIS method. *Journal of Manufacturing Systems, 50*, 9–24. https://doi.org/10.1016/j.jmsy.2018.11.002

Meng, W., & Qu, R. (2021). Automated design of search algorithms: Learning on algorithmic components. *Expert Systems with Applications, 185*, 115493. https://doi.org/10.1016/j.eswa.2021.115493

Mia, I., Estrada, J., & Geiger, T. (2007). Benchmarking National attractiveness for private investment in Latin American infrastructure. In *World Economic Forum, REF: 200407.*

Moller, B. (2006). Changing wind-power landscapes: Regional assessment of visual impact on land use and population in Northern Jutland, Denmark. *Applied Energy, 83*, 477–494.

Montandon, J. E., Politowski, C., Silva, L. L., Valente, M. T., Petrillo, F., & Gueheneuc, Y.-G. (2021). What skills do IT companies look for in new developers? A study with Stack Overflow jobs. *Information and Software Technology, 129*, 106429. https://doi.org/10.1016/j.infsof.2020.106429

Mora, J. C., Calero Baron, J. M., Riquelme Santos, J. M., & Burgos Payan, M. (2007). An evolutive algorithm for wind farm optimal design. *Neurocomputing, 70*, 2651–2658. https://doi.org/10.1016/j.neucom.2006.05.017

Mustakerov, I., & Borissova, D. (2014). A Web application for group decision-making based on combinatorial optimization. In *Proceedings of International Conference on Information Systems and Technologies* (pp. 46–56).

Mustakerov, I., & Borissova, D. (2013). Investments attractiveness via combinatorial optimization ranking. *WASET, International Scholarly and Scientific Research & Innovation, 7*(10). https://doi.org/10.5281/ZENODO.1088218

Paraschiv, S., Paraschiv, L. S., Serban, A., & Cristea, A. G. (2022). Assessment of onshore wind energy potential under temperate continental climate conditions. *Energy Reports, 8*(5), 251–258. https://doi.org/10.1016/j.egyr.2022.02.118

Peneva, V., & Popchev, I. (2007). Aggregation of fuzzy relations using weighting function. *Comptes Rendus de l'Académie Bulgarie des Science, 60*(10), 1047–1052.

Peneva, V., & Popchev, I. (2009). Models for fuzzy multicriteria decision making based on fuzzy relations. *Comptes Rendus de l'Académie Bulgarie des Science, 62*(5), 551–558.

Petrova, P., Kostadinova, I., & Alsulami, M. H. (2022). Embedded intelligence in a system for automatic test generation for smoothly digital transformation in Higher Education. In: V. Sgurev, V. Jotsov, & J. Kacprzyk (Eds.), *Advances in intelligent systems research and innovation. Studies in systems, decision and control* (Vol. 379, pp. 441–461). https://doi.org/10.1007/978-3-030-78124-8_20

Poddaeva, O., & Fedosova, A. (2022). Extreme wind speeds analysis using extended wind rose based on statistic methods. *Energy Reports, 8*(9), 1177–1184. https://doi.org/10.1016/j.egyr.2022.07.103

Popchev, I., & Georgieva, P. (2008). A fuzzy approach to solving multicriteria investment problems. In M. Iskander (Eds.), *Innovative techniques in instruction technology, E-learning, E-assessment, and education* (pp. 427–431). https://doi.org/10.1007/978-1-4020-8739-4_75

Rao, R. V. (2007). Introduction to multiple attribute decision-making (MADM) methods. In *Decision making in the manufacturing environment* (pp. 27–41). Springer. https://doi.org/10.1007/978-1-84628-819-7_3

Rediske, G., Burin, H. P., Rigo, P. D., Rosa, C. B., Michels, L., & Siluk, J. C. M. (2021). Wind power plant site selection: A systematic review. *Renewable and Sustainable Energy Reviews, 148*, 111293. https://doi.org/10.1016/j.rser.2021.111293

Rockmann, C., Lagerveld, S., & Stavenuiter, J. (2017). Operation and maintenance costs of offshore wind farms and potential multi-use platforms in the Dutch North Sea. In B. Buck & R. Langan (Eds.), *Aquaculture Perspective of Multi-Use Sites in the Open Ocean* (pp. 97–113). https://doi.org/10.1007/978-3-319-51159-7_4

Saeed, K. (2021). Limits to growth concepts in classical economics. In R. Y. Cavana, B. C. Dangerfield, O. V. Pavlov, M. J. Radzicki, & I. D. Wheat (Eds.), *Feedback economics. Contemporary systems thinking* (pp. 217–246). https://doi.org/10.1007/978-3-030-67190-7_9

Sang, X., Liu, X., & Qin, J. (2015). An analytical solution to fuzzy TOPSIS and its application in personnel selection for knowledge-intensive enterprise. *Applied Soft Computing, 30*, 190–204. https://doi.org/10.1016/j.asoc.2015.01.002

Sarkar, S., Pratihar, D. K., & Sarkar, B. (2018). An integrated fuzzy multiple criteria supplier selection approach and its application in a welding company. *Journal of Manufacturing Systems, 46*, 163–178. https://doi.org/10.1016/j.jmsy.2017.12.004

Schlichter, B. R., & Buchynska, T. (2021). Soft skills of delivery managers in a co-sourced software project. *Procedia Computer Science, 181*, 905–912. https://doi.org/10.1016/j.procs.2021.01.246

Sgurev, V. (2014). Decision Markov logic. *Communication Technologies – Bulgarian Academy of Sciences, 67*(2), 181–188.

Shih, H.-S., Shyur, H.-J., & Lee, E. S. (2007). An extension of TOPSIS for group decision making. *Mathematical and Computer Modelling, 45*, 801–813. https://doi.org/10.1016/j.mcm.2006.03.023

Sørensen, J. D. (2006). Optimal reliability-based design of offshore wind turbine parks. In *2nd IFED Forum*, April 26–29, Lake Louise.

Szumilas-Kowalczyk, H., Pevzner, N., & Giedych, R. (2020). Long-term visual impacts of aging infrastructure: Challenges of decommissioning wind power infrastructure and a survey of alternative strategies. *Renewable Energy, 150*, 550–560. https://doi.org/10.1016/j.renene.2019.12.143

Tao, Z., Liu, X., Chen, H., & Chen, Z. (2015). Group decision making with fuzzy linguistic preference relations via cooperative games method. *Computers & Industrial Engineering, 83*, 184–192. https://doi.org/10.1016/j.cie.2015.02.016

Tasci, T., Parlak, Z., Kibar, A., Tasbasi, N., & Cebeci, H. I. (2014). A novel agent-supported academic online examination system. *Educational Technology & Society, 17*(1), 154–168. https://www.jstor.org/stable/jeductechsoci.17.1.154

Tavana, M., Fallahpour, A., Di Caprio, D., & Santos-Arteaga, F. J. (2016). A hybrid intelligent fuzzy predictive model with simulation for supplier evaluation and selection. *Expert Systems with Applications, 61*, 129–144. https://doi.org/10.1016/j.eswa.2016.05.027

Tougaard, J., Hermannsen, L., & Madsen, P. T. (2020). How loud is the underwater noise from operating offshore wind turbines? *Journal of the Acoustical Society of America, 148*(5), 2885. https://doi.org/10.1121/10.0002453

Wetzstein, A., Hartmann, E., Benton, W. C., Jr., & Hohenstein, N. O. (2016). A systematic assessment of supplier selection literature – State-of-the-art and future scope. *International Journal of Production Economics, 182*, 304–323. https://doi.org/10.1016/j.ijpe.2016.06.022

White, G., Garrad, A., & Tindal, A. (1997). Integrated design methodology for wind farms. *IPENZ Transactions, 24*(1), 62–68.

Zhang, S., Zhu, J., Liu, X., & Chen, Y. (2016). Regret theory-based group decision-making with multidimensional preference and incomplete weight information. *Information Fusion, 31*, 1–13. https://doi.org/10.1016/j.inffus.2015.12.001

Zhang, S.-F., & Liu, S.-Y. M. (2011). A GRA-based intuitionistic fuzzy multi-criteria group decision making method for personnel selection. *Expert Systems with Applications, 38*, 11401–11405. https://doi.org/10.1016/j.eswa.2011.03.012

Chapter 3
Decision-Making in Wind Farm Design

Abstract The second stage of the life cycle of wind farms is related to the planning and design of them including the layout design of the wind turbines. Wind farm layout has become increasingly important in recent decades, on the one hand, due to the promotion of wind energy development, and on the other hand, due to the differences in different geographical locations and conditions. By optimizing the wind farm layout, the micrositing of wind turbines aims to achieve a good balance between power output and wind farm costs. Micrositing of wind turbines is the process by which the specific location of wind turbines is determined. Each position must meet several requirements in terms of existing wind resources, distances from other wind turbines and neighbors, etc. Therefore, the optimal layout of the wind farm aims at the optimal distribution of the wind turbines in the wind farm, so as to achieve the maximum energy production according to the design parameters of the wind turbine used.

This chapter describes various single-criteria and multicriteria models applicable to the layout design of wind farms with different specific characteristics. It is shown how multi-attribute decision-making techniques and multi-objective optimizations can be integrated into a framework to support the design of wind farms. The use of this integrated approach makes it possible both to generate alternatives (different layouts of wind farm design) and to realize a choice of one of them. Two models of wind farm layout design are also presented, in which the site area has forbidden zones with regular and irregular geometry for turbine placement. A model for optimizing the location of wind turbines, taking into account the location of the common connection point, is also described. Finally, a combinatorial decision-making model for the selection of an optimal site for the hybrid wind-solar power plant is proposed.

Keywords Wind farm design · Mixed-integer optimization · Multi-objective optimization · Forbidden zones · Location of common coupling · Hybrid wind-solar site selection

D. Borissova, *Decision-Making in Design, Maintenance, Planning, and Investment of Wind Energy*, International Series in Operations Research & Management Science 355, https://doi.org/10.1007/978-3-031-52219-2_3

3.1 Concept for Wind Farm Layout Design

As a result of technological advances, there exist a number of different types and models of wind turbines that can be used for wind farm design. The different wind turbines have different technical characteristics reflecting on their power production and price and that defines a necessity of a proper choice. Because of the fact that purchasing wind turbines is one of the most essential investments for wind farm development, the choice of optimal wind turbine type and the number is a critical wind farm design step. It is assumed from the technological and maintenance point of view that all wind farm turbines are to be of the same type.

Gathering the maximum amount of energy from wind energy is directly related to the location of the wind turbines in the wind farm. Designing a wind farm layout is about finding the optimal placement of each individual turbine relative to the others within the boundaries of the particular wind farm. The layout of wind farms is challenging due to the limitations due to the wave phenomenon. Calculation of the wake effect is one of the significant problems in wind farms, and its proper modeling and use will reduce power loss due to near and far wake effects (Shakoor et al., 2016). When designing the turbines' micrositing layout of a wind power plant, two main approaches can be distinguished: (1) by using heuristic methods (Pérez-Aracil et al., 2022; Grady et al., 2005; Mosetti et al., 1994) and (2) by continuous representation (Charhouni et al., 2019). In heuristic methods, the wind farm area is discretized into equal square grids and the center of each square represents a turbine's placement opportunity. In the second approach, it is assumed that the turbines can be placed at any position within the defined wild farm limits. With a continuous representation, there is an opportunity to not limit the possible solutions for turbine implementation, unlike predefined networks. Therefore, using a continuous representation approach is considered a more appropriate approach to determining the layout of a wind farm. Therefore, the subject of this section is the models of the second approach, which considers the continuous representation of wind turbines' micro-siting.

The proposed modeling approach assumes that any geographical area that is designated for the construction of a wind farm can be geometrically simplified to a rectangular area. In this regard, there are two typical cases of wind direction distributions: (a) uniform direction and (b) predominant direction within an area with a rectangular shape as shown in Fig. 3.1.

Determination of wind turbine type, number, and placement should conform to the given geographical area with known parameters—wind conditions and area size. The main goal is to determine the optimal wind turbine type, number, and placement to get maximal power output while minimizing investment costs and considering different practical requirements and restrictions. Two typical cases for wind direction are considered—uniform wind direction and known predominant wind direction. The spacing of wind turbines within a farm depends on the terrain, the wind direction and speed, and the rotor diameter. There exist some published recommendations for the turbine separation distances depending on the wind directions and rotor diameter

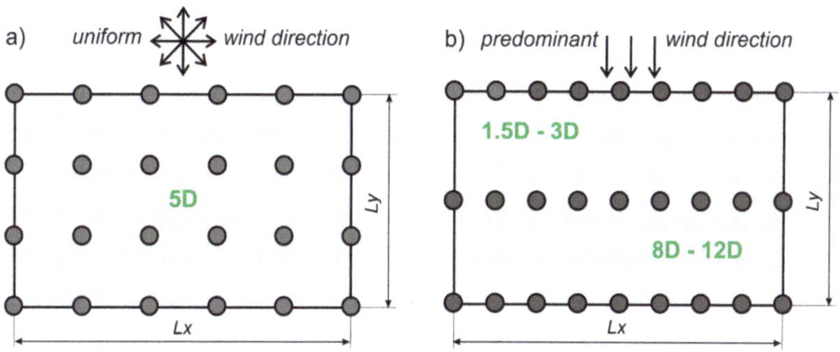

Fig. 3.1 Samples of wind turbines placement: (**a**) uniform wind direction; (**b**) for predominant wind direction

sizes. In the case of uniform wind direction, the required distance between turbines in the rows and columns is accepted to be approximately equal to 5 rotor diameters (Grady et al., 2005; Marmidis et al., 2008) (Fig. 3.1a). If constant predominant wind direction exists, the recommended wind turbines spacing is 8 to 12 rotor diameters in rows apart in the windward direction and 1.5 to 3 (Grady et al., 2005; Marmidis et al., 2008; Donovan, 2005) or 2 to 4 rotor diameters (Johnson, 2006) apart in the crosswind direction (Fig. 3.1b). Spacing the turbines further apart will decrease the wake influence and presumably will capture more wind power but at the expense of land wasting.

The optimal choice of wind turbine type, their number, and placement, using in the best way available area, should be done according to some optimization criteria. The widely accepted and used criterion for wind farm design includes wind farm costs per year as a function of the turbine number N (Grady et al., 2005; Marmidis et al., 2008; Mosetti et al., 1994):

$$\text{costs} = N\left(\frac{2}{3} + \frac{1}{3}\,exp\left(-0.00174N^2\right)\right) \tag{3.1}$$

where N is number of wind turbines within wind farm.

Another criterion is the wind farm installed power P based on the rated power of a single wind turbine P_{wt}.

$$P = NP_{wt} \tag{3.2}$$

The single wind turbine power production depends on the wind conditions (wind speed, direction, intensity, and probability) and on the turbine's type (rotor and tower size, installed power, thrust coefficient, etc.). The wind turbine electrical energy produced per year can be calculated as it is described in Kongnam et al., 2009; Marmidis et al. (2008). It can also be estimated approximately using some practically determined nominal power utilization coefficient η of the total wind farm installed power (Lepa et al., 2008).

As the third criterion could be used for the annual energy production from the wind farm. For the goal of the current modeling, the wind farm's annual energy production (*AEP*) is estimated as:

$$AEP = h_y \eta N P_{wt} \tag{3.3}$$

where h_y is the number of the hours over the year, η is nominal power utilization coefficient, N is the number of wind farm turbines, and P_{wt} is wind turbine rating power.

It should be mentioned that it is possible to use other criteria such as levelized cost of energy when designing the wind farm layout (Yang et al., 2021).

The total number of turbines N is determined by multiplying the number of turbines in rows and columns:

$$N = N_{\text{row}} N_{\text{col}} \tag{3.4}$$

The number of wind turbines in a row N_{row} is determined taking into consideration wind farm dimension length L_x and separation distances between turbines SD_x as in (Johnson, 2006):

$$N_{\text{row}} = \frac{L_x}{SD_x} + 1 \tag{3.5}$$

The existing recommendations about the turbines separation distances depending on the prevailing wind directions are taken into account by introducing coefficients k_{row} and k_{col} for wind turbines placement in rows and columns, respectively. SD_x can be expressed by using k_{row} and turbine rotor diameter D as:

$$SD_{\text{x}} = k_{\text{row}} D \tag{3.6}$$

and (3.5) is transformed to

$$N_{\text{row}} = \frac{L_x}{k_{\text{row}} D} + 1 \tag{3.7}$$

Analogically, the number of wind turbines in a column N_{col} is determined as:

$$N_{\text{col}} = \frac{L_y}{k_{\text{col}} D} + 1 \tag{3.8}$$

In the current modeling approach, the coefficients k_{row} and k_{col} are considered to be continuous variables within given upper and lower limits:

$$k_{\mathrm{row}}^{\min} \leq k_{row} \leq k_{\mathrm{row}}^{\max} \tag{3.9}$$

$$k_{\mathrm{col}}^{\min} \leq k_{col} \leq k_{\mathrm{col}}^{\max} \tag{3.10}$$

This gives some flexibility to the optimization model, thereby increasing the number of possible solution combinations.

The modeling of the wind turbine choice is realized by introducing and using binary integer variables $x_i \in \{0, 1\}$, satisfying the condition $\sum_i^m x_i = 1$. The choice itself is based on the simultaneous choice of multiple essential wind turbine parameters used as decision variables. For example, the rated power P_{wt} of the wind turbine is an important parameter influencing the overall wind farm electrical energy production. The choice of a turbine with particular rated power among the set of available turbines with known power P_{wt}^i is realized as:

$$P_{wt} = \sum_i^m x_i P_{wt}^i, \text{ where } \sum_i^m x_i = 1 \tag{3.11}$$

the rotor diameter D is another important wind turbine parameter related to the wind turbine effectiveness which also influences the separation distance between turbines and the number of turbines in a given limited wind farm area. The choice of particular wind turbine rotor diameter among the given set of m different types of turbines with known rotor diameters D_{wt}^i can be performed analogically as:

$$D = \sum_i^m x_i D_{wt}^i, \text{ where } \sum_i^m x_i = 1 \tag{3.12}$$

the nominal turbines wind speed S is a parameter reflecting the given particular wind conditions and can also be considered when choosing among the set of available turbines with known values for turbines' nominal speed S_{wt}^i:

$$S = \sum_i^m x_i S_{wt}^i, \text{ where } \sum_i^m x_i = 1 \tag{3.13}$$

the height of the wind turbine tower H is a parameter defining the turbine effectiveness (wind speed increases with the height) and can also be used to choose among the set of available turbines with known heights H_{wt}^i:

$$H = \sum_i^m x_i SH_{wt}^i, \text{ where } \sum_i^m x_i = 1 \tag{3.14}$$

Using the same technique other essential technological or ecological wind turbine parameters (electrical output voltage, frequency, noise, etc.) could be included in the wind turbine type choice introducing similar relations.

Some existing practical requirements about the turbines' parameters values can be additionally included by introducing lower and upper limits for them, for example:

$$D_{\min} \leq D \leq D_{\max} \text{–for the rotor diameter} \tag{3.15}$$

$$S_{\min} \leq S \leq S_{\max} \text{–for the nominal speed} \tag{3.16}$$

$$H_{\min} \leq H \leq H_{\max} \text{–for the tower height} \tag{3.17}$$

$$P_{wt}^{\min} \leq P_{wt} \leq P_{wt}^{\max} \text{–for the turbine generator power} \tag{3.18}$$

The advantage of the above-defined modeling approach is the possibility to introduce different criteria, relations, and restrictions to satisfy more precisely different practical requirements about the designed wind farm.

3.2 Single-Criterion Combinatorial Optimization Model for Wind Farm Design for Choice of Wind Turbines Type, Number, and Placement

Taking into account the above-described modeling considerations, a single-criterion mixed-integer nonlinear optimization model is formulated as follows (Mustakerov & Borissova, 2010):

$$\min\left(\frac{\text{costs}}{\text{AEP}}\right) = \frac{N\left(\frac{2}{3} + \frac{1}{3}\, exp\left(-0.00174N^2\right)\right)}{h_y \eta N P_{wt}} \tag{3.19}$$

subject to

$$\mathrm{N} = N_{\text{row}} N_{\text{col}} = \left(\frac{L_x}{k_{\text{row}} D} + 1\right)\left(\frac{L_y}{k_{\text{col}} D} + 1\right) \tag{3.20}$$

$$k_{\text{row}}^{\min} \leq k_{\text{row}} \leq k_{\text{row}}^{\max} \tag{3.21}$$

$$k_{\text{col}}^{\min} \leq k_{\text{col}} \leq k_{\text{col}}^{\max} \tag{3.22}$$

$$P_{wt} = \sum_{i}^{m} x_i P_{wt}^i \tag{3.23}$$

$$D = \sum_{i}^{m} x_i D_{wt}^{i} \tag{3.24}$$

$$H = \sum_{i}^{m} x_i H_{wt}^{i} \tag{3.25}$$

$$D_{in} \leq D \leq D_{\max} \tag{3.26}$$

$$\sum_{i}^{m} x_i = 1, x_i \in \{0, 1\} \tag{3.27}$$

$$H > D/2 \tag{3.28}$$

$$N_{\text{row}} \geq 0, N_{\text{col}} \geq 0, N > 0, \text{integers} \tag{3.29}$$

The last constraint (3.29) expresses the specific features of combinatorial optimization, namely the need to specify the number of turbines as an integer. Constraint (3.27) requires the selection of at least one turbine so the non-negativity constraints can be omitted ($P > 0$; $D > 0$; $L_x > 0$; $L_y > 0$; $k_{\text{row}} > 0$; $k_{\text{col}} > 0$).

The objective function expresses the ratio between costs and produced energy, i.e., the profitability of the wind farm. That means minimizing the investment while maximizing the generation of wind farm energy, and therefore, the optimum solution will be a compromise. Both depend on the particular wind turbine type, price, and number. The number of wind turbines is directly connected with the wind farm dimension area and also depends on the wind direction—predominant or uniform direction. On the other hand, generated wind farm energy depends on the number of installed turbines and their rated power and wind conditions. The wind conditions for a specific wind farm area are taken into account by the nominal utilization coefficient η and recommended values of the coefficients k_{row} and k_{col} used to determine the separation distances between turbines.

3.2.1 Numerical Results

The proposed single-criterion combinatorial optimization model for the selection of wind turbine type, number, and placement is numerically tested using real data sheets parameter values of commercial wind turbines.[1] The type and parameters of wind turbines are given in Table 3.1 where each row represents a particular turbine type and each column represents the corresponding parameter.

The AEP from wind farm is calculated as the number of hours over the year $h_y = 8760$ multiplied by utilization coefficient η for particular wind conditions assumed equal to 0.3.

[1] http://www.wind-energy-market.com/; http://www.enercon.de/en/; http://www.vestas.com/en/

Table 3.1 Commercial wind turbines parameters

No	Wind turbine Type	Model	Rated power output, MW	Rotor diameter, m	Tower height, m
1	1.1	ENERCON's E-33	0.330	33.4	37
2	1.2				44
3	1.3				50
4	2.1	ENERCON's E-48	0.800	48.0	50
5	2.2				60
6	2.3				76
7	3.1	ENERCON's E-53	0.800	52.9	60
8	3.2				73
9	4.1	Vestas V52	0.850	52.0	49
10	4.2				65
11	4.3				74
12	5.1	ENERCON's E-44	0.900	44.0	45
13	5.2				55
14	6.1	Vestas V82	1.650	82.0	78
15	7.1	Vestas V80	2.000	80.0	60
16	7.2				78
17	7.3				100
18	8.1	Enercon's E-82	2.000	82.0	78
19	8.2				138
20	9.1	ENERCON's E-70	2.300	71.0	58
21	9.2				113
22	10.1	Vestas V90	3.000	90.0	95
23	10.2				105
24	10.3				125

Using the formulated combinatorial optimization model (3.19)–(3.29) and input data from Table 3.1, several scenarios have been realized.

3.2.1.1 Scenario with Uniform Wind Direction and Wind Farm Area with a Square Shape

This scenario considers a 4 km^2 square-shaped windy area with dimensions $L_x = L_y = 2\ km$ and uniform wind direction. Two upper and lower limits used for the separation coefficients are as follows:

- Case-1: $k_{\text{row}}^{\min} = k_{\text{col}}^{\min} = 4.8$ and $k_{\text{row}}^{\max} = k_{\text{col}}^{\max} = 5.2$.
- Case-2: $k_{\text{row}}^{\min} = k_{\text{col}}^{\min} = 4.5$ and $k_{\text{row}}^{\max} = k_{\text{ol}}^{\max} = 5.5$

The obtained results are visualized in Fig. 3.2.

The turbine ENERCON's E-82 is the best selection when the limits in Case-1 are used and the separation distances between turbines are equal to 400 m. The number of turbines in this case is equal to 36. For Case-2, the best selection is ENERCON's

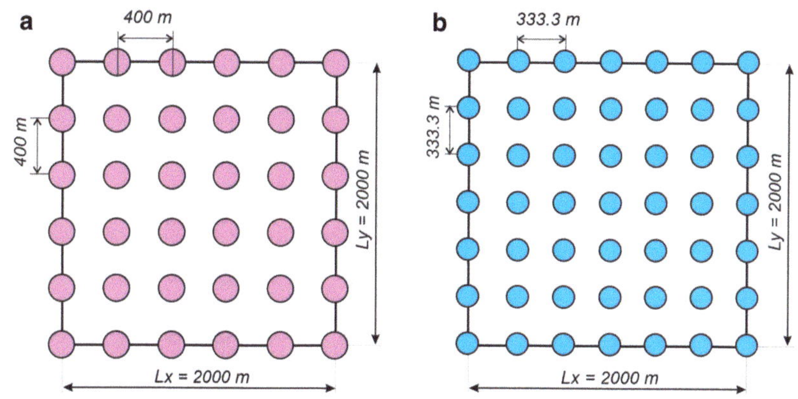

Fig. 3.2 Arrangement of wind turbines in a square shape of a wind farm, considering uniform wind direction: (**a**) selected ENERCON's E-82; (**b**) selected ENERCON's E-70

E-70 with a separation distance of 333.3 *m*. The number of turbines in this case is equal to 49.

3.2.1.2 Scenario with Uniform Wind Direction and Wind Farm Area with a Rectangular Shape

This scenario also considers a 4 km^2 windy area but with rectangular shaped and dimensions $L_x = 4\ km$, $L_y = 2\ km$ and uniform wind direction. Two upper and lower limits used for the separation coefficients are as follows:

- Case-1: $k_{\text{row}}^{\min} = k_{\text{col}}^{\min} = 4.8$ and $k_{\text{row}}^{\max} = k_{\text{col}}^{\max} = 5.2$
- Case-2: $k_{\text{row}}^{\min} = k_{\text{col}}^{\min} = 4.5$ and $k_{\text{row}}^{\max} = k_{\text{col}}^{\max} = 5.5$

The obtained results are visualized in Fig. 3.3.

The turbine Vestas V52 is the best selection when the limits in Case-1 are used and the separation distances between turbines are equal to 250 *m*. The number of turbines in this case is equal to 85. For Case-2, the best selection is ENERCON's E-70 with a separation distance of 333.3 *m*. The number of turbines in this case is equal to 52.

3.2.1.3 Scenario with Predominant Wind Direction and Wind Farm Area with a Square Shape

This scenario considers a 4 km^2 square-shaped windy area with dimensions $L_x = L_y = 2\ km$ and predominant wind direction. The difference is in the used upper and lower limits for the separation coefficients:

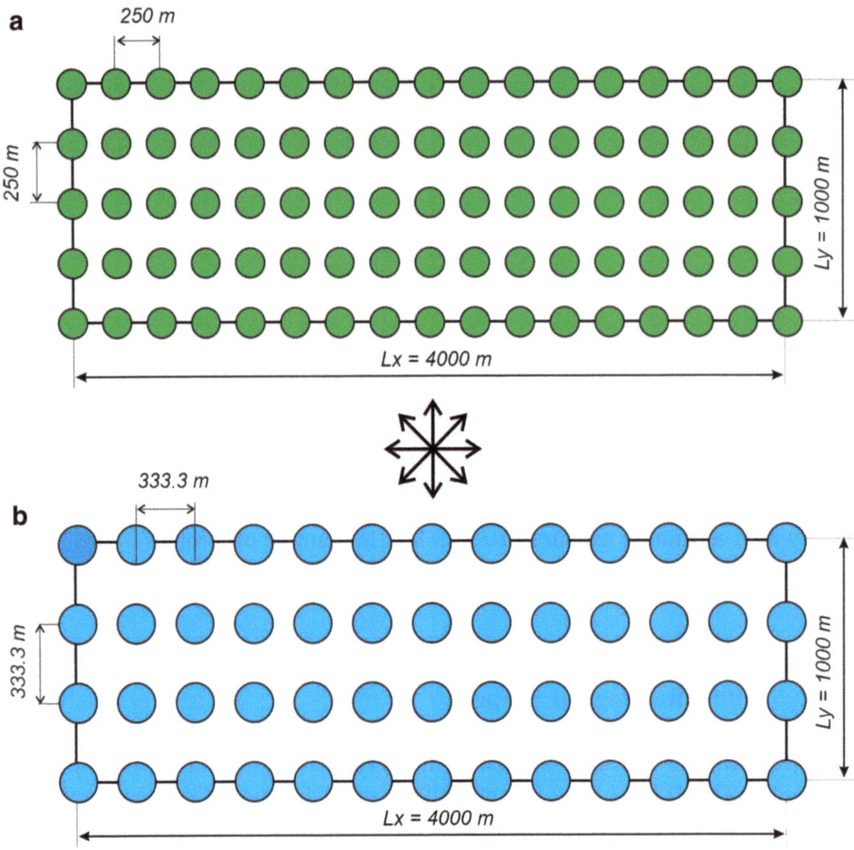

Fig. 3.3 Arrangement of wind turbines in a rectangular shape of a wind farm, considering uniform wind direction: (**a**) selected Vestas V52; (**b**) selected ENERCON's E-70

- Case-1: $k_{\text{row}}^{\min} = 1.5$, $k_{\text{row}}^{\max} = 3$; $k_{\text{col}}^{\min} = 9$, $k_{\text{col}}^{\max} = 11$
- Case-2: $k_{\text{row}}^{\min} = 2$, $k_{\text{row}}^{\max} = 4$; $k_{\text{col}}^{\min} = 8$, $k_{\text{col}}^{\max} = 12$

The obtained results under these two types of upper and lower limits are visualized in Fig. 3.4.

The turbine ENERCON's E-70 is the best selection when the limits in Case-1 are used and the separation distances between turbines in wind direction are equal to 111.1 *m* and 666.6 *m* for the opposite direction. The number of turbines in this case is equal to 76. For Case-2, the best selection is Vestas V90 with a separation distance of 181.8 *m* in the wind direction and 1000 *m* in the opposite direction. The number of turbines in this case is equal to 36.

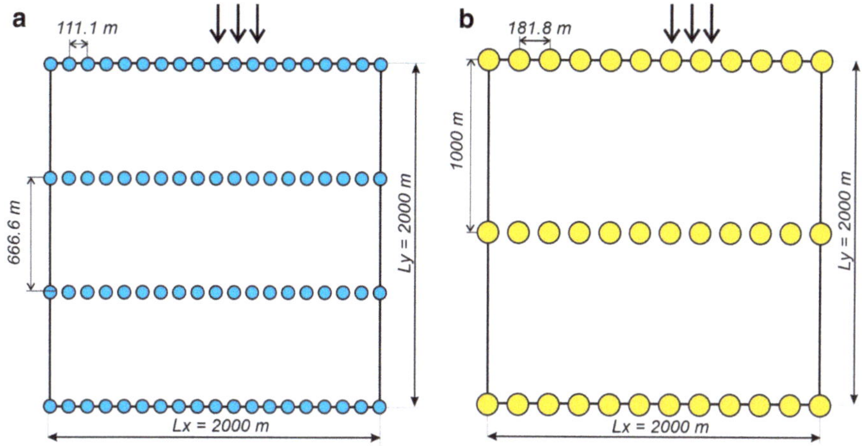

Fig. 3.4 Arrangement of wind turbines in a rectangular shape of a wind farm, considering uniform wind direction: (**a**) selected ENERCON's E-70; (**b**) selected Vestas V90

3.2.1.4 Scenario with a Predominant Wind Direction (Perpendicular to the Longer Side) and a Wind Farm Area with a Rectangular Shape

This scenario also considers a 4 km^2 windy area with a rectangular shape and dimensions $L_x = 4\ km$, $L_y = 2\ km$ and a predominant wind direction that is perpendicular to the longer side.

- Case-1: $k_{\text{row}}^{\min} = 1.5$, $k_{\text{row}}^{\max} = 3$; $k_{\text{col}}^{\min} = 9$, $k_{\text{col}}^{\max} = 11$
- Case-2: $k_{\text{row}}^{\min} = 2$, $k_{\text{row}}^{\max} = 4$; $k_{\text{col}}^{\min} = 8$, $k_{\text{col}}^{\max} = 12$

The corresponding results for this scenario are visualized in Fig. 3.5.

The turbine Vestas V52 is the best selection when the limits in Case-1 are used and the separation distances between turbines in wind direction are equal to 78.4 *m* and 500 *m* for the opposite direction. The number of turbines in this case is equal to 96. For Case-2, the best selection is Vestas V90 with a separation distance of 181.8 *m* in the wind direction and 1000 *m* in the opposite direction. The number of turbines in this case is equal to 44.

3.2.1.5 Scenario with a Predominant Wind Direction (Perpendicular to the Short Side) and a Wind Farm Area with a Rectangular Shape

This scenario differs from the previous one only in that the wind direction is perpendicular to the shorter side.

The obtained results for this scenario are visualized in Fig. 3.6.

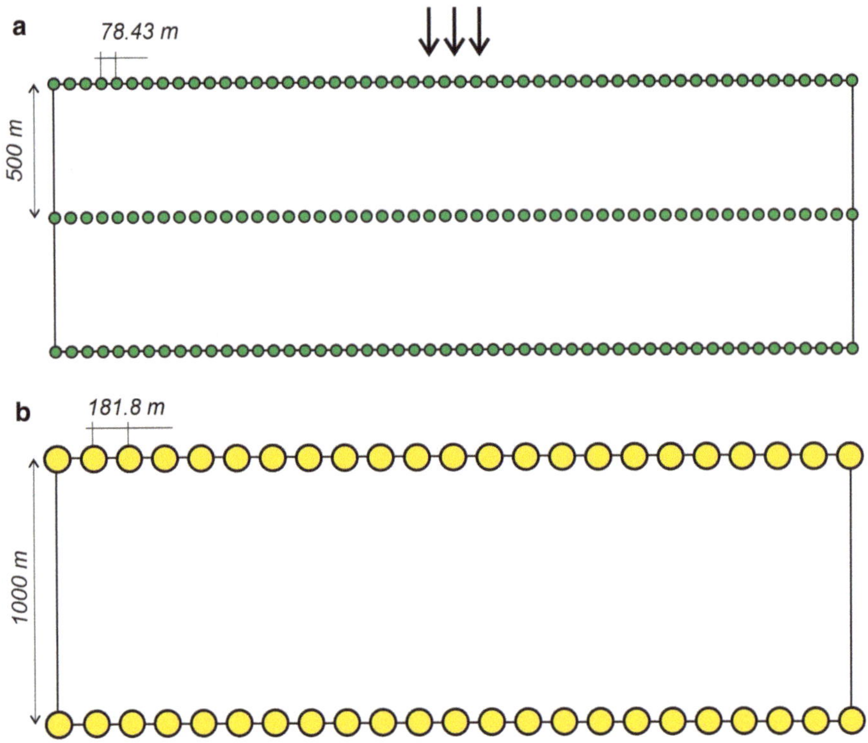

Fig. 3.5 Arrangement of wind turbines in a rectangular shape of a wind farm, considering uniform wind direction: (**a**) selected Vestas V52; (**b**) selected Vestas V90

In this scenario, the turbine Vestas V90 is the best selection when the limits in Case-1 are used and the separation distances between turbines in wind direction are equal to 200 *m* and 800 *m* for the opposite direction. The number of turbines in this case is equal to 36. For Case-2, the best selection is ENERCON's E-70 with a separation distance of 111.1 *m* in wind direction and 363.6 *m* in the opposite direction. The number of turbines in this case is equal to 120.

Considering all of these scenarios and the obtained results described above, a motivated decision could be made about the selection of a suitable type of wind turbine according to the wind site dimensions and wind direction using the criterion for minimization of the ratio of investment cost as a function of turbine number and corresponding annual energy production expressed by (3.19).

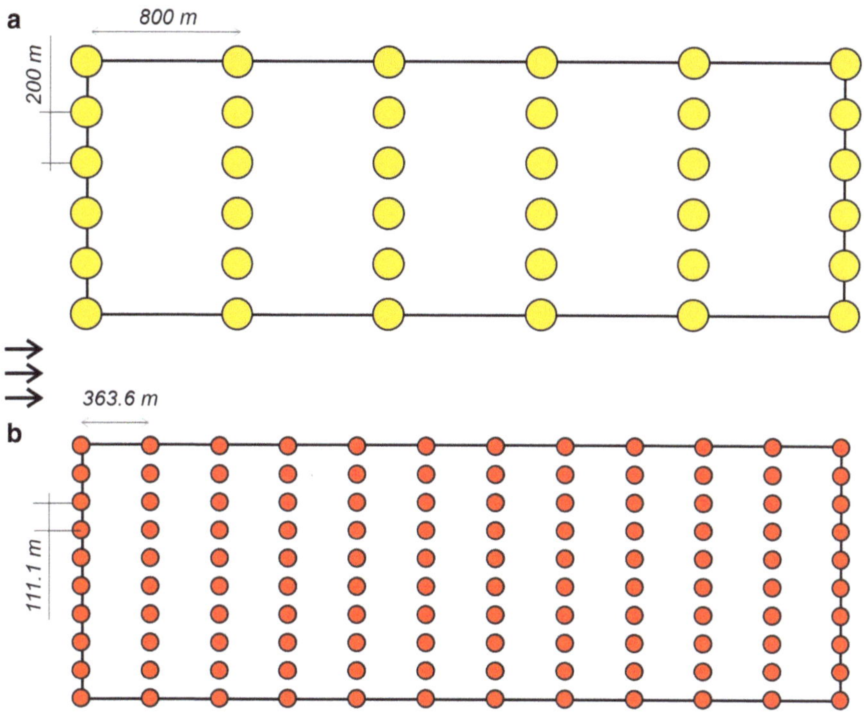

Fig. 3.6 Arrangement of wind turbines in a rectangular shape of a wind farm, considering uniform wind direction: (**a**) selected Vestas V90; (**b**) selected ENERCON's E-70

3.3 Multicriteria Combinatorial Optimization Model for Wind Farm Layout Design for Choice of Wind Turbines Type, Number, and Placement

The multicriteria model for designing wind farm layout takes into account two of the most important criteria—energy output and costs. Based on the assumptions in this chapter, the following multicriteria combinatorial optimization model is formulated (Borissova & Mustakerov, 2017b):

$$\begin{cases} \max\left(h_y \eta N P_{wt}\right) \\ \min N\left(\frac{2}{3}+\frac{1}{3} exp\left(-0.00174N^2\right)\right) \end{cases} \tag{3.30}$$

subject to:

$$N = \left(\frac{L_x}{k_{\text{row}} D} + 1\right)\left(\frac{L_y}{k_{\text{col}} D} + 1\right) \tag{3.31}$$

$$k_{\text{row}}^{\min} \le k_{\text{row}} \le k_{\text{row}}^{max} \tag{3.32}$$

$$k_{\text{col}}^{\min} \le k_{\text{col}} \le k_{\text{col}}^{\max} \tag{3.33}$$

$$P_{wt} = \sum_{i}^{m} x_i P_{wt}^{i} \tag{3.34}$$

$$D = \sum_{i}^{m} x_i D_{wt}^{i} \tag{3.35}$$

$$H = \sum_{i}^{m} x_i H_{wt}^{i} \tag{3.36}$$

$$D_{\min} \le D \le D_{\max} \tag{3.37}$$

$$\sum_{i}^{m} x_i = 1, x_i \in \{0, 1\} \tag{3.38}$$

$$N_{\text{row}} \ge 0, N_{\text{col}} \ge 0, N > 0, \text{integers} \tag{3.39}$$

The multicriteria optimization model (3.30)–(3.39) is used to formulate optimization problems solved using a scalarization technique.

3.3.1 Using of Weighted Sum Method for Wind Farm Layout Design

The weighted sum method as the most common multi-objective optimization method based on scalarization techniques aggregates different objectives as a weighted linear sum of their normalization (Marler & Arora, 2005; Shakoor et al., 2016). The DM preferences about importance of objectives are expressed by assigning corresponding weight coefficients w_j, which results in transformed optimization task:

$$\max\left(w_e\left(\frac{\text{AEP} - AEP^{\min}}{\text{AEP}^{\max} - AEP^{\min}}\right) + w_c\left(\frac{E^{\max} - C}{C^{\max} - C^{\min}}\right)\right) \tag{3.40}$$

subject to (3.31)–(3.39) with additional restriction:

$$w_e + w_c = 1 \tag{3.41}$$

Table 3.2 List of wind turbines parameters

#	WT	Rated power, kW	WT rotor diameter, m
1	Vestas V80	2000	80
2	ENERCON's E-82	2000	82
3	ENERCON's E-70	2300	71
4	SWT-2.3-82 VS	2300	82.4
5	ENERCON's E-82 E2	2300	82
6	SWT-2.3-113	2300	113
7	SWT-2.3-108	2300	108
8	SWT-2.3-93	2300	93
9	ENERCON's E-92	2350	92
10	Clipper Windpower	2500	96
11	Vestas V100	2600	100
12	ENERCON's E-82	3000	82
13	ENERCON's E-101	3050	101
14	Vestas V90	3000	90
15	Vestas V112	3000	112
16	SWT-3.6-120	3600	120
17	SWT-3.6-107	3600	107
18	SWT-6.0-154	6000	154
19	Vestas V164–7.0	7000	164
20	ENERCON's E-126	7580	127

where $\mathrm{AEP}^{\max}$, $\mathrm{AEP}^{\min}$, $C^{\max}$, $C^{\min}$ are the minimal and maximal values for energy and costs obtained by solution of (3.30)–(3.39) as single-objective optimization problems.

To investigate the possibility of using multi-objective optimization as a tool for design and assessment of wind farm layout, the transformed task is solved for different preferences of DM toward importance of objectives. In the first case, more important is the output energy ($w_e = 0.9$ and $w_c = 0.1$), while the second case illustrates the prevailing importance of costs ($w_e = 0.1$ and $w_c = 0.9$).

A predefined set of wind turbines given in Table 3.2 is used as input data for optimization task solving.

Using of parameters of wind turbines from Table 3.2 and the transformed optimization task, the following wind farm layout designs are obtained as shown in Fig. 3.7.

The results show that the described multi-objective modeling approach could be useful in determining an acceptable wind farm layout design. It is worth mentioning that the used weights in the weighted sum method toward criteria importance influence the wind farm layout design and the expected costs and energy production. The obtained results are illustrated graphically in Fig. 3.8.

The usage of the weighted sum method is capable of providing the Pareto-optimal wind farm layouts. In addition, due to the type of objective functions, it is possible to

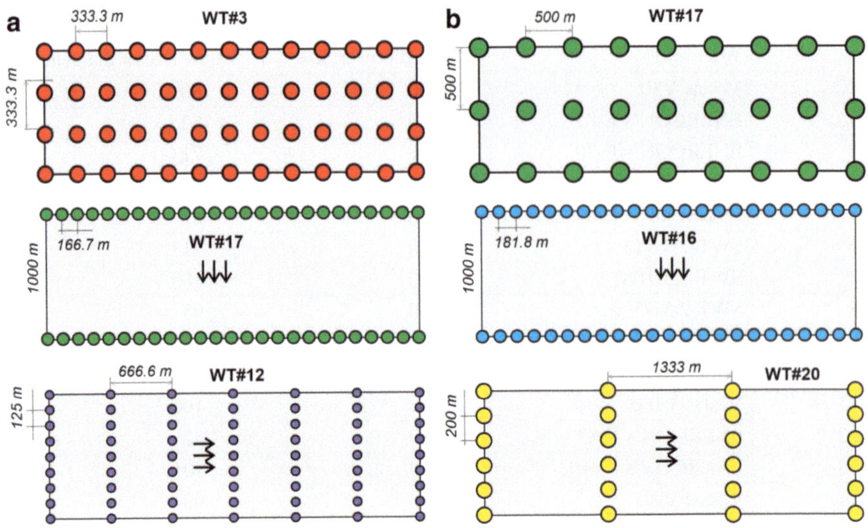

Fig. 3.7 Wind farm layout design using weighted sum method: (**a**) the preference is about output energy ($w_e = 0.9$, $w_c = 0.1$); (**b**) the preference is about costs ($w_e = 0.1$, $w_c = 0.9$)

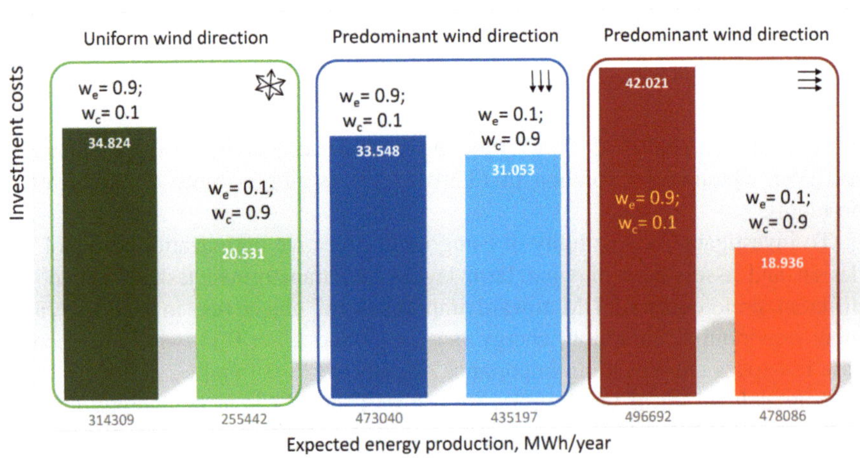

Fig. 3.8 Estimation of energy and costs under different wind directions and different preferences using weighted sum method

get an estimation of the expected energy production and costs under different wind directions and different preferences toward objective functions.

Based on the obtained results (Fig. 3.8), a reasoned decision could be made about the selection of a suitable type of wind turbine according to the available dimensions of the site and the direction of the wind and, accordingly, the obtained assessment of the investment costs and the energy produced by this type of turbine.

3.3.2 *Using Lexicographic Method for Wind Farm Layout Design*

The lexicographic method requires ranking of objectives according to the DM preferences and optimizing them one at a time. The solution of each single-objective problem gives a limiting measure used to define a proper restriction on the next step when the next objective is optimized and so on (Marler & Arora, 2004; Marler & Arora, 2005).

Two lexicographic orders were used to investigate their impact on wind farm layout. In the first case, the energy output has been chosen as more important than the costs. The corresponding optimization procedure is as follows:

Step 1: Solving of the optimization task:

$$\max\ \left(\mathrm{AEP} = h_y \eta N P_{wt}\right) \tag{3.42}$$

subject to restrictions (3.31)–(3.39)

Step 2: Solving of the optimization task:

$$\min N\left(\frac{2}{3} + \frac{1}{3}\exp\left(-0.00174N^2\right)\right) \tag{3.43}$$

subject to restrictions (3.31)–(3.39) and additional restriction about the energy output:

$$h_y \eta N P_{wt} \geq \varepsilon \left(h_y \eta N P_{wt}\right)^{\max} \tag{3.44}$$

where the coefficient ε determines the degree of proximity to the optimal value of $(h_y \eta N P_{wt})^{\max}$ determined in the first step. To explore the possibilities of this method, two cases for this coefficient were used to express the preferences of the DM regarding the importance of the energy objective and to show their influence on the final decision.

In the second case, the costs objective is chosen as more important, which leads to the following computational procedure:

Step 1: Solving of the optimization task:

$$\min N\left(\frac{2}{3} + \frac{1}{3}\exp\left(-0.00174N^2\right)\right) \tag{3.45}$$

subject to restrictions (3.31)–(3.39).

Step 2: Solving the optimization task:

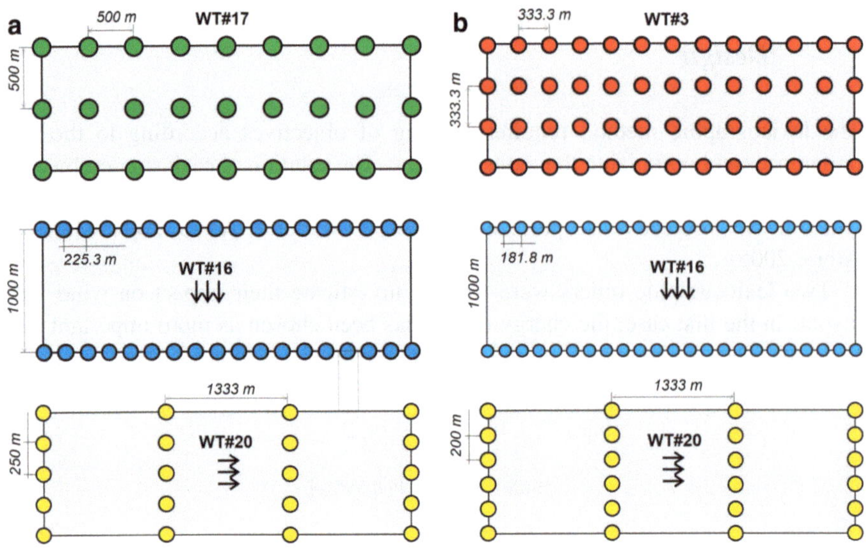

Fig. 3.9 Wind farm layout by lexicographic method: (**a**) energy, cost,s and $\varepsilon = 0.7$; (**b**) energy, costs, and $\varepsilon = 0.9$

$$\max\ \left(\mathrm{AEP} = h_y \eta N P_{wt}\right) \tag{3.46}$$

subject to restrictions (3.31)–(3.39) and additional restriction about the costs:

$$N\left(\frac{2}{3} + \frac{1}{3}\,exp\left(-0.00174N^2\right)\right) \leq \varepsilon\left(N\left(\frac{2}{3} + \frac{1}{3}\,exp\left(-0.00174N^2\right)\right)\right)^{\min} \tag{3.47}$$

Here, the coefficient ε determines the degree of proximity to the optimum value $N\left(\frac{2}{3} + \frac{1}{3}\exp\left(-0.00174N^2\right)\right)^{min}$ from the first step. Again two values of this coefficient are used for the DM preferences regarding the importance of costs objective and to show their influence on the final decision.

Using parameters of wind turbines from Table 3.2 and the transformed optimization tasks, the following wind farm layout designs are obtained as shown in Figs. 3.9 and 3.10.

The results show that the described multi-objective modeling approach could be useful in determining an acceptable wind farm layout design. It should be mentioned that the lexicographic ordering of criteria influences the wind farm layout design and the expected costs and energy production. The obtained results are illustrated graphically in Fig. 3.11.

The usage of the lexicographic method is capable of providing the Pareto-optimal wind farm layouts. In addition, due to the type of objective functions, it is possible to get an estimation of the expected energy production and costs under different wind directions and different lexicographic ordering of the objective functions.

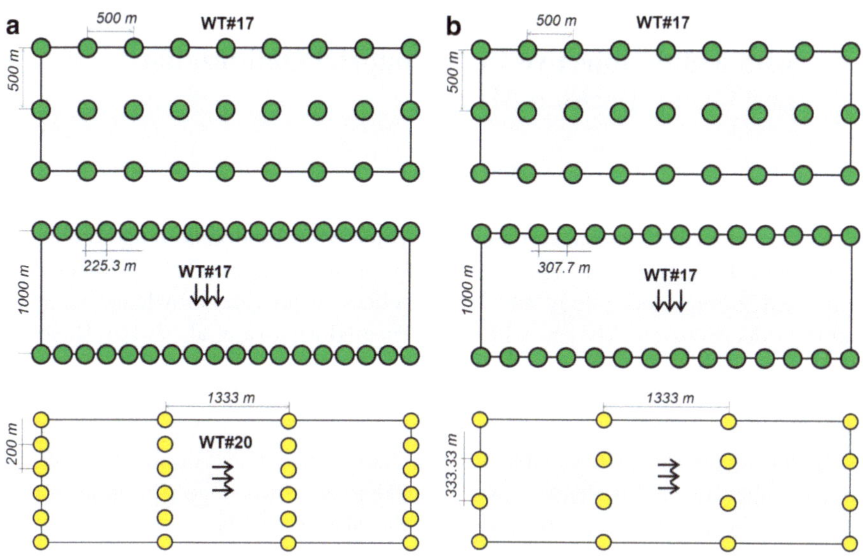

Fig. 3.10 Wind farm layout by lexicographic method: (**a**) costs, energy, and $\varepsilon = 1.3$; (**b**) costs, energy, and $\varepsilon = 1.1$

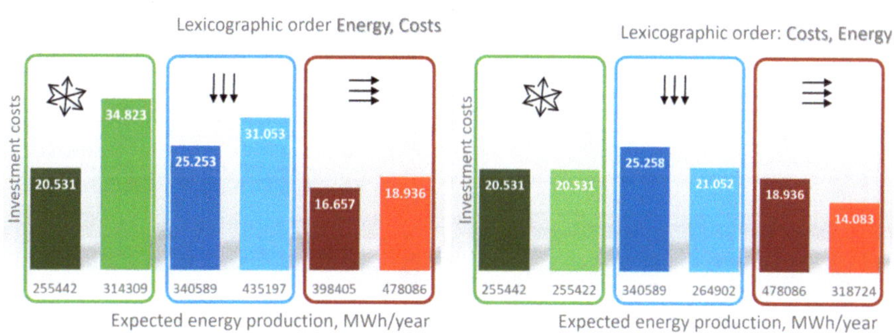

Fig. 3.11 Estimation of costs and expected energy under different wind directions and different preferences using lexicographic method

Based on the obtained results (Fig. 3.11), it is possible to make a reasoned decision to choose the most suitable type of wind turbine according to the available dimensions of the site and the direction of the wind and, accordingly, the obtained assessment of the investment costs and the energy produced by this type of turbine.

3.4 Two-Stage Algorithm for Wind Farm Layout Design with Multi-Objective Combinatorial Optimization and Group Decision-Making

The final measure of decision quality expresses the degree to which the decision led to the best business outcome. A modern direction is the use of business intelligence to access and analyze information to improve and optimize decisions and efficiency (Larson & Chang, 2016). The complexity of the decision-making process is addressed through technology and best practices to provide knowledge to make the right decisions quickly and with confidence (Borissova et al., 2016). Business decisions are made based on system-wide information, including financial and environmental aspects, and rely on experts with the ability to evaluate decision alternatives.

In this regard, to realize suitable business intelligence in decision-making for the determination of design alternatives for wind farms, a two-stage algorithm is proposed as shown in Fig. 3.12 (Borissova & Mustakerov, 2017a).

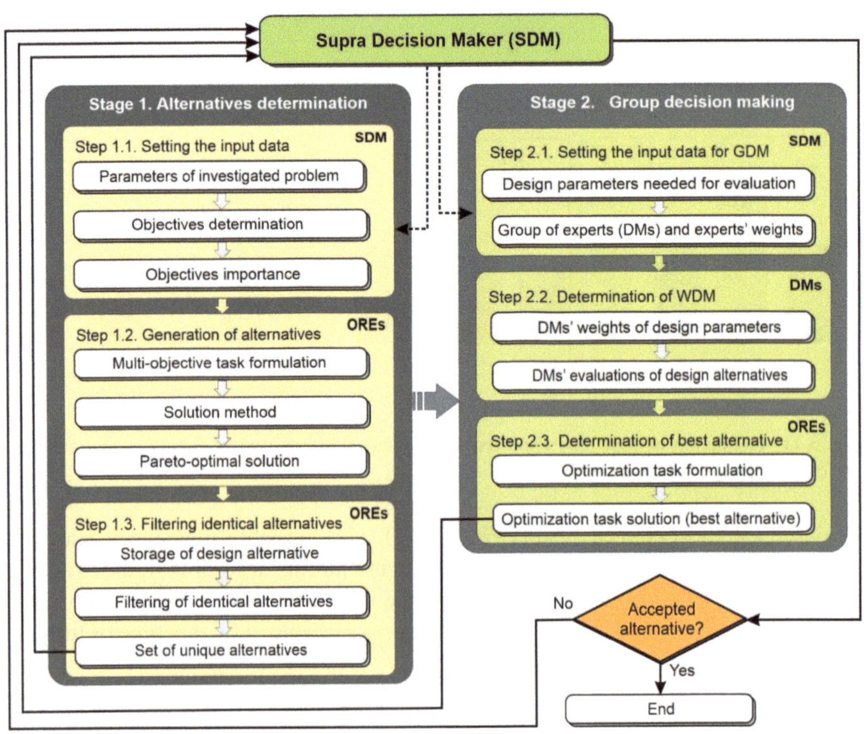

Fig. 3.12 Flowchart of two-stage decision-making algorithm for the determination of design alternatives for wind farms

The two-stage algorithm for wind farm layout design consists of:

- Determination of a set of design alternatives conforming to different design scenarios
- Choice of the best alternative by GDM

It is assumed that overall project management is done by a supra decision-maker (SDM). The SDM is authorized to organize the overall decision-making process, to gather a group of different experts (DMs) relevant to the project design aspects and to collect experts' evaluations of different alternatives. The SDM is usually a manager who is responsible to propose a final decision at a higher level of management or has the right to make the final decision by himself. The responsibilities for input data about the parameters, objectives, and their importance are assigned to SDM in Step 1.1. The goal of multi-objective optimization problem is to identify Pareto-optimal solution using additional information from DM about the objectives importance. In Step 1.2 the experts in operation research (OREs) have to formulate and solve multi-objective optimization tasks. They choose and implement proper multi-objective solution method and express objectives importance.

A common approach for dealing with multi-objective optimization problems is to use scalarization techniques (Ehrgott, 2006; Genova et al., 2013; Marler & Arora, 2004) due to their simplicity and effectiveness. As a result of multi-objective task solution, a Pareto-optimal design alternative is determined. Then, in Step 1.3, design alternatives are filtered to get a set of unique alternatives.

The input data for GDM are implemented in Step 2.1. SDM determines the evaluation criteria and group of experts (DMs) from different knowledge domains and defines some weight coefficients for each of DMs. Weights of DMs play a very important role in MAGDM because they reflect the expertise of each particular expert. The range of weights for experts could be limited within interval of 0–1, where 1 means the most experienced and trusted level of expertise (Fulop, 2005).

The essence of Step 2.2 is based on MADM described by a weighted decision matrix that indicates both the set of alternatives and the set of criteria subject to evaluation. The proposed structure of WDM includes DMs' weights for the importance of the design parameters (criteria), DMs' scores toward design alternatives' parameters, and weighted coefficients for DMs. Each of the DMs estimates alternatives independently, without negotiations with other DMs and accordingly to his specific point of view. For example, a financial expert probably will put more weight on financial parameters of the design while engineers would be more interested in wind farm energy production. The SDM assigns weighting coefficients for each DM according to their hierarchy positions in the management of wind farm design.

In Step 2.3, a single-criterion optimization model is to be formulated to select the best alternative. Then, the solution is to be presented to SDM for a decision—to accept the determined alternative or to repeat stages of algorithm.

The proposed single-criterion optimization and combinatorial model is as follows (Borissova & Mustakerov, 2017a):

$$\max \left(\sum_{i}^{M} \sum_{k}^{K} \alpha^k \omega_i^k A_i^k \right) \tag{3.48}$$

subject to

$$\forall i = 1, 2, \ldots, M : \left(\forall k = 1, 2, \ldots, K : A_i^k = \sum\nolimits_{j}^{J} a_{i,j}^k x_j \right) \tag{3.49}$$

$$\forall i = 1, 2, \ldots, M : \left(\forall k = 1, 2, \ldots, K : A_i^k = \sum\nolimits_{j=1}^{J} a_{i,j}^k x_j \right) \tag{3.50}$$

$$\sum_{j}^{J} x_j = 1, x_j \in \{0, 1\} \tag{3.51}$$

$$\alpha^k \in (0, 1) \tag{3.52}$$

where i = 1, 2, ..., M are indexes of design parameters against which the design alternatives are to be evaluated, j = 1, 2, ..., J are indexes of design parameters to be evaluated, k = 1, 2, ..., K are indexes of group of DMs involved into the decision-making process, w_i^k are weighting coefficients representing relative importance of design parameters as evaluated by different DMs, a_{ij}^k is evaluation score of k-th DM for performance of alternative j against parameter p_i, coefficients α^k represent the weight of opinion of the *k-th* DM, and x_j are binary integer decision variables used to perform choice of a single alternative.

3.4.1 Numerical Results

To test the applicability of the proposed two-stage algorithm for wind farm layout design, a set of 30 wind turbines is used to choose from and determine the most appropriate type of turbine and corresponding layout (Table 3.3).

The weighted sum method and lexicographic method are used to generate possible alternatives for wind farm layout design considering multi-objective combinatorial model (3.30)–(3.39). Once the alternatives are determined, it is needed to be filtered as it is possible that some of alternatives are the same.

The identical design alternatives are considered as one unique alternative as shown in Table 3.4.

Three wind farm parameters are subject of evaluation by a group of three DMs as shown in Table 3.5.

—

The WDM for GDM considers the case Pre-2 data which is characterized by 4 unique design alternatives (Table 3.5). Three different experts—engineer (DM-1), a financial expert (DM-2), and ecologist (DM-3)—are involved in the process of alternatives evaluation toward 3 wind farm design parameters (energy output, costs,

Table 3.3 List of wind turbines parameters

#	Wind turbine type	Rated power, kW	Rotor diameter, m
1	ENERCON's E-33	330	33.4
2	ENERCON's E-48	800	48.0
3	ENERCON's E-53	800	52.9
4	Vestas V52	850	52.0
5	Vestas V60	850	60.0
6	ENERCON's E-44	900	44.0
7	GAMESA AE-61-1320	1320	61.0
8	Vestas V82	1650	82.0
9	M Torres	1650	77.0
10	Vestas V100	1800	100.0
11	Vestas V80	2000	80.0
12	ENERCON's E-82	2000	82.0
13	ENERCON's E-70	2300	71.0
14	SWT-2.3-82 VS	2300	82.4
15	ENERCON's E2–82	2300	82.0
16	SWT-2.3-113	2300	113.0
17	SWT-2.3-108	2300	108.0
18	SWT-2.3-93	2300	93.0
19	ENERCON's E-92	2350	92.0
20	C96 clipper Windpower	2500	96.0
21	Vestas V100	2600	100.0
22	ENERCON's E-82	3000	82.0
24	Vestas V90	3000	90.0
25	Vestas V112	3000	112.0
23	ENERCON's E-101	3050	101.0
26	SWT-3.6-120	3600	120.0
27	SWT-3.6-107	3600	107.0
28	SWT-6.0-154	6000	154.0
29	Vestas V164–7.0	7000	164.0
30	ENERCON's E-126	7580	127.0

and number of turbines). For each of these experts, a corresponding weighting coefficient is assigned to represent its opinion importance.

The obtained results based on the solution of single-objective combinatorial optimization model (3.48)–(3.52) for 4 sets of values of DMs' weights are shown in Table 3.6.

The graphical illustration of the optimal layout design alternatives obtained as a result of execution of Stage 2 is shown in Fig. 3.13.

The main idea of a two-stage placement algorithm with multi-objective optimization and group decision-making is to generate multiple alternatives and select one of them via a group of experts to make the most suitable business decision. Using multi-objective combinatorial optimization, in the first stage of the algorithm,

Table 3.4 Unique design alternatives as a result of execution of Stage 1

Unique alternatives	Turbine #	Number of turbines	SDx, m	SDy, m	AEP, MWh/year	Costs
Uniform wind direction						
A-1*	# 13	52	333.4	333.4	314,309	34.824
A-2*	# 27	27	500.0	500.0	255,442	20.531
Predominant wind direction (Pre-1)						
A-1′	# 27	50	166.7	1000.0	473,040	33.55
A-2′	# 26	46	181.8	1000.0	435,197	31.05
A-3'	# 26	36	235.3	1000.0	340,589	25.25
A-4'	# 27	28	307.7	1000.0	264,902	21.05
Predominant wind direction (Pre-2)						
A-1"	# 22	63	666.7	125.0	496,692	42.02
A-2"	# 30	24	1333.4	200.0	478,086	18.94
A-3"	# 30	20	1333.4	250.0	398,405	16.66
A-4"	# 30	16	1333.4	333.4	318,724	14.08

provides different Pareto-optimal alternatives and can be considered as a flexible analytical tool to simulate different wind farm design scenarios. The different DMs' preferences toward the importance of used criteria are also used to simulate different wind farm design alternatives. All of these alternatives are used in the second stage of the proposed algorithm where it is needed to select the most suitable alternative by group decision-making.

The obtained results using the described two-stage algorithm for the design of wind farm layouts show its practical applicability. The runtimes for tested examples are in the range of a few seconds. This demonstrates the effectiveness of the use of combinatorial optimization modeling to solve decision-making problems related to the optimization of wind farm layouts.

3.5 Integrated Decision Support Framework by Combining MADM and MODM to Select and Evaluate the Effectiveness of the Selected Wind Turbine

To get more transparent and objective decisions, the proposed integrated framework for wind farm design could be used. This framework relies on the effective integration of MADM/MAGDM models and mathematical modeling based on single- or multi-objective optimization as shown in Fig. 3.14 (Borissova et al., 2021).

The user interface (UI) should be designed to provide easy access to all resources that require the active participation of DMs. It should provide the required information for entering the needed data about the parameters relevant to the considered domain area. All of the entered data are to be stored in a predefined database. The UI

Table 3.5 WDM for GDM

	Parameters' weights			A1 evaluations			A2 evaluations			A3 evaluations			A4 evaluations		
Design Parameters	DM-1	DM-2	DM-3	DM-1	DM-2	DM-3	DM-1	DM-2	DM-3	DM-1	DM-2	DM-3	DM-1	DM-2	DM-3
AEP MWh/ year	10	4	3	10	8	4	8	6	7	6	5	6	4	3	5
Costs	4	10	3	5	2	4	6	7	5	7	6	6	8	5	7
Turbines number	5	8	10	3	2	2	4	6	5	5	7	7	6	6	8

Table 3.6 GDM results for different weights of DMs

Tasks	Weighting coefficients of DMs			Best alternative
	α_1	α_2	α_3	
1	0.5	0.5	0.5	A-3
2	0.9	0.2	0.2	A-2
3	0.2	0.9	0.2	A-2
4	0.2	0.2	0.9	A-4

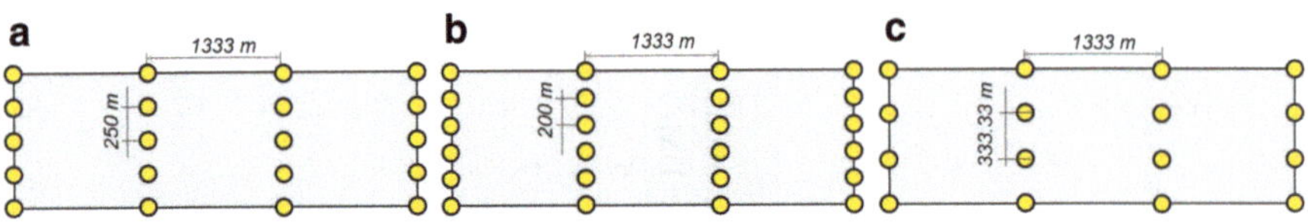

Fig. 3.13 Graphical representation of (**a**) Alternative-3 (20 of WT#30); (**b**) Alternative-2 (24 of WT#30); (**c**) Alternative-4 (16 of WT#30)

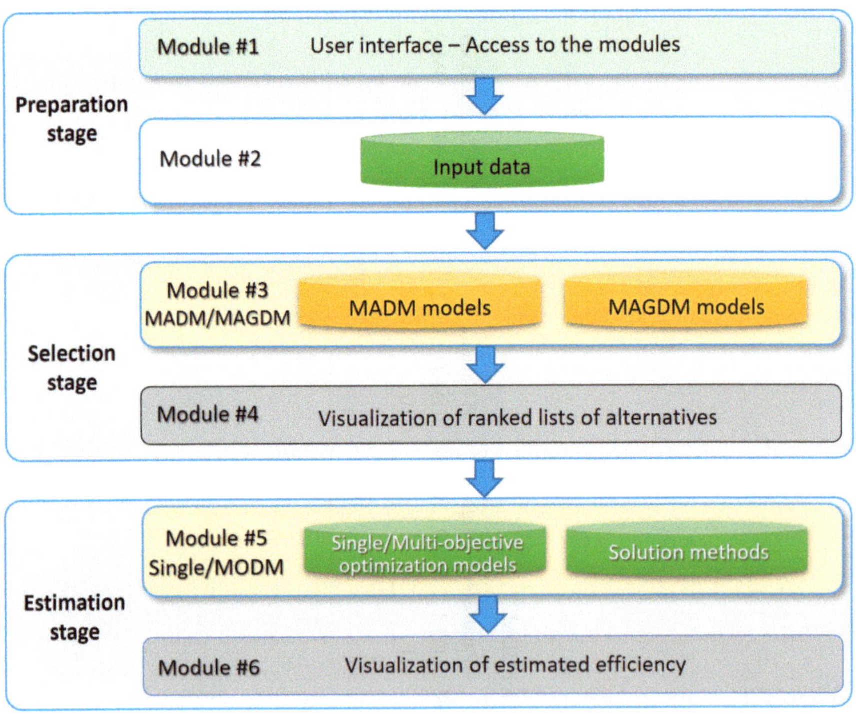

Fig. 3.14 A framework of an intelligent system to support decision-making by effective integration of MADM/MAGDM and MODM

and the created database with the input information about the particular problem are two essential parts of the preparation stage of the proposed intelligent system.

The next essential stage is the selection stage that is composed of module #3 and module #4. Module #3 is intended to keep different databases, related to the MADM models and MAGDM models. These MADM/MAGDM models could be used to provide the ranked lists of alternatives considering single or group decision-making.

There are different and well-known models for MADM/MAGDM that could be utilized to rank a predefined set of alternatives in accordance with the DM preferences' or group of experts. The usage of module #4 will provide a proper interface to visualize the ranked list of alternatives. Once the ranked list is known, the best alternative in the ranked list should be estimated toward determined criteria for effectiveness.

The third stage of the proposed intelligent system to support decision-making is the estimation stage and involves two additional modules (module #5 and module #6). Module #5 is composed of additional two databases. The first database contains single- and multi-objective optimization models, while the second is composed of different suitable methods for their solving. The database with single- and multi-objective optimization models relies on the formulation of different optimization criteria expressed as objective function/s under given restrictions. The second database with different solution methods is intended to maintain a variety of exact or heuristic approaches to solve the transformed multi-objective problem into a single-objective one. Using one of these solution methods will provide a Pareto-optimal solution of the investigated problem. The advantage of this approach is the fact that the obtained solution is surely optimal or Pareto-optimal for the formulated objective function and used restrictions.

In such a way, it is possible to estimate the ranked alternative from the selection stage using module #3 with respect to its effectiveness expressed via formulated single- or multi-objective problems via module #5. Module #6 is intended to provide a proper interface to visualize the effectiveness of the alternative.

3.5.1 MAGDM Models for Wind Turbine Type Selection

One of the essential stages of the proposed framework in Fig. 3.14 is the formulation of a mathematical model for the selection of turbine type via group decision-making considering differences in DMs' opinions importance. For the goal, the classical SAW model is modified by introducing of binary integer variables corresponding to the number of given alternative and using additional weighted coefficients to reflect the background experience and knowledge of the within-group experts (DMs). The resulting formulation leads to the determination of aggregated group decision by modified simple additive weighting model (MSAW) as follows:

$$\max\left(\sum_{i=1}^{M}\sum_{k=1}^{K}x_i\lambda^k A_i^k\right) \tag{3.53}$$

subject to

$$\forall i=1,2,\ldots,M:\left(\forall k=1,2,\ldots,K:A_i^k=\sum_{j=1}^{N}w_j^k e_{i,j}^k\right) \tag{3.54}$$

$$\sum\nolimits_{j=1}^{N}w_j^k=1,\forall k=1,2,\ldots,K \tag{3.55}$$

$$\sum_{i=1}^{M}x_i=1,x_i\in\{0,1\} \tag{3.56}$$

$$\sum_{k=1}^{K}\lambda^k=1,\lambda^k\in(0,1) \tag{3.57}$$

The aggregate assessment of the i-th alternative against evaluation criteria, according to the k-th expert (DM) in the group, is expressed by A_i^k. The weighted coefficient w_j^k represents the relative importance among criteria from DMs' point of view. Each DM determines corresponding evaluation scores $e_{i,j}^k$ to express his opinion on the performance of i-th alternative toward j-th criterion. The binary integer variables x_i is used to perform a selection of the single alternative (turbines' type). To distinguish the importance of alternatives assessment between DMs, another type of coefficients λ^k is involved to indicate the differences in the expertise of each group's members. The values of these coefficients make it possible to differentiate the point of view of the experts within the group. The sum of corresponding coefficients λ^k should be equal to 1. It should be noted that acceptable values of coefficients λ^k are from the open interval (0, 1). This restriction guarantees that all DMs' opinions are taken into account. The relation (3.53) expresses the assessment of each alternative as a linear combination of the relative weight of criteria importance multiplied by corresponding evaluation scores for each criterion.

The assessment of alternatives via classical WPM uses multiplication of evaluation scores for each criterion on the degree of the relative weight of criteria importance. The proposed modified model for the selection of turbines type via group decision-making based on WPM uses binary integer variables assigned to each alternative and a weighted coefficient for background experience and knowledge of each DM. The resulting formulation leads to the determination of aggregated group decision by modified weighted product model (MWPM):

$$\max\left(\sum_{i=1}^{M}\sum_{k=1}^{K}x_i\lambda^k R(A_i)^k\right) \tag{3.58}$$

subject to (3.55)–(3.57) and additional restriction

$$\forall i=1,2,\ldots,M:\left(\forall k=1,2,\ldots,K: R(A_i)^k=\prod_{j=1}^{N}\left(e_{ij}^k\right)^{w_j^k}\right) \tag{3.59}$$

The overall assessment of the *i*-th alternative against a set of the evaluation criteria, according to the *k*-th DM, is denoted by $R(A_i)^k$. The rest of the notations are the same as described in (3.53)–(3.57) restrictions of GDM-SAW model.

3.5.2 *Optimization Model for Determining the Efficiency of the Selected Wind Turbine Type for Building Wind Farm*

The efficiency of a particular turbine for the building of a wind farm in accordance with the last stage of the proposed methodology is determined by the following single-objective optimization model:

$$\min\left(\frac{\text{Costs}}{\text{Energy}}\right)=\frac{N\left(\frac{2}{3}+\frac{1}{3}e^{-0.00174N^2}\right)}{NP_{wt}h_y\eta} \tag{3.60}$$

subject to

$$SD_x=k_xD_{wt} \tag{3.61}$$

$$SD_y=k_yD_{wt} \tag{3.62}$$

$$N=N_xN_y \tag{3.63}$$

$$N_x=\frac{L_x}{SD_x}+1 \tag{3.64}$$

$$N_y=\frac{L_y}{SD_y}+1 \tag{3.65}$$

$$k_x^{\min}\le k_x\le k_x^{\max} \tag{3.66}$$

$$k_y^{\min}\le k_y\le k_y^{\max} \tag{3.67}$$

3.5.3 Numerical Results

For numerical testing, six types of wind turbines with rated power equal to 2 MW are used and their corresponding parameters are shown in Table 3.7 (Sagbansua & Balo, 2017).

The capacity factor of a particular turbine represents the actual annual energy output divided by the theoretical maximum output if the turbine is running at its maximum during certain hours of the year. If this relation is bigger, more energy output will be produced. The parameter of hub height is related to the temperature, air density, and air pressure. The wind speed is increasing with heights above ground level. This is due to the decrease of the temperature and the pressure of the air when the altitude is increased. The usage of turbines with higher hub increases the extracted wind power and turbines energy output. Therefore, this is better if its value is greater. Cut-out wind speed is the highest wind speed value at which wind turbines should stop in order to prevent overloading and stresses of the blades (Hansen et al., 2004) to avoid damaging the turbine. If turbines are able to work at a bigger range of wind velocities, more energy output will be produced. Output energy depends on the rotor swept area, i.e., on rotor diameter. The bigger rotor diameter will generate more energy. The nominal wind speed represents the lowest speed at which the nominal (rated) power output of the turbine is reached.

Following all of the above considerations, the DMs from the group should evaluate turbines types by their parameters and should determine corresponding weights for each of these parameters. According to the proposed methodology, the most preferred wind turbine type will be determined as a result of the aggregating of all DMs' opinions by using the proposed GDM-SAW and GDM-WPM models.

The group of experts is formed by four DMs with different expertise and background who should make a decision for selection of the best from 6 wind turbine types (Table 3.7). Each DM evaluates a given set of turbines toward their parameters (performance criteria). The alternatives of turbines types are evaluated toward these criteria by direct scoring using values in the range between 0 and 10.

Table 3.7 Wind turbines' parameters with rated power of 2.0 MW

Turbines	Annual output (at 8 m/s) in MWh	Capacity factor	Rotor diameter (m)	Hub height (m)	Cut-out wind speed (m/s)	Nominal wind speed (m/s)
T-1	9269	32.4	114	80/93/125/	25	12.0
T-2	9327	32.9	116	80/94	25	10.0
T-3	8498	30.2	103	80/90	22	10.0
T-4	9074	32.0	110	80/95	20	11.5–12
T-5	8825	31.1	98.0	60/80/98	25	15.0
T-6	8676	30.7	92.5	64/80/100	24	12.5

Table 3.8 The summarized evaluations from each DM

#	Evaluation criteria and their weights						DMs
	C1	C2	C3	C4	C5	C6	
	$w_1^1=0.1$	$w_2^1=0.21$	$w_3^1=0.25$	$w_4^1=0.12$	$w_5^1=0.22$	$w_6^1=0.10$	DM-1
T-1	9	9.8	6	10	10	8.4	
T-2	9.7	9.8	5	7	9.9	9.6	
T-3	4.5	5.8	8.3	5.3	8	10	
T-4	7.8	8.5	8.9	8	8.2	8.6	
T-5	6.6	7.3	9	7.3	10	6	
T-6	5.9	6	10	9	9.4	7.3	
	$w_1^2=0.16$	$w_2^2=0.18$	$w_3^2=0.22$	$w_4^2=0.12$	$w_5^2=0.1$	$w_6^2=0.22$	DM-2
T-1	8.4	7	8.4	9	6.3	8.1	
T-2	9.5	7.8	9.7	9.8	7.4	9	
T-3	5.8	5.3	9.6	5.6	8	8.2	
T-4	8.4	8.3	9.2	8.4	7.3	7.5	
T-5	8.2	7.8	6.5	7	9	9.1	
T-6	6.3	9.1	8.4	7.6	10	8.9	
	$w_1^3=0.25$	$w_2^3=0.1$	$w_3^3=0.1$	$w_4^3=0.1$	$w_5^3=0.23$	$w_6^3=0.22$	DM-3
T-1	9.3	9.9	8	6.4	9	9.3	
T-2	9.4	10	9.5	5	10	9.8	
T-3	5.4	8	9.8	8.3	5	5.8	
T-4	7.7	8.2	9.1	7.4	8.7	8	
T-5	8.8	10	7.4	9.1	8.8	7	
T-6	7.1	9	9.3	9.8	7.9	6.7	
	$w_1^4=0.15$	$w_2^4=0.20$	$w_3^4=0.20$	$w_4^4=0.18$	$w_5^4=0.15$	$w_6^4=0.12$	DM-4
T-1	8.8	8.9	9.5	6.7	9.2	9.5	
T-2	9.7	9.1	7.9	5	9.3	9.2	
T-3	8.9	8.1	5.4	8.2	5.8	6.2	
T-4	9.2	9.7	8.9	7.9	8.4	8.6	
T-5	7.4	9.2	9.7	9	8.6	7	
T-6	7.2	9	10	9.8	6.7	7.9	

Value 0 means that the criterion in question (alternative) does not entail any benefit, while the value of 10 means that the alternative best meets the efficiency requirements for a wind farm. The sum of the weights of all turbines' parameters should be equal to 1. The summarized results of DMs' evaluation scores of different turbine type alternatives and weights of the criteria of each expert from the group are shown in Table 3.8.

Using the results from Table 2.8 and MSAW model (3.53)–(3.57), and MWPM with objective function (3.58) subject to (3.55)–(3.57) and (3.59), the best wind turbine types considering 3 different sets of weighting coefficients reflecting the members' expertise are obtained and shown in Table 3.9.

Table 3.9 Selected turbine types by using of MSAW and MWPM models

	Weighted coefficients for DMs' expertise				Optimal selection of turbine	
Cases	DM-1	DM-2	DM-3	DM-4	MSAW	MWPM
Case-1	0.25	0.25	0.25	0.25	T-2	T-2
Case-2	0.25	0.10	0.35	0.30	T-1	T-1
Case-3	0.20	0.30	0.15	0.35	T-2	T-4

The results of using both modified MSAW and MWPM models show that the T-1 and T-2 turbines are good enough alternatives for the building of wind farms. For example, in Case-1 where experts are considered to have equal background and knowledge (i.e., equal weighted coefficients) the objective function values for both models (MSAW and MWPM) determines T-2 as the most preferred turbine type. In Case-2, where evaluation scores of DM-3 for turbines' performance are given more importance, followed by DM-4 and DM-1, and less importance of DM-2 evaluation scores, the preferable turbine type is T-1 also for both models (MSAW and MWPM). Case-3 expresses the importance of evaluation scores of the experts in the following order DM-4, DM-2, DM-1, and DM-3. For this case, modified MSAW determines turbine T-2 as the most preferred while modified GDM-WPM defines turbine T-4 as the best one.

The ultimate goal of using the optimal choices of turbines types is to build an efficient wind farm. To investigate the wind farm efficiency, a wind farm with rectangular shape and dimensions $4km \times 2km$ is considered. The efficiency of the wind farm is defined by objective function (3.60) that expresses the costs per energy ratio over a year for 3 possible wind directions: (1) predominant wind direction perpendicular to the long side of the wind farm (predominant-1); (2) wind direction predominant perpendicularly to the short side of the wind farm (predominant-2); and (3) uniform wind directions. The value for the wind power utilization coefficient η is considered to be equal to 0.32. The separation coefficients k_x and k_y (used for determination of turbines separations) are chosen as number of rotor diameters limited as:

- $1.5 \leq k_x \leq 3$ and $8 \leq k_y \leq 11$ for predominant-1 wind direction
- $8 \leq k_x \leq 11$ and $1.5 \leq k_y \leq 3$ for predominant-2 wind direction
- $4.5 \leq k_x = k_y \leq 5.8$ for uniform wind directions

The wind farm efficiency is estimated by using the optimization model (3.60)–(3.67) for the determined wind turbine types in accordance with Tables 3.8 and 3.9. The obtained results for the efficiency of the wind farm and for the number of turbines for its building under different wind directions are shown in Table 3.10.

The optimization model (3.60)–(3.67) along with estimation for the efficiency of wind farm also determines the number of turbines and their separation distances needed to define the position of each turbine within the designed wind farm. Using turbine T-1 or T-2 determines the overall turbines number as 45 for the predominant wind direction perpendicular to the longest farm site, 54 turbines for opposite wind direction, and 28 in case of uniform wind directions. The number of turbines for

Table 3.10 Wind farm efficiency for different wind directions

	Predominant-1 wind direction		Predominant-2 wind direction		Uniform wind directions	
Turbine type	Wind farm efficiency	Turbine number	Wind farm efficiency	Turbine number	Wind farm efficiency	Turbine number
T-1	0.1206654E-03	45	0.1192838E-03	54	0.1341084E-03	28
T-2	0.1206654E-03	45	0.1192838E-03	54	0.1341084E-03	28
T-4	0.1196791E-03	50	0.1191202E-03	57	0.1206654E-03	45

wind farm building using turbine T-4 for predominant-1 wind direction is 50 turbines, for predominant-2, wind direction the farm could contain 57 turbines, and for uniform wind directions, the number of turbines is equal to 45.

The determined best turbine types by using of MSAW are illustrated in Fig. 3.15a and corresponding selections when using of MWPM are shown in Fig 3.15b.

In Case-1, where all DMs have equal weighted coefficients show that turbine T-2 as the best selection for using both models MSAW and MWPM (Fig. 3.15). Based on the same input data from Table 1, other authors show that the optimal technical parameters are observed at wind turbine T-1 using AHP techniques (Sagbansua & Balo, 2017). In contrast to the AHP, where the problem is decomposed hierarchically of more comprehended sub-problems which are analyzed independently, the proposed approach integrates all evaluations, coefficients, and DMs competency in a single utility function. The most preferred alternative is obtained by a single run of optimizations task, based on using binary integer variables for selection.

In Case-2, both models determine as the best selection turbine T-1, while in Case-3 MSAW model determines T-2 as the most preferable one unlike the MWPM model where T-4 is determined as the most preferable turbine. This difference is due to the used different utility function and aggregation of evaluations weights for criteria and weighed coefficients for DMs. Regardless of the model used to determine the best turbine, the final stage of the proposed framework of an intelligent system to support decision-making requires evaluating the effectiveness of a selected turbine for wind farm building. The results for determining wind farm efficiency based on the optimization model (3.60)–(3.67) are illustrated in Fig. 3.16.

The selection of the most appropriate wind turbines is an important step in the building of wind farms, but the evaluation of the effectiveness of the designed wind farm should not be underestimated. Using the parameters of turbines T-1 and T-2 in the proposed model (3.60)–(3.67), obtained effectiveness for wind farm building is less than effective when parameters of turbine T-4 are used (Fig. 3.16). The number of turbines for the building of wind farms depends on available resources like wind speeds, wind directions, and determined site dimensions for building the farm. In this regard, the contemporary trend is to make turbines with a higher power while preserving the rotor diameter or make it smaller. For example, the 2 *MW* platform

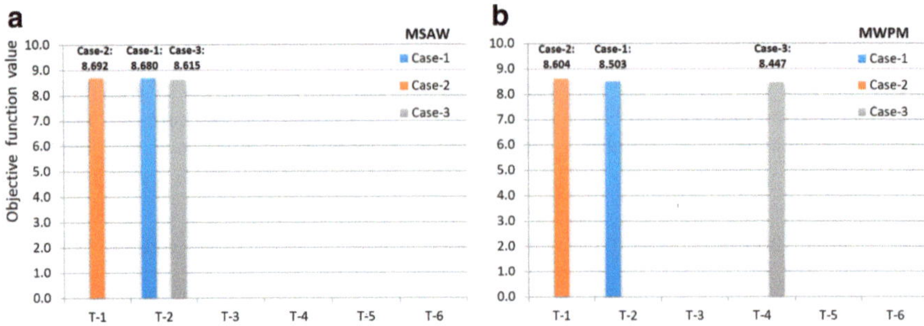

Fig. 3.15 (**a**) The best turbine selection under different DMs weights and MSAW; (**b**) the best turbine selection under different DMs weights and MWPM

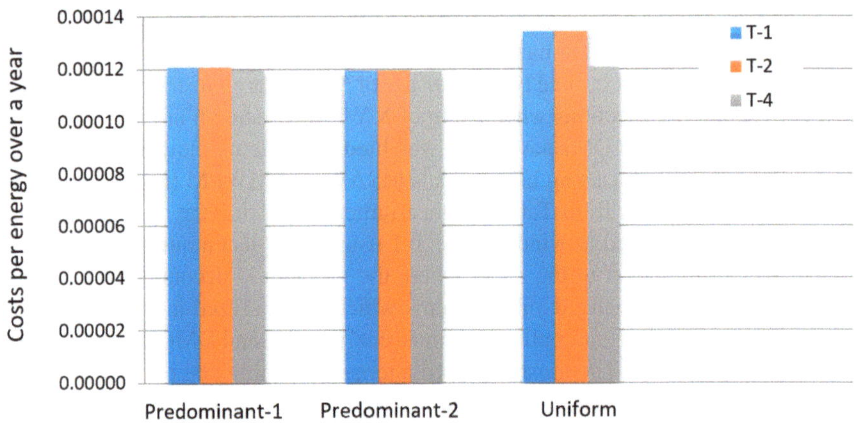

Fig. 3.16 Wind farm efficiency as function of selected turbine type

of Vestas Company offers wind turbines with rotor diameters of 90 *m*, 100 *m*, and 110 *m*. Using smaller rotor diameter allows building more compact wind farms and more efficient.

As it is shown in Fig. 3.16, not always the best choice of the turbine (alternative) determines the best efficiency of the project for a wind farm building. Therefore, it is important to estimate somehow the expected efficiency before starting the wind farm building.

The proposed intelligent framework integrates techniques of MADM/MAGDM and MODM to evaluate the selected alternative with respect to its efficiency. Two new utility functions suitable for group decision-making are proposed based on classical SAW and WPM models. The essence of the proposed MSAW and MWPM models is the fact that they are capable of aggregating not only evaluations and weights for evaluation criteria given by each DM but also incorporating the importance of the opinion of each group member into a final group decision. It is shown that using different weighted coefficients for DM opinions' importance influences a group decision. Therefore, it is important carefully to identify the group members and properly determine their coefficients expressing the importance of a formed group decision. The described approach could integrate other models based on the formulation of different utility functions. The second part of the described intelligent framework emphasizes the formulation of single- or multi-objective optimization models. The idea of this part is to estimate the effectiveness of the obtained selected alternative from the previous stage. These models should correspond to the goal of the company.

3.6 Wind Farm Design Avoiding Forbidden Zones with Regular Geometric Shape

Many real-world engineering problems are related to the layout design of some objects within the given area where exist some forbidden object placement zones. Locations of stations of mobile communication, placement of modules on reconfigurable resources under the constraints of field-programmable gate array, wireless sensor networks (for example of forest fires detection), architectural floor plan layout design, placement of wind turbines, etc., are just some examples that could be mentioned. Due to the complexity of those types of problems, the researchers propose different approaches, models, and algorithms to tackle fuzziness in relations and restrictions (Peneva & Popchev, 2010; Zadeh, 1999; Zadeh, 1975;).

Reasonable decision-making for any technical system design relies on the selection of important parameters that influence the final choice and different preliminary requirements need to be considered. In the case of layout design problems, the restrictions are related to given dimensions of the area, integer number of objects for placement, and forbidden for placement zones. In many real problems, there also exist restrictions for the distance between objects which is an important species of layout design problems. For example, construction site layout planning requires the determination of reasonable approximations to the path followed by workers, machinery, etc., between existing facilities (Hammad et al., 2015). A typical example of applying restrictions about the distance between objects is wind farm layout design where the wake effect between turbines should be considered (Shakoor et al., 2016). The wind farm design is essential in the contemporary context of renewable energy usage. The existence of forbidden zones in wind site areas is an important problem of wind farms' optimal layout design. Such zones could be the result of physical obstacles due to archaeological ruins, natural areas, visual impact, etc.

3.6.1 Combinatorial Optimization Model for Placement of Objects Avoiding Forbidden Zones with Regular Geometric Shape

The basic problem is to determine the optimal layout placement of certain objects within the area with predefined dimensions and with the existence of forbidden placement zones with different shapes. Two different shapes of forbidden zones are considered—a trapeze shape (*A*) and a circular shape (*B*) as shown in Fig. 3.17 (Hou et al., 2016).

Zone *A* can be described by two equations of a straight line:

$$y_1 = ax + C_1$$

$$y_2 = ax + C_2$$

where a, C_1, and C_2 are constants for each line description. The circular shape of zone *B* is described by the equation

$$(x - x_c)^2 + (y - y_c)^2 = R^2$$

where R is radius of the circle and x_c, y_c are center coordinates.

An important feature of the described modeling approach is using relative coordinates i and j for each object O_{ij} defined by means of required inter-objects distances d_x and d_y as $i = x/d_x$ and $j = y/d_y$.

The proposed generalized optimization model for the placement of objects O_{ij} avoiding forbidden zones and which is also suitable for the placement of wind turbines is as follows (Borissova & Mustakerov, 2017c):

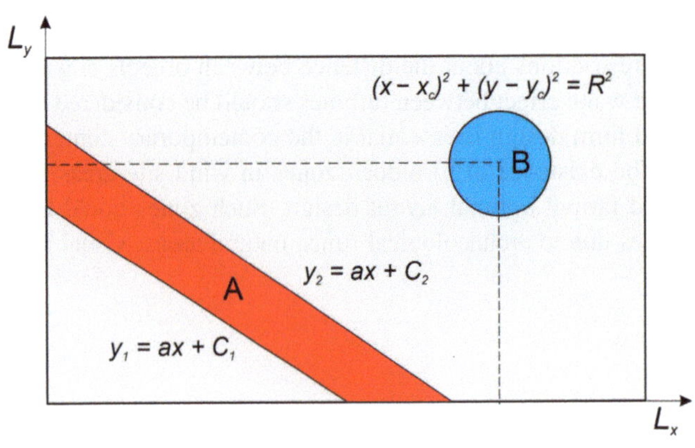

Fig. 3.17 Area with two forbidden zones A and B

$$\max\{F(N^O)\} \tag{3.68}$$

subject to

$$N^O = \sum_i^{N_x} \sum_j^{N_y} O_{ij} \tag{3.69}$$

$$N_x = \frac{L_x}{d_x} - 1, \text{integer} \tag{3.70}$$

$$N_y = \frac{L_y}{d_y} - 1, \text{integer} \tag{3.71}$$

$$\forall i \in \{1, \ldots, N_x\} : J_1 = a_i + C_1 \tag{3.72}$$

$$\forall i \in \{1, \ldots, N_x\} : J_2 = a_i + C_2 \tag{3.73}$$

$$\forall i \in \{1, \ldots, N_x\} : Z_1 = \frac{2j_c - \sqrt{(2j_c)^2 - 4(R^2 - i^2 - i2i_c + i_c^2)}}{2} \tag{3.74}$$

$$\forall i \in \{1, \ldots, N_x\} : Z_2 = \frac{2j_c - \sqrt{(2j_c)^2 - 4(R^2 - i^2 - i2i_c + i_c^2)}}{2} \tag{3.75}$$

$$\forall i \in \{1, \ldots, N_x\} : O_{ij} = \begin{cases} 0, \forall j \in \{1, .., N_y\} : J_1 \le j \le J_2 \text{ or } Z_1 \le j \le Z_2 \\ 1, \text{otherwise} \end{cases} \tag{3.76}$$

In most cases, objective function (3.68) aims to maximize some function depending on the number of objects N^O. The inter-objects distances (d_x and d_y) define the number of nodes and their relative coordinates excluding nodes on the boundaries by relations (3.70) and (3.71). The site area dimensions where objects are to be placed are denoted by L_x and L_y, respectively. The forbidden zones are described by relations of a straight line (3.72) and (3.73) for zone A, and (3.74) and (3.75) for zone B. These equations also do not allow placement of objects on the area boundaries. The relation (3.76) expresses the objects O_{ij} that are considered as binary integer variables equal to 1 if an object is placed at optimal location outside forbidden zones or 0 if it is placed in the areas of forbidden zones (including their boundaries).

3.6.2 Numerical Results

The applicability of the proposed model for optimal layout design avoiding forbidden zones is numerically tested for the design of offshore wind farm. This problem is adopted from Hou et al. (2016) where the number of certain type of wind turbine with rated power $P = 10\ MW$ and rotor diameter $D_r = 0.1783\ km$ has to be placed within rectangular area $L_x = 10.5\ km$ and $L_y = 13\ km$. The prevailing wind direction (east to west) is given together with shapes and dimensions of forbidden zones.

The coefficients for equations describing zone (A) are:

$$a = -1; C_1 = 6.5; C_2 = 7.8;$$

and for zone (B):

$$x_c = 8\ km; y_c = 6\ km; R = 0.7\ km.$$

The examples for forbidden zones that are to be avoided for placement of wind turbines could be] (A)—installed gas pipe (or Marine Traffic Line) and (B)—oil well availability (Fig. 3.18).

Based on the generalized optimization model (3.68)–(3.76) and a utility function that expresses the ratio of annual energy per costs as function of number of turbines (Marmidis et al., 2008) and their rated power (Mustakerov & Borissova, 2011), the following combinatorial optimization task for determination number and placement of given type of turbines and considering forbidden zones is formulated as:

$$\max\left(\frac{\eta h_y N^O P_{wt}}{N^O\left(\frac{2}{3}+\frac{1}{3}\,exp\left(-0.00174\left(N^O\right)^2\right)\right)}\right) \tag{3.77}$$

subject to

$$N^O = \sum\nolimits_i^{N_x}\sum\nolimits_j^{N_y} O_{ij} \tag{3.78}$$

$$N_x = \frac{10.5}{0.1783k_x} - 1, \text{integer} \tag{3.79}$$

$$N_y = \frac{13}{0.1783k_y} - 1, \text{integer} \tag{3.80}$$

$$9 \le k_x \le 11 \tag{3.81}$$

$$1.5 \le k_y \le 3 \tag{3.82}$$

$$\forall i \in \{1, \ldots, N_x\} : J_1 = -i + \frac{6.5}{0.1783k_x} \tag{3.83}$$

$$\forall i \in \{1, \ldots, N_x\} : J_2 = -i + \frac{7.8}{0.1783k_y} \tag{3.84}$$

$$\forall i \in \{1, \ldots, N_x\} : Z_1 = \frac{2j_c - \sqrt{(2j_c)^2 - 4\left(R^2 - i^2 - i2i_c + i_c^2\right)}}{2} \tag{3.85}$$

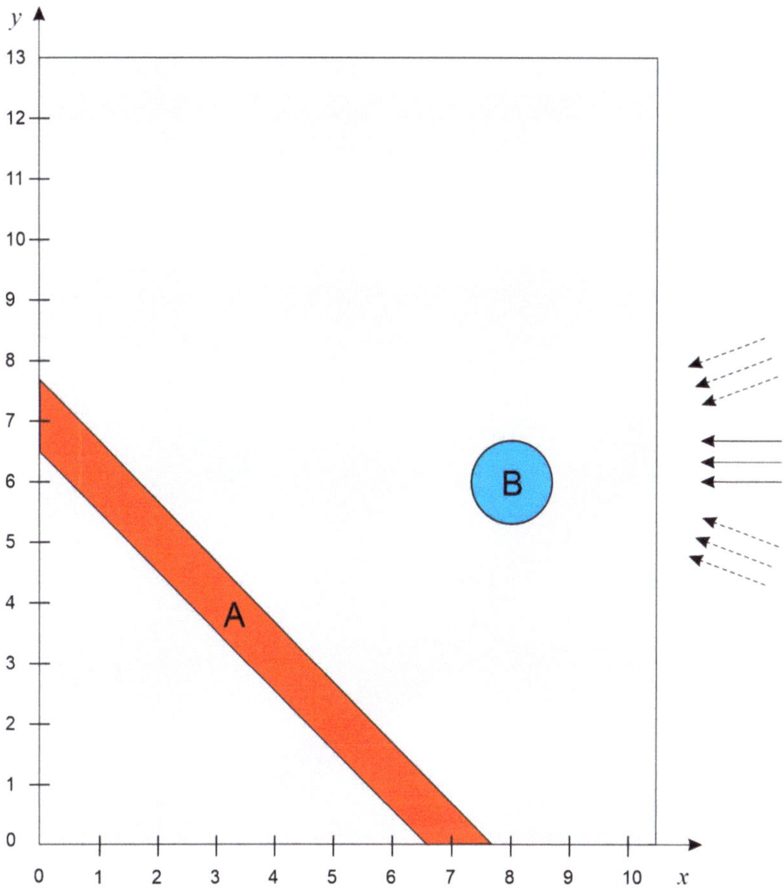

Fig. 3.18 Offshore wind farm area with forbidden zones A and B and prevailing wind direction

$$\forall i \in \{1, \ldots, N_x\} : Z_2 = \frac{2j_c - \sqrt{(2j_c)^2 - 4\left(R^2 - i^2 - i2i_c + i_c^2\right)}}{2} \tag{3.86}$$

$$i_c = \frac{8}{0.1783k_x} \tag{3.87}$$

$$j_c = \frac{6}{0.1783k_y} \tag{3.88}$$

$$\forall i \in \{1, \ldots, N_x\} : O_{ij} = \begin{cases} 0, \forall j \in \{1, .., N_y\} : J_1 \le j \le J_2 \ or \ Z_1 \le j \le Z_2 \\ 1, \text{otherwise} \end{cases} \tag{3.89}$$

The optimization task was solved by means of LNGO software on PC with Intel Core i3 @ 2.93 GHz under MS Windows. The branch-and-bound algorithm is used and solution is found after 127 iterations for about a few seconds. The heuristic

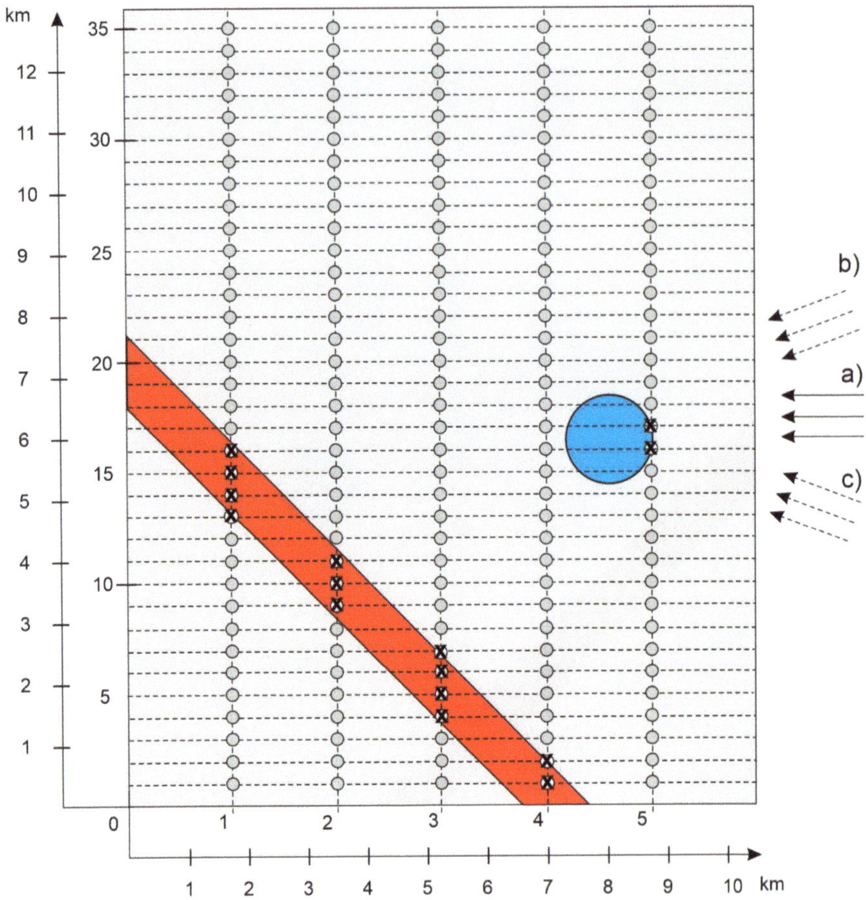

Fig. 3.19 Offshore wind farm optimal layout avoiding forbidden zones

algorithms take more solution time without guarantee to get to optimal solution. For example, 1180 iterations are needed to get stable (not optimal) solution in (Hou et al., 2016). The results of optimization task define 160 turbines of given type to be installed outside of forbidden zones using grid layout with separation distances $d_x = 1.75\ km$ and $d_y = 0.361\ km$. The energy output per unit of costs for particular offshore wind farm example is equal to 32850.00 *MW/h*. The result of optimization task solution is graphically illustrated in Fig. 3.19.

The obtained values for separation distances between turbines satisfy requirement to overcome the wake effect and guarantee the optimal placement of turbines even in the presence of small displacement of predominant wind direction shown in Fig. 3.19 as (b) and (c). The advantage of the proposed combinatorial optimization model in contrast to the heuristic optimization method proposed in Hou et al. (2016) is the optimal and compact utilization of a given area using grid layout for the placement of turbines. The solution of the formulated combinatorial optimization task defines a considerably larger number of turbines, which reflect in increasing installed wind farm capacity and in extracted output energy. Another advantage is a computational time of order of seconds which is far better than heuristic approaches.

The integer variables for turbines number combined with narrow limits for separations coefficients could result in an infeasible solution for some fixed dimensions of the wind farm area. This typical drawback for mixed-integer optimization can be tackled by widening limits for separation coefficients.

The essence in the described modeling approach is using grid mesh relative coordinates of optimal objects locations instead of metric units coordinates defined by dimensions of the site. This reduces the number of locations that have to be checked if they belong to forbidden zones.

The described model can be used with different objective functions and different shapes of forbidden zones. These zones could be taken into account by decomposing them to quadrangles and circles (with some approximations) and using proper equations to describe them. The applied approximations for complex forbidden zone shapes are not critical for practical applications. Therefore, the presence of restricted access areas is not an obstacle to installing a wind power plant in a specific location that has shown excellent wind potential indicators.

3.7 Wind Farm Design Avoiding Forbidden Zones with Irregular Geometric Shape

In this section, a problem with the design and assessment of wind farm potential on restricted terrain is discussed. The existence of forbidden areas can be due to physical obstacles, archaeological ruins, natural areas, visual impact, or any kind of specific constraints (Serrano-Gonzalez et al., 2011). The investigated problem aims to assess the most appropriate design of wind plant layout taking into account the wind power plant parameters and considering forbidden zones to place turbines. The illustration of terrain with forbidden zones that is used as an example in the chapter is shown in Fig. 3.20.

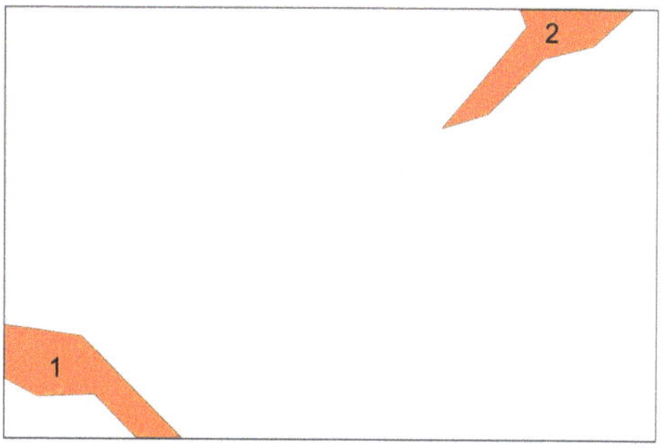

Fig. 3.20 Wind site area with two forbidden zones (1) and (2)

Given several types of turbines, the problem is to choose the most suitable type and number, as well as their location, taking into account restricted areas and estimating the costs and energy output of a wind farm.

3.7.1 Algorithm for Assessment of Wind Power Plant Layout Design Considering the Forbidden Zones for Location of Turbines

The proposed algorithm is based on combinatorial optimization model for theoretical estimation of the annual wind power plant energy per unit of costs. It is adapted from Mustakerov and Borissova (2011) and considers wind farm layout design for given turbines types and known site conditions. This model is used to define a single-criterion mixed-integer nonlinear problem as follows:

$$\max\left(\frac{\eta 8760NP}{N\left(\frac{2}{3}+\frac{1}{3}\,exp\left(-0.00174(N)^2\right)\right)}\right) \tag{3.90}$$

subject to

$$N=\left(\frac{L_x}{k_xD}+1\right)\left(\frac{L_y}{k_yD}+1\right) \tag{3.91}$$

$$k_x^{\min}\leq k_x\leq k_x^{\max} \tag{3.92}$$

$$k_y^{\min}\leq k_y\leq k_y^{\max} \tag{3.93}$$

$$SD_x=k_xD \tag{3.94}$$

$$SD_y=k_yD \tag{3.95}$$

The solution of optimization problem (3.90)–(3.95) will define the number of turbines and their layout for given dimensions of wind site and wind conditions. The obtained layouts for turbines placement are optimal in sense of maximizing energy per unit of costs.

The diagram of the proposed algorithm is shown in Fig. 3.21 (Borissova & Mustakerov, 2017e).

It is assumed that wind speed and direction and the probability of their occurrence are known. This information is used in the first step of the algorithm to select a number of appropriate wind turbine types. The second step uses combinatorial optimization model (3.90)–(3.95) for the determination of different wind power plant layouts for each of the given turbine types. The designated locations of the turbines of each type are investigated in the third step whether some of them fall into the forbidden zones. If this is true, the corresponding turbines are excluded from the

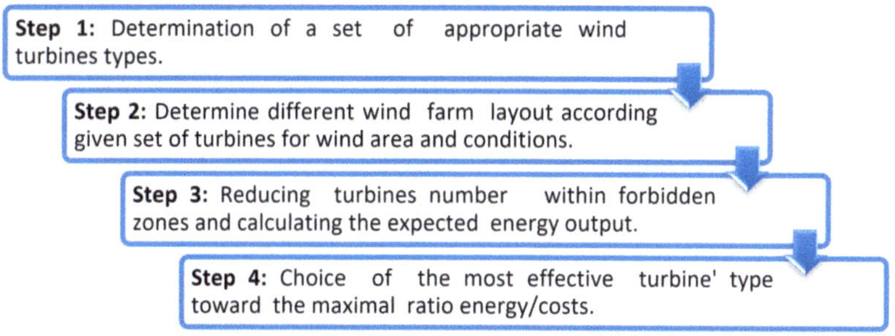

Fig. 3.21 Algorithm for design and assessment of wind power plant with forbidden zones

layout. This results in reducing in the overall number of turbines. Then the ratio energy output/costs for each of the turbines types and corresponding layouts is calculated by the expression used in the objective function (3.90).

The results for the energy output/costs ratio are used in the fourth step of the algorithm to select the most effective turbine type considering the restriction about forbidden zones.

3.7.2 Numerical Results

The applicability of the proposed algorithm for design and assessment of wind power plant considering terrain with forbidden zones is numerically tested. It is assumed that by some practical investor's recommendations three types of wind turbines are used with parameters shown in Table 3.11.

Wind site area with dimensions 3 *km x* 1 *km* and 2 prohibited zones (Fig. 3.20) for uniform and predominant wind directions is used for testing. In the case of uniform wind direction, the separation coefficients between columns and rows in wind site are within the limits of 4.5–5.5 rotor diameters as recommended by Patel (1999). For nonuniform wind, the separation coefficients for predominant wind direction are limited between 9 and 11 rotor diameters and are limited between 1.5 and rotor diameters for perpendicular wind direction (Marmidis et al., 2008; Donovan, 2005). The obtained results for each turbine type from Table 1 by using the optimization problem (3.90)–(3.95) are summarized in Table 3.12.

For uniform wind direction, the numerical testing determines equal layouts for all three turbine types from Table 3.7 (Fig. 3.22). The places where turbines cannot be installed due to the existence of forbidden areas are marked as **x**.

Figure 3.23 illustrates nonuniform wind direction case for two opposite predominant wind directions for the same wind site.

The performance of the investigated wind site with regard to maximization of energy output and investment costs ratio is illustrated in Fig. 3.24.

Table 3.11 Wind turbines' parameters

#	Turbine type	Rated power, MW	Rotor diameter, m
WT#1	ENERCON's E-92	2350	92
WT#2	Vestas V100	2600	100
WT#3	SWT-3.6-107	3600	107

Table 3.12 Numerical results

WT#	Wind direction	Turbines number without forbidden zones	Turbines number with forbidden zones	k_x	k_y	Energy/ Costs without forbidden zones	Energy/ Costs with forbidden zones
1	Uniform	35	32	5.35	5.35	8744.870	8544.502
	Pre-1	69	62	1.53	10.60	9262.531	9257.937
	Pre-2	64	60	10.84	1.50	9259.982	9254.891
2	Uniform	35	32	5.02	5.02	9675.176	9453.492
	Pre-1	63	57	1.55	9.75	10244.07	10231.26
	Pre-2	60	57	10.33	1.50	10239.45	10231.26
3	Uniform	35	32	4.83	4.83	13396.40	13089.45
	Pre-1	60	54	1.52	9.11	14177.71	14146.93
	Pre-2	56	54	9.66	1.51	14160.98	14146.93

The solution of formulated optimization tasks on step 2 of the algorithm is obtained by means of LINGO v. 12 solver system. The solution times took a few seconds on PC with Intel Core i3 CPU at 2.93 GHz, 3.37 GB RAM under MS Windows OS.

The analysis of testing results for given wind site and turbines shows that the maximum of energy production per unit of costs can be achieved if turbine WT#3 is used (Fig. 3.24a, b, c). Due to small differences in used turbine rotor diameters (less than 10%) the number and layouts of turbines are equal in case of uniform wind direction. Nevertheless, the differences in the turbines' rated power have a significant impact on the energy production as it is seen in Fig. 3.24a.

For nonuniform wind direction, the number of turbines is different as a result of the existence of forbidden zones. The analysis of results shows that the turbines with smaller rotor diameters are less effective. This is because they require the placement of a greater number of turbines which in turn decreases the energy/costs ratio. Along with this, the experiments show that the turbines with larger rotor diameters have less probability to fall into forbidden zones.

The proposed algorithm in Fig. 3.21 aims to assist the process of the development of a wind farm project. The innovative idea in the described algorithm is using combinatorial optimization for the determination of the best wind power plant layout taking into account the existence of forbidden zones in wind sites. The numerical testing shows the practical applicability of the algorithm. Using the proposed

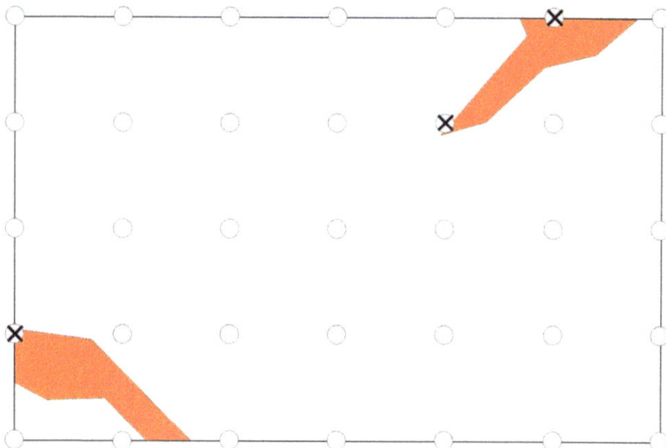

Fig. 3.22 Wind turbine layout for uniform wind direction and forbidden zones

algorithm allows preliminary assessment of wind farm projects for different numbers and types of turbines toward energy/costs ratio while forbidden zones are available.

The proposed algorithm for the design and assessment of the wind farm parameters with forbidden zones could be used with other single- or multi-objective mathematical models for the determination of wind turbine placement.

3.8 Wind Farm Design Considering Location of Point of Common Coupling

The third stage of the wind farm life cycle is related to the commissioning, i.e., actual construction and inclusion in the transmission network. In this regard, after the wind farm layout design is determined it is needed to decide how this power plant will be integrated into the electrical grid. The challenging problem is the connection of this power plant with the electrical grid in a such way as to minimize the distance between the wind farm and the grid, thus decreasing the transmission costs. Various studies and reports on various industrial wind farms show that about 60–80% of the total investment costs are related to the cost of the installed wind turbines. This means that 20–40% of the investment is invested in infrastructure and any possible reduction in connection costs will be beneficial. A review of state-of-the-art research related to wind farm layout optimization as well as electrical system design, including cable connection scheme optimization, is presented (Hou et al., 2019).

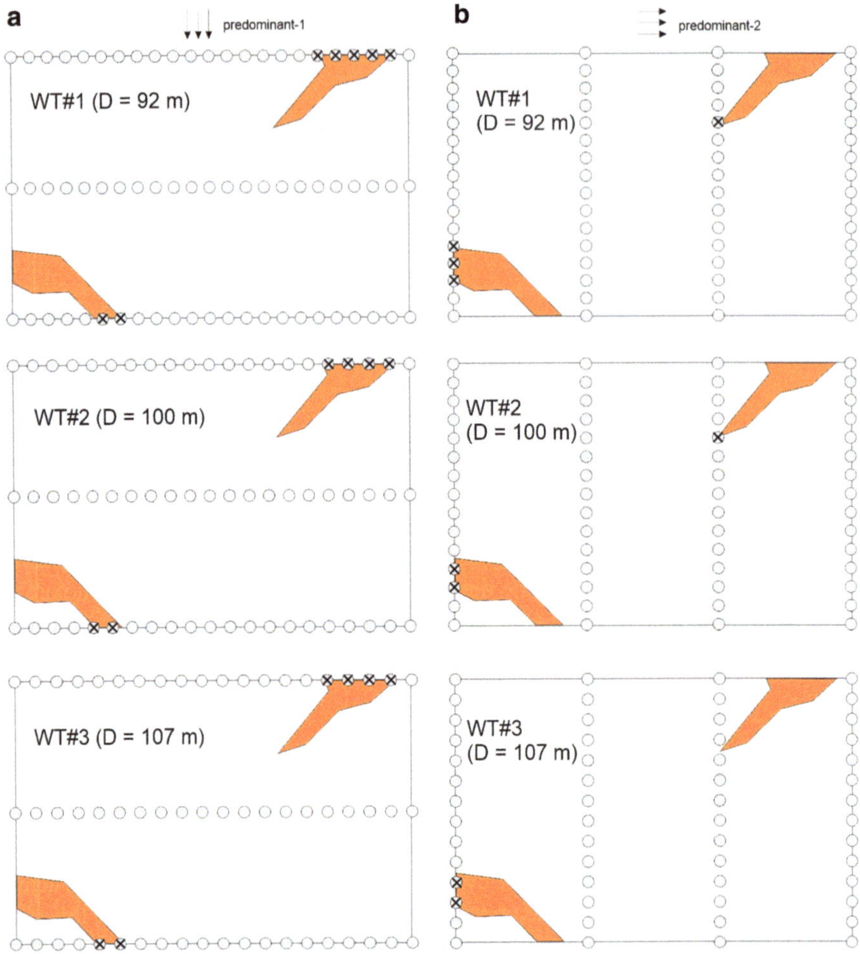

Fig. 3.23 Wind turbine layout for 2 opposite predominant wind directions and forbidden zones (**a**) predominant-1 (left), (**b**) predominant-2 (right)

3.8.1 Model for the Optimal Layout of a Wind Farm Design Considering the Location of the Common Connection Point

The main task in designing the placement of turbines in the wind farm is the optimal use of the wind site area, taking into account the existing wind conditions. Therefore, the practical task of determining the location of the common point of connection of the packer to the electric field can be divided into several stages as shown in Fig. 3.25.

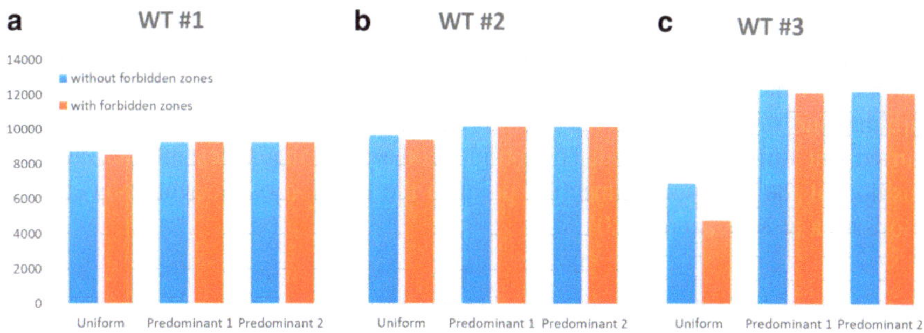

Fig. 3.24 The wind site performance toward turbine type and energy/costs ratio for (**a**) uniform direction; (**b**) predominant direction north-south; (**c**) predominant direction west-east

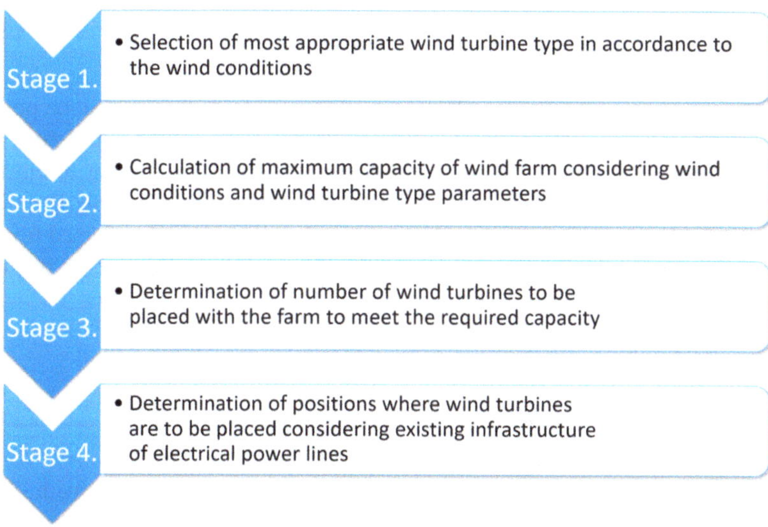

Fig. 3.25 Optimal wind farm layout considering the location of point of common coupling

The selection of the appropriate type of wind turbine depends on the specific wind conditions in the wind site and the selected supplier (stage 1). The local authorities are an important factor in determining the number of wind turbines to be installed on the power plant and in the implementation of the project. Depending on the selected type of wind turbine and based on the formulated optimization models described in this chapter, the maximum allowable number of turbines that can be located in the wind power plant can be determined. All these models are able to determine the optimal power of the wind power plant and the number of turbines by determining their optimal placement according to the specifics of the given wind site (stage 2). Depending on the approved project for the construction of a wind power plant, it is also necessary to determine the number of turbines that will make up the wind farm, so that the planned capacity of the wind farm corresponds to the installed capacity—

stage 3. Of interest is stage 4, where it is necessary to determine which of the already identified turbine locations to skip and which to build turbine foundations on. The selection of these turbine locations should be such that the distances between the turbines and the common connection point are minimal. Solving this new task is carried out through a formulated optimization model that takes into account the position of the common connection point to the location of the wind power plant (Borissova & Mustakerov, 2017d):

$$\min\left(\sum_{i,j} x_{ij} r_i\right) \tag{3.96}$$

subject to

$$\sum x_{ij} = N_{wt} \tag{3.97}$$

where x_{ij} are binary integer variables, r_{ij} are distances from turbines to the point of common coupling, and N_{wt} is the number of turbines to be placed with farm.

The objective function (3.96) seeks to determine such positions of turbines for which the distances r_{ij} to the point of common coupling are minimal. The binary integer variables x_{ij} are assigned to all possible turbine locations determined from the solution of optimization problem (3.19)–(3.29). The number of turbines to be placed is expressed by N_{wt}.

3.8.2 Numerical Results

The considered wind farm site area is restricted by a rectangular shape with dimensions $Lx = 4\ km$ and $Ly = 2\ km$. This area is with wind conditions that could be expressed by known prevailing wind direction parallel to the longer side of the rectangular shape. The separation coefficients needed to overcome the negative wake effect according to the given wind conditions are limited within the range of $7.8 \le k_x \le 9.5$ for the prevailing wind direction and $1.3 \le k_y \le 2.2$ for the opposite wind direction. The selected turbine type for the investigated wind farm has a rated power of 3.6 MW and a rotor diameter of 107 m. The nominal utilization coefficient is considered equal to $\eta = 0.32$ and the hours over the year are equal to $h_y = 8760$. Using these input data determine the following wind farm parameters with theoretically possible maximum capacity as shown in Table 3.13.

The positions of turbines of the same type that ensure the maximum capacity of wind area under minimum cost per unit of annual energy production are illustrated in Fig. 3.26.

The number of turbines for the new wind farm is required to be $N_{wt} = 44$ turbines. This means that out of the specified 50 maximum possible turbine locations, only

Table 3.13 Parameters of the wind farm for optimal maximum capacity

Input data				Solution				
P_{wt}, W	D_{wt}, m	Lx, km	Ly, km	N	Sx, m	Sy, m	k_x	k_y
3.6	107	4	2	50	1000	222.2	9.35	2.07

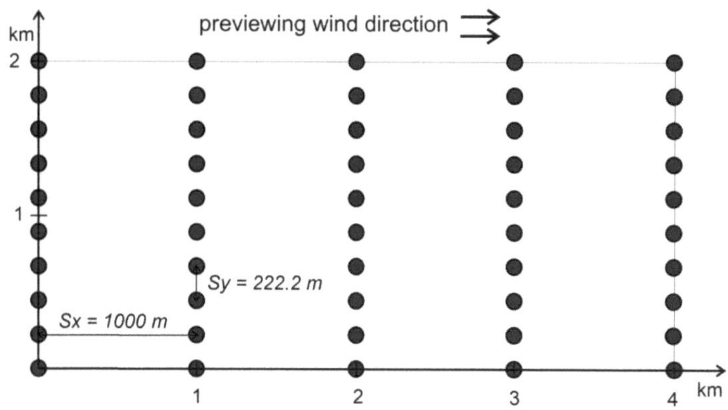

Fig. 3.26 Layout of turbines for maximum wind farm capacity under minimum cost per unit of annual energy

44 of these turbine locations need to be selected. The selected positions should conform to the optimization model (3.96)–(3.97), where the distances from each turbines to the common connection point are minimal (see Fig. 3.27).

In this case, the solution determines zero values for 6 variables and the rest 44 variables are equal to 1; i.e., these variables define the best locations of wind turbines according to minimal distances to point of common coupling.

If the location of the point of common coupling has another position as shown in Fig. 3.28, then the optimal locations for the placement of 44 turbines are different.

The formulated combinatorial optimization model (3.96)–(3.97) determines only these locations of turbines for which the overall distance to the point of common coupling is minimal. This means that the position of the point of common coupling must be carefully considered in advance because it influences the selection of locations of the turbines in the farm (see Figs. 3.27 and 3.28). In both cases, the results show the applicability of the proposed approach based on the used optimization models.

The transmission of wind energy from the wind power plants to the point of connection to the electrical grid predetermines the length of the transmission lines. The more distant these farms are, the cabling is longer. In addition, the amount of power system loss depends on the length and type of the transmission or distribution line (Sadovskaia et al., 2019). Moreover, the distance to the power grid heavily impacts the cost of the project by adding losses in the annual energy production and additional costs associated with longer cabling. Therefore, it is important at the design stage of the wind farm to determine not only the location of the turbines in the

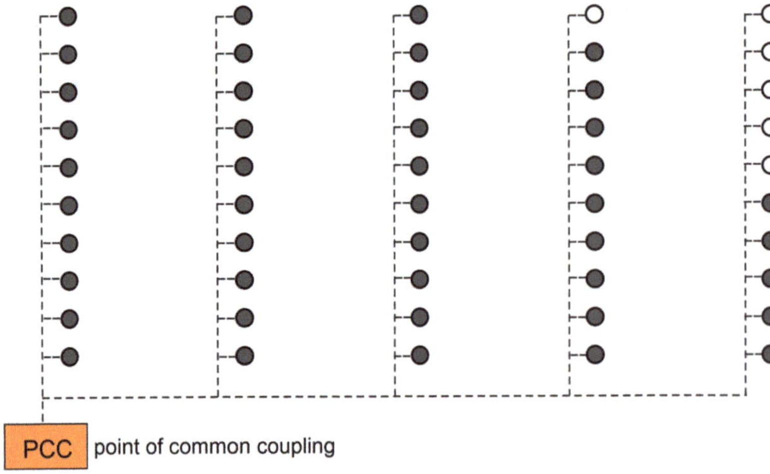

Fig. 3.27 Optimal positions for 44 wind turbines considering location of point of common coupling

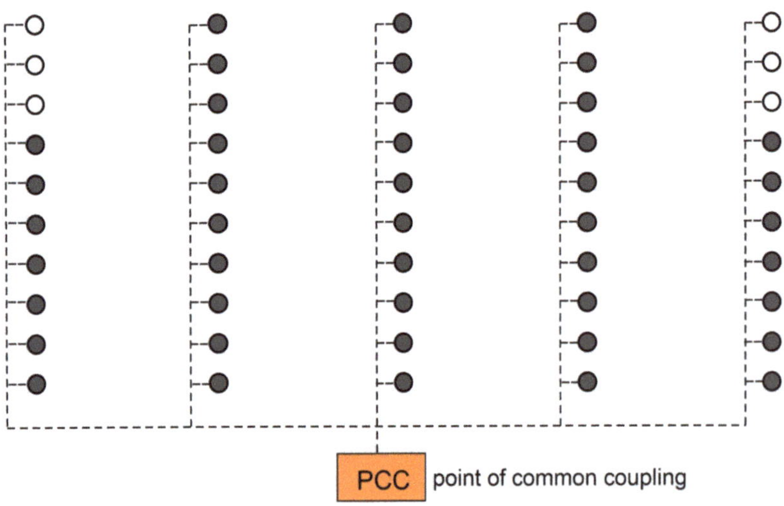

Fig. 3.28 Optimal positions for 44 wind turbines considering location of point of common coupling

farm but also their connection to the electrical network. The proposed combinatorial optimization model clearly shows that the connection point of the wind power plant to the electrical network is essential whether it is onshore or offshore.

3.9 Decision-Making in Selection of Site for Hybrid Wind-Solar Power Plant

Another promising direction related to wind energy is the construction of hybrid wind-solar power plants. On the one hand, this is related to the nature of these two types of renewable energy (wind and solar energy), namely its impermanent nature. Wind turbines only produce power when the wind blows, while solar panels only generate electricity when the sun is shining. Consequently, output from these renewable energy sources can be significantly reduced on cloudy or quiet days. To overcome this challenge, researchers and engineers are working on innovative solutions related to the development of hybrid renewable energy systems that integrate solar panels and wind turbines into a single power generation unit. These hybrid systems rely on the complementary nature of solar and wind energy. Typically, solar panels generate the most electricity in the middle of the day, when the sun is at its highest point in the sky, and for wind turbines, this occurs in the early morning and late evening, when wind speeds are highest. Therefore, combining these two sources of renewable energy is seen as a more reliable solution for electricity supply. Another benefit of building hybrid wind-solar power plants is the shared installation location, which on the one hand is related to the lease or purchase of land, and on the other hand, the potential for reduced environmental impact related to habitat loss and other environmental problems. The integration of solar and wind energy into one system contributes to reducing the amount of energy that needs to be stored, thereby reducing the overall cost of the system. Therefore, combining solar and wind energy enables the creation of a more sustainable energy system that reduces dependence on fossil fuels and helps combat climate change. This combination of solar and wind power represents a significant step forward in our quest for a cleaner and greener future.

The first and most fundamental step toward the sustainable deployment of renewable energy resources in the energy sector is finding suitable locations for the construction of power plants, whether wind or solar. A key requirement for identifying these sites is the presence of sufficient potential to maximize energy production while seeking to minimize project-related investment. Choosing the optimal site for solar photovoltaic power plants and wind power plants is a complex process for energy planners and policymakers due to many conflicting criteria, making decisions difficult. The process of site selection for building solar photovoltaic and wind farms involves many factors, including some constraints. The factors are to be analyzed from an economic and technological point of view, including from a political and environmental point of view.

3.9.1 Criteria for Selection of Site for Hybrid Wind-Solar Power Plant

Using hybrid wind-solar systems contributes to increasing the total production of the hybrid installation. It should be mentioned that the generation of electric energy by wind and solar installations in different months of the year is different. In the summer months, mainly photovoltaic batteries operate, and in the winter months, wind turbines make the main contribution to energy generation (Abd Ali et al., 2020). As a result, the joint work of wind and solar installations to generate electrical energy helps to establish more uniform generation throughout the year. Using such hybrid systems increases the reliability of the energy complex based on renewable energy sources. All of these motivated the researchers to propose different approaches to deal with such complex problems. For example, a multiple linear regression model is used, with the help of the combination of geographical information system (GIS) and AHP methods, to model the siting of wind and solar power plants (Asadi et al., 2023). It is shown that wind-photovoltaic-shared energy storage systems can improve the utilization efficiency of renewable energy resources while reducing the idle rate of energy storage resources. Using the GIS and the MCDM method, a two-stage evaluation model is developed for site selection of wind-photovoltaic-shared energy storage projects using a wide range of constraints and evaluation criteria (Gao et al., 2022). GIS and MCDM are also used to determine the suitable regions for wind-solar hybrid energy systems in Kayseri Province/Turkey (Genc & Karipoglu, 2021). Other research shows that it is possible to determine a suitable site for wind-solar energy by a GIS-AHP-based approach (Koc et al., 2019). Practical evaluation criteria for the site selection under the fuzzy environment are established, which consist of 15 criteria from resource and technical to environmental, social and policy, and economic (Wu et al., 2022).

A photovoltaic power station, also known as a solar power plant, is a large-scale photovoltaic system designed to supply power to the national electricity grid. The selection of a suitable location for a solar photovoltaic farm is influenced by several factors, divided into several main categories (Hasti et al., 2023; Fard et al., 2022; Deveci et al., 2021; Guaita-Pradas et al., 2019): climatic, technical, economic, environmental, and social/political.

A brief description and influence of the indicators from these categories are shown below.

3.9.1.1 Climatic Criteria

The climatic criteria describe the factors related to meteorological parameters such as solar radiation, temperature, sunshine hours, slope, elevation, wind speed and average rainfall, humidity, hydrographic areas, and lines.

a) Solar irradiance is the most impactful parameter to determine the energy generation potential of the solar photovoltaic power plant.
b) Ambient temperature also has an impact on the performance of solar photovoltaic panels (Ebhota & Tabakov, 2023). These panels usually operate between 15 °C and 35 °C, during which solar cells harvest maximum or close to maximum efficiency.
c) Sunshine hours are used as a measure to quantify the duration of sunshine for a specific period (year or day) for a specific location.
d) In addition to solar radiation and temperature other meteorological parameters such as wind speed and average rainfall are to be taken into account, to ensure safer design and operation of the solar photovoltaic power plant. During the maintenance and upkeep of photovoltaic power plants, water is needed to clean the surface of photovoltaic equipment. Therefore, water is indispensable including annual mean relative humidity, annual mean temperature, and annual mean precipitation. On the other hand, the amount of photovoltaic power generation decreases as the annual mean precipitation increases.
e) Ambient humidity has some negative impact on performance in terms of the annual energy production of the project and the degradation rate of the photovoltaic panels in the long run. Short-wave radiation is absorbed by aerosol particles like carbon dioxide and water vapor, and solar energy received by photovoltaic panels decreases as relative humidity increases.
f) The slope of the surface where the solar photovoltaic panels will be located has an impact on the annual energy production. The tilt angle of the slope of the photovoltaic panels is to be determined depending on the location of the project site, geographic topography, and the type of elements under construction. The slope is a critical criterion for construction as it is essential in evaluating the economic feasibility of the project and is largely influenced by the accessibility of the construction site and ease of transportation during construction.
g) Location elevation and hydrographic areas and lines have a very limited impact on the performance of solar photovoltaic power plants if the photovoltaic project is not a floating photovoltaic solar-type project.

3.9.1.2 Technical Criteria

The technical criteria describe the technical indicators related to the design, construction, and operation phases of the solar photovoltaic power plant. They envelop the following sub-criteria such as topological characteristics, distance to network connection, distance to residential areas, distance to roads and logistic works, service life, strength of the existing grid, and solar photovoltaic material technology and efficiency.

a) Topological characteristics of a particular site for a solar power plant have an impact on the activities related to planning, construction, and investment. Topographical surveys are required to determine soil relief requirements in the planning phase of a solar photovoltaic plant project.

b) Distance to the power grid is another important indicator in terms of identifying technical challenges. This is due to the fact that the amount of power system loss depends on the length and type of the transmission or distribution line (Sadovskaia et al., 2019). On the other hand, this parameter heavily impacts the economic metrics of the project by adding losses in the annual energy production and revenue of the project due to additional electrical power losses and additional costs associated with longer cabling.
c) Distance to residential areas is another parameter to consider when choosing a site for solar power plant projects. The more remote the location, the greater will be the losses due to transmission. On the other hand, the proximity to residential areas and especially tall buildings can shade part of the wind farm and this can negatively affect the energy produced. Therefore, it is desirable to place them in an optimal place if possible.
d) Distance to roads is related to logistics. During the construction of an energy power plant, logistics could be a challenging task for the project investors. The site location is preferable to be easily accessible to transfer the needed construction elements, photovoltaic panels, and other needed power systems-related equipment. The site location needs to be also accessible during the operation phase so that the technical team is able to reach the site in the case of planned and unplanned technical problems. Besides this, the solar panels are to be located more than 100 m from roadways to minimize the amount of non-natural dust.
e) The service life of power plant components such as photovoltaic panels, structural elements, and electrical equipment affects the technical and economic viability of the investment. The industry standard for the life of most solar panels is 25–30 years. Most reputable manufacturers offer manufacturing warranties of 25 years or more. The breakeven point for energy savings from solar panels occurs 6–10 years after installation.
f) The strength of the existing grid is another parameter that has to be considered. Power grids are designed and operated to satisfy some set of technical requirements in terms of voltage stability and quality. The fluctuating behavior of renewable energy generation often poses additional challenges to the power grids in terms of satisfying the required quality metrics. It is therefore vital to perform detailed power systems integration analyses, especially if the project scale of the solar photovoltaic power plant is large.
g) The efficiency of standard solar photovoltaic panels depends on the quality of the material of the solar cells on them. The most commercial panels have efficiencies from 15% to 20%.

3.9.1.3 Economic Criteria

The *economic criteria* of solar photovoltaic power plant involve the following indicators such as initial investment cost, annual income, operation and maintenance cost, construction/infrastructure cost, land cost, levelized cost of energy evolution, economic performance indicators (NPV, IRR, RoI), local government subsidies, impact on regional development and local economies, impact on agriculture, and utility fee of electrical energy (electricity price for consumers) (Deveci et al., 2021).

a) Costs are related to the investments made during the initial stage of the project, such as the cost of solar panels, construction, and electrical and civil works related to the project. These are capital costs associated with initial investment costs.
b) The annual energy production and the annual income that could be generated from the solar photovoltaic power plant investment primarily depend on the solar radiation parameter.
c) O&M cost is related to the cost incurred during the operational life of the project, such as the cost of the services related to maintenance, overhead costs of the company that operates the project, insurance premiums, and similar recurring cost elements. Operational expenditures are related to operation and maintenance costs.
d) Depending on the type of land use, and land cover characteristics of the site location, the land where the solar photovoltaic power plant will be built can be bought or leased. If the land needs to be bought, then this type of land cost can be counted as capital costs; otherwise, it can be considered a recurring cost.
e) Levelized cost of energy (LCOE) is one of the most effective metrics in energy economics, which uses the per unit cost of the electrical energy harvested from the given solar photovoltaic power plant. It is the average cost of electricity per kWh generated by a system, considering all the expected costs, including construction, maintenance, and market price of electricity.
f) Local government subsidies or national governments may utilize some policy instruments to help promote the selected renewable energy resources with additional subsidies. Such support can be given directly during the investment phase or can be associated with the produced electricity, or can be applied as a tax credit depending on the country, state, and region.
g) Utility fees for electrical energy express the electricity price for consumers. This is an excellent indicator comparable to LCOE to show the competitiveness of project investment compared to the regular electricity prices from the utility. Ideally, the LCOE of the project will be lower than the utility fee if the project is built in a cost-effective manner.

3.9.1.4 Environmental Criteria

The next type of criteria is related with environmental indicators such as land use and land cover, soil structure and geology, impact on agriculture, urban planning, impact on ecological system and natural habitat, visual impact and light pollution, impact on wildlife, sand/dust risk, protected areas, and carbon emission savings.

a) The current land usage and future plans are among the main factors of solar PV energy project development. The investment decisions concerning land use are important, as they relate to the preservation of the ecological balance, and the availability of land for agricultural purposes. Land cover criterion is closely related to the potential for creating interrelationships between solar power plant development efforts and ecosystem preservation.

b) Soil structure and geological surveys are essential to investigate before the project construction phase. The outcomes of such surveys will help to make better decisions when identifying the best available type of construction where the solar power plant will be installed. Geological surveys are also used within the scope of other civil works, such as road construction to the project site.
c) Investments in solar power plants can potentially occupy huge amounts of land; on the other hand, land-related issues need to be investigated very carefully. Such an investigation should focus on the impact on local agricultural economies for the proposed project locations (Pascaris et al., 2021).
d) Urban planning activities for cities and counties are challenging undertakings since such analysis and surveys require well-trained interdisciplinary teams of engineers, urban planners, and environmental experts to analyze the contribution of solar energy to their urban plans (Akrofi & Okitasari, 2022).
e) Depending on the country where the solar power plant project is planned, the developers are to be informed about current regulations and problems that could impact the ecological system and natural habitat.
f) Visual impact of solar power plant dependent on the altitude difference between the solar panels and the observer. Light pollution from solar power plants is related to the polarization characteristics of the light. This could impose a negative impact on some insects and water birds whose habitat is situated close to the solar power plant (Fritz et al., 2020). This type of solar power plant impact should be taken into account during the planning phase.
g) Solar energy can have both positive and negative impacts on wildlife depending on various factors such as the location and design of solar energy facilities. The construction and operation of solar power plants can lead to the loss and fragmentation of wildlife habitats. This can lead to the displacement and disruption of wildlife populations, and in some cases can even lead to the extinction of certain species. There is a range of regulations in place to protect biodiversity; according to this, landowners and developers need to obtain permits or carry out assessments before carrying out any activities that may affect protected habitats or species.
h) Environmental conditions such as sand/dust accumulation, dust storms, hail, wind speed, and air pollution all influence the power output of PV modules and systems. In addition, in the design phase, it is recommended to buy PV modules that have successfully passed the certification tests for sand and dust (Shi et al., 2020). The same factors also reduce the degradation rates of the PV panels, which thus yields lower service lives.
i) In some regions, there could be protected areas, such as national parks, nature reserves, or wildlife sanctuaries, where development is restricted or prohibited. The project developers need to be aware of protected areas before they start their first project development activities, in order not to make faulty decisions. So,

they should obtain a permit from the relevant authorities to carry out activities within these areas including permission in various historical and environmentally protected areas (Lucchi et al., 2022).

j) Solar energy as a renewable source has significantly lower carbon emission levels than traditional energy from fossil. The majority of the energy for the entire life cycle of solar power plants is consumed during the extraction of raw materials, and the production and assembly of solar panels, which together represent 53% of the total consumption. Energy related to battery production is also an important contributor, around 28% of total consumption (Mehedi et al., 2022).

3.9.1.5 Social and Political Criteria

The last criteria type is related to social and political aspects of the investments such as public acceptance and support, national policy and local regulations, legal constraint, impact on the regional and local economies, and skilled manpower availability (Deveci et al., 2021).

a) Public acceptance and support of sustainable energy innovations and policies are of crucial importance in enabling the transition to decarbonized energy systems. Perhaps, this is the most critical component for the project realization. Distance to urban areas is a variable that is highly controlled by policy and public acceptance. The acceptance and support of the broader public are fundamental preconditions for further increasing the share of renewable energies. A lack of such support might for example lead to a slowing down of the planning and permitting processes for renewable energy installations, due to time-consuming participation processes and stakeholder consultations. Strong opposition from local citizens can risk the entire investment, and support from local citizens can help the project develop quickly. Therefore in the design and planning phase surveys, interviews, and meetings, to understand and measure public acceptance and support is a vital part of the project preparations. It is worth pointing out that public acceptance of renewable energies is fundamental to further increasing their share in the overall energy supply.
b) National policy and local regulatory frameworks should be examined to identify regulatory policy boundaries and associated legislative documents, for example, whether the state provides a subsidy to support investments in renewable sources, for what period, and under what conditions, to be reflected in the financial models of the project.
c) Legal constraints related to regulatory, legislative, and contractual issues with respect to land use, protection areas, land leases, engineering procurement construction contracts, and similar domains should be carefully handled by a professional internal or outsourced legal team before making any investment decision. The legal services should ideally continue until the end of the project's life cycle.

d) In practice, renewable energy investments usually have a positive impact on the regional development and local economies where they are built. This is due to the fact that they contribute to creating additional jobs during the phase of construction and operation of the project. It is also possible that some of the project equipment and material can be ordered from local producers or suppliers, to impact local economies positively.
e) The lack of sufficient manpower to develop and operate the hybrid wind-solar power plant may threaten the viability of the entire project. Therefore, it is vital to have sustainable HR planning activities. Training of the manpower on different levels is usually a national or regional task undertaken by universities, vocational schools, and companies that have internal learning-by-doing processes. Therefore, the local and national governments need to use the expansion of renewable energy to promote jobs, training, and reskilling opportunities and create a skilled workforce for the energy transition.

Finding optimal sites for the construction of solar farms is a complex task with many factors to be taken into account (environmental, social, legal and political, technical-economic, etc.), which classic site selection models do not address efficiently (Guaita-Pradas et al., 2019). It is verified that the determining factors for choosing the best locations are (1) solar irradiation, (2) substation distance, (3) slope, (4) distance of roads, (5) distance from urban areas, and (6) land use (Rediske et al., 2019). Solar photovoltaic site suitability studies considered solar irradiation amount as the highest important decision criterion followed by the proximity to power lines and land slope, whereas the protected lands and watercourses were considered the highest restriction factors (Al Garni & Awasthi, 2018). Therefore, solar energy can be a sustainable economic activity as long as a balance is struck between the benefits it brings to mitigate climate change and the damage it can cause to biodiversity and ecosystems.

Taking into account the division of the groups of criteria described above, the selection process can be divided into separate stages, each of which leads to the determination of a suitable and acceptable site from a climatic, technical, economic, environmental, and social-political point of view for the construction of a hybrid wind-solar power plant. In this regard, the task of choosing a site for the construction of a hybrid wind-solar power plant can be divided into several tasks according to the groups of criteria, as shown in Fig. 3.29.

The process starts with the determination of climatically acceptable sites, after which these sites are evaluated according to their specific technical parameters, followed by the relevant economic assessment; the next assessment is against the environmental criteria; and finally, the assessment is about the sociopolitical acceptance criteria.

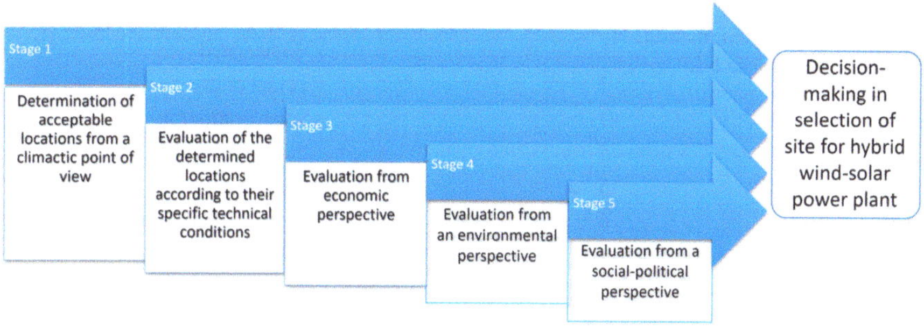

Fig. 3.29 Decision-making stages in selection of site for hybrid wind-solar power plant

3.9.2 Basic Criteria for Assessing the Potential of the Site for Construction of a Hybrid Wind-Solar Power Plant

It was found that several sets of criteria should be considered together to make the best decision for hybrid wind-solar site selection, as shown in Gao et al. (2022). In practice, to determine the best location for the construction of a hybrid wind-solar farm, the potential for renewable energy production is to be first determined. In this regard, a restricted number of indicators could be used to realize such decision problems related to the determination of sites with best suited potential. The reduced number of these indicators is shown in Table 3.14.

The climatology criteria involve sub-criteria such as wind speed, temperature, and solar irradiation. The annual average wind speed is a key criterion in measuring the economy and technology of wind farm installation sites. High wind speeds indicate plentiful wind resources, which is beneficial to increase power generation. Global horizontal irradiance, diffuse horizontal irradiation, and direct normal irradiance are the three components of solar radiation, each of which is a measure of the amount of available solar resources for photovoltaic plants. Their higher values influence the generation of more photovoltaic energy. Areas with lower average temperatures will improve the efficiency of renewable energy production from photovoltaics and wind generators. This is due to the fact that the photovoltaic panels are designed to work in the range of 15–35 °C; on the other hand, as the temperature decreases, the air becomes more humid and its density increases accordingly, which leads to a decrease in the wind energy produced. It should be pointed out that the amount of wind and photovoltaic power generation decreases as the annual mean precipitation increases, and therefore, the power generation is affected by annual mean precipitation. In this criteria group, the elevation and slope are considered the most important for hybrid wind-solar power plant (Noorollahi et al., 2022). As the altitude increases, the air becomes thinner, which is a prerequisite for receiving more solar energy and higher wind speed, which makes the terrain more suitable for the construction of photovoltaic and wind power plants. The slope reflects the smoothness of the terrain in the area, and the larger it is the higher

Table 3.14 Groups of criteria and sub-criteria for estimation of the potential for hybrid wind-solar power plant building

No	Sub-criteria (indicators)
1	Wind speed
2	Temperature
3	Solar irradiation
4	Elevation
5	Slope
6	Distance to power line
7	Distance to railways lines / road

construction costs, energy transfer, and equipment transportation. Large power plants work better in flat locations than in areas with steep slopes.

The *technical* criteria are reduced to distance to power line and distance to the road or railway. Distance to power lines is directly related to the transfer of energy to transmission lines and transformers, and the smaller this distance, the less energy will be transformed into losses. On the other hand, shorter distances also mean lower network connection costs. The authors show that the most significant theoretical technical wind energy potential loss results from missing infrastructure and soft consideration criteria (Jung & Schindler, 2021). During the project's construction, a lot of transportation work is required, and being close to the major road can save cost on transportation while also avoiding unnecessary environmental damage.

3.9.3 Combinatorial Optimization Model for Site Selection for Hybrid Wind-Solar Power Plant

The main criteria described in the previous section are used to formulate a combinatorial decision-making model for choosing a site for the building of a hybrid wind-solar power plant. The objective function of this model is to select such a site that simultaneously maximizes the performance of criteria for wind power plants and solar photovoltaic power plants as follows:

$$\max\left\{\sum_{i}^{N} x_i \sum_{j}^{M} w_{ij} + \sum_{i}^{N} x_i \sum_{q}^{Q} s_{iq}\right\} \tag{3.98}$$

subject to

$$\sum_{i}^{N} x_i = 1, or\ (N-1) \tag{3.99}$$

Table 3.15 Parameters of potential sites

No	Criteria	S-1	S-2	S-3	S-4	S-5
1	Wind speed, m/s	9.3	7.5	9.1	7.9	8.3
2	Temperature, °C	15.2	18.7	17.3	20.4	14.8
3	Solar irradiation, kWh/m2/year	1680	1570	1590	1600	1610
4	Elevation, m	977	525	582	612	492
5	Slope, %	3	0	1.7	1	0.65
6	Distance to power line, km	1.98	3.72	2.37	2.95	2.34
7	Distance to railways lines / road, km	4.8	3.77	1.7	1.6	2.8

$$w_{ij}^{\min} \leq w_{ij} \leq w_{ij}^{\max} \tag{3.100}$$

$$s_{iq}^{\min} \leq s_{iq} \leq s_{iq}^{\max} \tag{3.101}$$

where x_i are binary integer variables associated with the set of sites to choose from and w_{ij} expresses the performance of i-th alternative (site) toward j-th indicator for the wind power plant while s_{iq} expresses the performance of i-th alternative (site) toward q-th indicator for solar power plant. The relation (3.99) expresses the possibility of choosing only one alternative (site) or $(N - 1)$ where N is the number of sites to choose from. It could be seen from (3.98) that the number of indicators for the selection of wind and solar power plants is different. The inequalities (3.100) and (3.101) determine some lower and upper acceptable limits for site selection.

3.9.4 Numerical Application

To demonstrate the applicability of the proposed combinatorial decision-making model for hybrid wind-solar power plant site selection, several sites denoted as S-1, S-2, S-3, S-4, and S-5 are considered as shown in Table 3.15.

These input data from Table 3.15 are used to formulate the following optimization task based on the proposed model (3.98)–(3.101):

$$\max \left\{ \sum_{i=1}^{5} x_i (sp_i - t_i + e_i - s_i - dpl_i - drl_i) + \sum_{i=1}^{5} x_i (- t_i + si_i - e_i - s_i - dpl_i - drl_i) \right\} \tag{3.102}$$

subject to

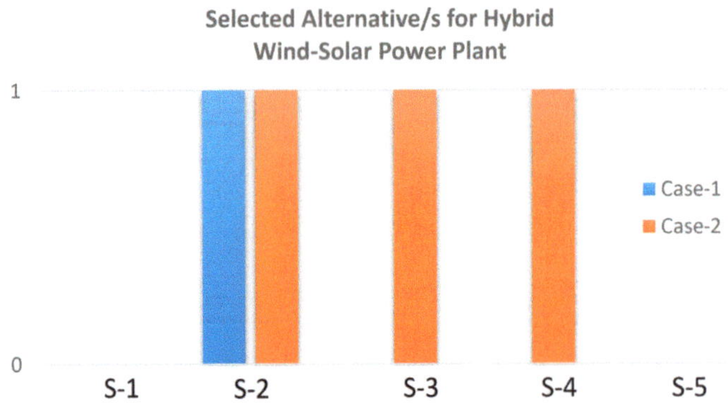

Fig. 3.30 Selected site/s for hybrid wind-solar power plant: Case-1—selection of one alternative (site); Case-2—simultaneous selection of 3 alternatives (sites)

$$\sum_{i=1}^{5} x_i = 1 \tag{3.103a}$$

$$\sum_{i=1}^{5} x_i = 3 \tag{3.103b}$$

All values about the sites' parameters are acceptable and additional restrictions are not required. Two cases of this task are investigated when it is needed to select only one site by using relation (3.103a) and for selection of 3 sites by (3.103b).

The tricky moment in solving the optimization problem (3.102)–(3.103) is the fact that it is impossible to sum different types of units of measurement. So before running it, some normalization should be done. This is necessary to make the ranges of parameters uniform. However, the objective function value will not have any physical dimension again. In this case, the objective function is used to direct the search to the best site according to its parameters, which will become clear from the determination of the values of the binary control variables.

The obtained results shown in Fig. 3.30 are based on the usage of the normalization scheme given by (1.15).

The obtained results prove that the proposed combinatorial model could be used to select the most appropriate site(s) for building hybrid wind-solar power plant. In addition, other parameters that can be expressed by values can be easily included in the objective function (3.98) to reflect in detail the specifics of such types of decision-making problems. When more of one selection is to be done, that means the selection of the site will be done later considering other selection criteria related to economics, environmental, or sociopolitical indicators.

Once the site is selected, the next stage (see Fig. 3.29) should perform the economic evaluation according to the described criteria. For example, the annual wind energy production can be estimated by the following relation:

$$\mathrm{AEP}^{\mathrm{wind}} = 8760.IC.CF \tag{3.104}$$

where *AEP* expresses the annual energy production in *GWh*, *IC* is the installed capacity, and *CF* is the capacity factor that depends on the technical performance of the wind turbine, speed distribution, etc., and varies in the range between 20 and 30% (McKenna et al., 2022).

The annual solar energy production can be estimated as follows (Rekik & El Alimi, 2023.):

$$\mathrm{AEP}^{\mathrm{solar}} = r.A.AF.\eta \tag{3.105}$$

where AEP expresses the annual generated power in *TWh/year*, *r* is the annual solar irradiation in *kWh/m2/year*, *A* is the total area of suitable location in km^2, *AF* is the area factor of total *A* that could be converted by solar panels in %, and η is the photovoltaic system effectiveness in %.

Since in numerical testing, the area of potential sites is not taken into account, the related costs for CAPEX and OPEX are not the focus of this decision problem. These types of costs are described in Chap. 6.

It should be noted that it is preferable to choose a site with a larger area, as more wind turbines and solar panels could be installed. This, in turn, is related to the production of more energy and wind-solar plants would be more profitable. For example, the produced solar energy can be calculated as a function of solar panel number (Ang et al., 2022):

$$\mathrm{Energy}^{\mathrm{solar}}(kWh) = N_s.r.\mu.p \tag{3.106}$$

where N_s is number of solar panels, *r* is solar irradiance (kWh/m^2), μ is solar panel yield (or efficiency), and *p* is performance ratio of the solar panel, including consideration for losses.

The produced wind energy can be calculated too as a function of wind turbines number (Ang et al., 2022):

$$\mathrm{Energy}^{\mathrm{wind}}(kWh) = 0.5\frac{N_w.C_p.\rho.A.v_w^3}{1000} \tag{3.107}$$

where N_w is number of wind turbines, C_p is power coefficient, ρ is air density, *A* is rotor swept area (m^2), and v_w is wind speed (m/s).

Using these practical calculations (3.106) and (3.107), it is possible to estimate the produced energy from the hybrid wind-solar power plant taking into account the number of installed solar panels and wind turbines that depend on the site area.

Unlike other approaches that rely on various MCDM methods with combinations of GIS (Rekik & El Alimi, 2023; Saraswat et al., 2021; Diaz-Cuevas et al., 2019), the proposed modeling approach unambiguously and categorically determines optimal site selection as a result of solving a single-criterion optimization problem. An advantage of the proposed combinatorial optimization model (3.98)–(3.101) is the possibility of easily including or excluding indicators based on which the site selection will be realized. A drawback of the proposed combinatorial optimization model (3.98)–(3.101) is the procedure for the normalization of indicators to obtain dimensionless units within the same range.

References

Abd Ali, L. M., Al-Rufaee, F. M., Kuvshinov, V. V., Krit, B. L., Al-Antaki, A. M., & Morozova, N. V. (2020). Study of hybrid wind-solar systems for the Iraq energy complex. *Applied Solar Energy, 56*, 284–290. https://doi.org/10.3103/S0003701X20040027

Akrofi, M. M., & Okitasari, M. (2022). Integrating solar energy considerations into urban planning for low carbon cities: A systematic review of the state-of-the-art. *Urban Governance, 2*(1), 157–172. https://doi.org/10.1016/j.ugj.2022.04.002

Al Garni, H. Z., & Awasthi, A. (2018). Chapter 2–Solar PV power plants site selection: A review. In I. Yahyaoui (Ed.), *Advances in renewable energies and power technologies* (pp. 57–75). https://doi.org/10.1016/B978-0-12-812959-3.00002-2

Ang, Y. Q., Polly, A., Kulkarni, A., Chambi, G. B., Hernandez, M., & Haji, M. N. (2022). Multi-objective optimization of hybrid renewable energy systems with urban building energy modeling for a prototypical coastal community. *Renewable Energy, 201*(Part 1), 72–84. https://doi.org/10.1016/j.renene.2022.09.126

Asadi, M., Pourhossein, K., Noorollahi, Y., Marzband, M., & Iglesias, G. (2023). A new decision framework for hybrid solar and wind power plant site selection using linear regression modeling based on GIS-AHP. *Sustainability, 15*(10), 8359. https://doi.org/10.3390/su15108359

Borissova, D., Dimitrova, Z., & Dimitrov, V. (2021). Intelligent system to support decision making using optimization business models for wind farm design. In D. Simian & L. F. Stoica (Eds.), *Modelling and development of intelligent systems. MDIS 2020* (Vol. 1341, pp. 287–301). Communications in Computer and Information Science. https://doi.org/10.1007/978-3-030-68527-0_18

Borissova, D., Mustakerov, I., & Korsemov, D. (2016). Business intelligence system via group decision making. *Cybernetics and Information Technologies, 16*(3), 219–229.

Borissova, D., & Mustakerov, I. (2017a). A two-stage placement algorithm with multi-objective optimization and group decision making. *Cybernetics and Information Technologies, 17*(1), 87–103. https://doi.org/10.1515/cait-2017-0007

Borissova, D., & Mustakerov, I. (2017b). Designing of wind farm layout by using of multi-objective optimization. *International Journal of Mathematical Models and Methods in Applied Sciences, 11*, 290–295.

Borissova, D., & Mustakerov, I. (2017c). Mixed-integer model for placement of objects avoiding forbidden zones. *Comptes rendus de l'Académie bulgare des Sciences., 70*(9), 1297–1304.

Borissova, D., & Mustakerov, I. (2017d). Optimal planning of wind farm layout and integration to electric grid infrastructure. *MAJLESI Journal of Electrical Engineering, 11*(3), 1–5.

Borissova, D., & Mustakerov, I. (2017e). Wind power plant layout design and assessment considering forbidden zones for location of turbines. *Advanced Modeling and Optimization, 19*(1), 29–38.

Charhouni, N., Sallaou, M., & Mansouri, K. (2019). Realistic wind farm design layout optimization with different wind turbines types. *International Journal of Energy and Environmental Engineering, 10*, 307–318. https://doi.org/10.1007/s40095-019-0303-2

Deveci, M., Cali, U., & Pamucar, D. (2021). Evaluation of criteria for site selection of solar photovoltaic (PV) projects using fuzzy logarithmic additive estimation of weight coefficients. *Energy Reports, 7*, 8805–8824. https://doi.org/10.1016/j.egyr.2021.10.104

Diaz-Cuevas, P., Dominguez-Bravo, J., & Prieto-Campos, A. (2019). Integrating MCDM and GIS for renewable energy spatial models: Assessing the individual and combined potential for wind, solar and biomass energy in southern Spain. *Clean Technologies and Environmental Policy, 21*, 1855–1869. https://doi.org/10.1007/s10098-019-01754-5

Donovan, S. (2005). *Wind farm optimization*. https://citeseerx.ist.psu.edu/document?repid=rep1&type=pdf&doi=389ffabaf08bb1c214ebd683e2f55038558ebd0b

Ebhota, W. S., & Tabakov, P. Y. (2023). Influence of photovoltaic cell technologies and elevated temperature on photovoltaic system performance. *Ain Shams Engineering Journal, 14*(7), 101984. https://doi.org/10.1016/j.asej.2022.101984

Ehrgott, M. (2006). A discussion of scalarization techniques for multiple objective integer programming. *Annals of Operations Research, 147*, 343–360. https://doi.org/10.1007/s10479-006-0074-z

Fard, M. B., Moradian, P., Emarati, M., Ebadi, M., Chofreh, A. G., & Klemes, J. J. (2022). Ground-mounted photovoltaic power station site selection and economic analysis based on a hybrid fuzzy best-worst method and geographic information system: A case study Guilan province. *Renewable and Sustainable Energy Reviews, 169*, 112923. https://doi.org/10.1016/j.rser.2022.112923

Fritz, B., Horvath, G., Hunig, R., Pereszlenyi, A., Egri, A., Guttmann, M., Schneider, M., Lemmer, U., Kriska, G., & Gomard, G. (2020). Bioreplicated coatings for photovoltaic solar panels nearly eliminate light pollution that harms polarotactic insects. *PLoS One, 15*(12), e0243296. https://doi.org/10.1371/journal.pone.0243296

Fulop, J. (2005). *Introduction to decision making methods*. pp. 1–15. http://citeseerx.ist.psu.edu/viewdoc/download?doi=10.1.1.86.6292&rep=rep1&type=pdf

Gao, J., Wang, Y., Huang, N., Wei, L., & Zhang, Z. (2022). Optimal site selection study of wind-photovoltaic-shared energy storage power stations based on GIS and multi-criteria decision making: A two-stage framework. *Renewable Energy, 201*, 1139–1162. https://doi.org/10.1016/j.renene.2022.11.012

Genc, M. S., & Karipoglu, F. (2021). Wind-solar site selection using a GIS-MCDM-based approach with an application in Kayseri Province/Turkey. In *7th Iran Wind Ener-gy Conference (IWEC2021)* (pp. 1–4). Shahrood. https://doi.org/10.1109/IWEC52400.2021.9467003

Genova, K., Kirilov, L., & Guljashki, V. (2013). New reference-neighbourhood scalarization problem for multiobjective integer programming. *Cybernetics and Information Technologies, 13*, 104–114.

Grady, S. A., Hussainia, M. Y., & Abdullah, M. M. (2005). Placement of wind turbines using genetic algorithms. *Renewable Energy, 30*, 259–270. https://doi.org/10.1016/j.renene.2004.05.007

Guaita-Pradas, I., Marques-Perez, I., Gallego, A., et al. (2019). Analyzing territory for the sustainable development of solar photovoltaic power using GIS databases. *Environmental Monitoring and Assessment, 191*, 764. https://doi.org/10.1007/s10661-019-7871-8

Hammad, A. W. A., Akbarnezhad, A., & Rey, D. (2015). Comparison of mixed integer programming models for the construction site layout problem. In *Proc. ISARC* (pp. 1–9). https://doi.org/10.22260/ISARC2015/0109

Hansen, A. D., Iov, F., Blaabjerg, F., & Hansen, L. H. (2004). Review of contemporary wind turbine concepts and their market penetration. *Wind Engineering, 28*(3), 247–263. https://doi.org/10.1260/0309524041590099

Hasti, F., Mamkhezri, J., McFerrin, R., & Pezhooli, N. (2023). Optimal solar photovoltaic site selection using geographic information system-based modeling techniques and assessing environmental and economic impacts: The case of Kurdistan. *Solar Energy, 262*, 111807. https://doi.org/10.1016/j.solener.2023.111807

Hou, P., Hu, W., Chen, C., Soltani, M., & Chen, Z. (2016). Optimization of offshore wind farm layout in restricted zones. *Energy, 113*, 487–496. https://doi.org/10.1016/j.energy.2016.07.062

Hou, P., Zhu, J., Ma, K., Yang, G., Hu, W., & Chen, Z. (2019). A review of offshore wind farm layout optimization and electrical system design methods. *Journal of Modern Power Systems and Clean Energy, 7*, 975–986. https://doi.org/10.1007/s40565-019-0550-5

Johnson, G. L. (2006, October 10). *Wind energy systems*. Electronic Edition, Manhattan, KS. https://www.e-booksdirectory.com/details.php?ebook=3536

Jung, C., & Schindler, D. (2021). Distance to power grids and consideration criteria reduce global wind energy potential the most. *Journal of Cleaner Production, 317*, 128472. https://doi.org/10.1016/j.jclepro.2021.128472

Koc, A., Turk, S., & Şahin, G. (2019). Multi-criteria of wind-solar site selection problem using a GIS-AHP-based approach with an application in Igdir Province/Turkey. *Environmental Science and Pollution Research, 26*, 32298–32310. https://doi.org/10.1007/s11356-019-06260-1

Kongnam, C., Nuchprayoon, S., Premrudeepreechacharn, S., & Uatrongjit, S. (2009). Decision analysis on generation capacity of a wind park. *Renewable and Sustainable Energy Reviews, 13*(8), 2126–2133. https://doi.org/10.1016/j.rser.2009.01.023

Larson, D., & Chang, V. (2016). A review and future direction of agile, business intelligence, analytics and data science. *International Journal Information Managementm, 36*, 700–710. https://doi.org/10.1016/j.ijinfomgt.2016.04.013

Lepa, J., Kokin, E., Annuk, A., Poder, V., & Jurjenson, K. (2008). Wind power stations performance analysis and power output prognosis. *Engineering for Rural Development, 92-96.*

Lucchi, E., Dall'Orto, I., Peluchetti, A., Toledo, L., Pelle, M., Lopez, C. P., & Guazzi, G. (2022). Photovoltaic technologies in historic buildings and protected ar-eas: Comprehensive legislative framework in Italy and Switzerland. *Energy Policy, 161*, 112772. https://doi.org/10.1016/j.enpol.2021.112772

Marler, R. T., & Arora, J. S. (2005). Function-transformation methods for multi-objective optimization. *Engineering Optimization, 37*(6), 551–570. https://doi.org/10.1080/03052150500114289

Marler, R. T., & Arora, J. S. (2004). Survey of multi-objective optimization methods for engineering. *Structural and Multidisciplinary Optimization, 26*, 369–395. https://doi.org/10.1007/s00158-003-0368-6

Marmidis, G., Lazarou, S., & Pyrgioti, E. (2008). Optimal placement of wind turbines in a wind park using Monte Carlo simulation. *Renewable Energy, 33*, 1455–1460. https://doi.org/10.1016/j.renene.2007.09.004

McKenna, R., Pfenninger, S., Heinrichs, H., Schmidt, J., Staffell, I., Bauer, C., Gruber, K., Hahmann, A. N., Jansen, M., Klingler, M., Landwehr, N., Guo Larsén, X., Lilliestam, J., Pickering, B., Robinius, M., Tröndle, T., Turkovska, O., Wehrle, S., Weinand, J. M., & Wohland, J. (2022). High-resolution large-scale onshore wind energy assessments: A review of potential definitions, methodologies and future research needs. *Renewable Energy, 182*, 659–684. https://doi.org/10.1016/j.renene.2021.10.027

Mehedi, T. H., Gemechu, E., & Kumar, A. (2022). Life cycle greenhouse gas emissions and energy footprints of utility-scale solar energy systems. *Applied Energy, 314*, 118918. https://doi.org/10.1016/j.apenergy.2022.118918

Mosetti, G., Poloni, C., & Diviacco, B. (1994). Optimization of wind turbine positioning in large wind farms by means of a genetic algorithm. *Journal of Wind Engineering and Industrial Aerodynamics, 51*(1), 105–116. https://doi.org/10.1016/0167-6105(94)90080-9

Mustakerov, I., & Borissova, D. (2011). Wind park layout design using combinatorial optimization. *Wind Turbines, 403–424*. https://doi.org/10.5772/15029

Mustakerov, I., & Borissova, D. (2010). Wind turbines type and number choice using combinatorial optimization. *Renewable Energy, 35*(9), 1887–1894. https://doi.org/10.1016/j.renene.2009.12.012

Noorollahi, Y., Senani, A. G., Fadaei, A., Simaee, M., & Moltames, R. (2022). A framework for GIS-based site selection and technical potential evaluation of PV solar farm using fuzzy-Boolean logic and AHP multi-criteria decision-making approach. *Renewable Energy, 186*, 89–104. https://doi.org/10.1016/j.renene.2021.12.124

Pascaris, A. S., Schelly, C., Burnham, L., & Pearce, J. M. (2021). Integrating solar energy with agriculture: Industry perspectives on the market, community, and socio-political dimensions of agrivoltaics. *Energy Research & Social Science, 75*, 102023. https://doi.org/10.1016/j.erss.2021.102023

Patel, M. R. (1999). *Wind and power solar systems*. CRC Press.

Peneva, V., & Popchev, I. (2010). Fuzzy multi-criteria decision making algorithms. *Comptes Rendus de L'Academie Bulgare des Sciences, 63*(7), 979–992.

Pérez-Aracil, J., Casillas-Pérez, D., Jiménez-Fernández, S., Prieto-Godino, L., & Salcedo-Sanz, S. (2022). A versatile multi-method ensemble for wind farm layout optimization. *Journal of Wind Engineering and Industrial Aerodynamics, 225*, 104991. https://doi.org/10.1016/j.jweia.2022.104991

Rediske, G., Mairesse Siluk, J. C., Gava Gastaldo, N., Donaduzzi Rigo, P., & Brum Rosa, C. (2019). Determinant factors in site selection for photovoltaic projects: A systematic review. *International Journal of Energy Research, 43*(5), 1689–1701. https://doi.org/10.1002/er.4321

Rekik, S., & El Alimi, S. (2023). Optimal wind-solar site selection using a GIS-AHP based approach: A case of Tunisia. *Energy Conversion and Management: X, 18*, 100355. https://doi.org/10.1016/j.ecmx.2023.100355

Sadovskaia, K., Bogdanov, D., Honkapuro, S., & Breyer, C. (2019). Power transmission and distribution losses–A model based on available empirical data and future trends for all countries globally. *International Journal of Electrical Power & Energy Systems, 107*, 98–109. https://doi.org/10.1016/j.ijepes.2018.11.012

Sagbansua, L., & Balo, F. (2017). Decision making model development in increasing wind farm energy efficiency. *Renewable Energy, 109*, 354–362.

Saraswat, S. K., Digalwar, A. K., Yadav, S. S., & Kumar, G. (2021). MCDM and GIS based modelling technique for assessment of solar and wind farm locations in India. *Renewable Energy, 169*, 865–884. https://doi.org/10.1016/j.renene.2021.01.056

Serrano-Gonzalez, J., Gonzalez-Rodriguez, A. G., Castro-Mora, J., Riquelme-Santos, J., & Burgos-Payan, M. (2011). Overall design optimization of wind farms. *Renewable Energy, 36*, 1973–1982. https://doi.org/10.1016/j.renene.2010.10.034

Shakoor, R., Hassan, M. Y., Raheem, A., & Wu, Y. K. (2016). Wake effect modeling: A review of wind farm layout optimization using Jensen's model. *Renewable and Sustainable Energy Reviews, 58*, 1048–1059.

Shi, C., Yu, B., Liu, D., Wu, Y., Li, P., Chen, G., & Wang, G. (2020). Effect of high-velocity sand and dust on the performance of crystalline silicon photovoltaic modules. *Solar Energy, 206*, 390–395. https://doi.org/10.1016/j.solener.2020.06.018

Wu, Y., Chu, H., Dong, H., Xu, C., Chen, W., & Wu, G. (2022). Optimal site selection of rural wind–photovoltaic-storage station from a sustainable development perspective. *Energy Reports, 8*, 8381–8395. https://doi.org/10.1016/j.egyr.2022.06.050

Yang, Q., Li, H., Li, T., & Zhou, X. (2021). Wind farm layout optimization for levelized cost of energy minimization with combined analytical wake model and hybrid optimization strategy. *Energy Conversion and Management, 248*, 114778. https://doi.org/10.1016/j.enconman.2021.114778

Zadeh, L. A. (1975). Calculus of fuzzy restrictions. In L. A. Zadeh, K.-S. Fu, K. Tanaka, & M. Shimura (Eds.), *Fuzzy sets and their applications to cognitive and decision processes* (pp. 1–39). https://doi.org/10.1016/B978-0-12-775260-0.50006-2

Zadeh, L. A. (1999). Fuzzy sets as a basis for a theory of possibility. *Fuzzy Sets and Systems, 100*(Supplement 1), 9–34. https://doi.org/10.1016/S0165-0114(99)80004-9

Chapter 4
Decision-Making in 1D Cutting of Blanks for Wind Turbine Manufacturing and Processing Planning

Abstract The third stage of the wind farm life cycle refers to the actual construction and commissioning. Activities related to this stage refer to logistics, civil engineering, foundations, turbine installation, grid connection, etc. Generally, wind turbines consist of three main components—blades, tower, and nacelle—each of which must be designed and manufactured separately. However, the heart of the wind turbine remains the nacelle, containing the gears, generator, and other mechanical components of the turbine. In addition to these components, there are also many high-tech electronic components that allow the turbine to monitor changes in wind speed and direction, turn on and off, or automatically change direction.

Wind turbines, as complex machines, are designed and manufactured by companies known as original equipment manufacturers. Such manufacturers are in most cases multinational corporations that rely on many smaller businesses to construct the individual components and systems that make up a wind turbine. Given the growing wind energy market, many original equipment manufacturers are looking to localize production to reduce transportation costs and potentially reduce the logistical challenges associated with transporting large turbines and components. That means the wind industry supports many small companies that manufacture specialized parts, including gears.

One major problem at the stage of construction and commissioning is the problem of stock-cutting. This problem arises when the available material has to be cut to fulfill certain goals such as cutting patterns with minimal material waste and cost-efficient production and higher customer satisfaction. Another important problem at this stage is personnel scheduling.

Therefore, the models described in this chapter are designed to optimize the processes for the required details processing and determine the optimal number of personnel for wind turbine installation.

Keywords 1D cutting · Job shop scheduling · Product configuration design · Staff scheduling

D. Borissova, *Decision-Making in Design, Maintenance, Planning, and Investment of Wind Energy*, International Series in Operations Research & Management Science 355, https://doi.org/10.1007/978-3-031-52219-2_4

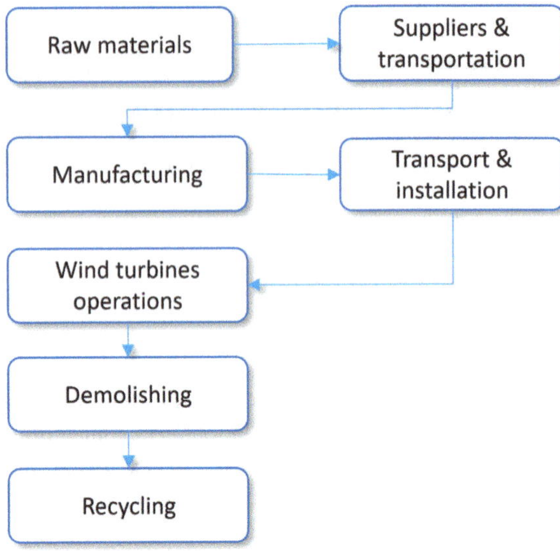

Fig. 4.1 Life cycle of wind turbines

4.1 Wind Turbines' Life Cycle

A comprehensive life cycle assessment for large onshore wind turbines including all phases of the turbine's life cycle separately (materials acquisition, manufacturing, transportation, installation, operation and maintenance, and end of life) and multiple impact categories (environmental, human health, resource consumption) is presented in (Alsaleh & Sattler, 2019). It should be pointed out that Vestas[1] works with life cycle assessments to develop increasingly energy-efficient products while mitigating the environmental impacts throughout the turbine's lifetime. The studies assess a wind turbine's entire bill-of-materials accounting for the approximately 25,000 parts that make up a wind turbine. In a life cycle assessment, a complete wind power plant is assessed up to the point of the electricity grid, including the wind turbine itself, foundation, site cabling, and the transformer station. This process of the wind turbines' life cycle is illustrated in Fig. 4.1.

There are several clear stages in the life cycle of wind turbines, following a step-by-step gradation, namely:

1. Supply of raw material
2. Supplier selection and transportation
3. Production of blanks, elements, and assembly (manufacturing)
4. Wind turbine installation and operation
5. Demolishing
6. Recycling

[1] https://www.vestas.com/en/sustainability/environment/lifecycle-assessments

For the production of wind turbines, it is needed to provide the supply of raw materials and the processing of the necessary details with subsequent assembly. When choosing suppliers for raw materials, the models described in Chap. 2 can be used.

Taking into account the need to minimize waste leading not only to the optimal utilization of resources but also to mitigate the environmental impacts, the purpose of this chapter is to show how combinatorial optimization models can help to make optimal decisions for one-dimensional cutting stock of blanks, their processing, and determination of the optimal product combination for the design of product family based on modules.

4.2 1D Cutting-Stock Model for Minimizing Waste

In general, cutting-stock problems consist in cutting large pieces (blanks), available in stock, into a set of smaller pieces (elements) according to the given requirements while optimizing a certain objective function. These problems are relevant in the production planning of many industries such as metallurgy, plastics, paper, glass, furniture, textile, joinery manufacturing, even in the elements of wind turbines, etc. In the last four decades cutting-stock problems have been studied by an increasing number of researchers (Mobasher & Ekici, 2013; Dikili et al., 2007; Cui & Lu, 2009; Cherri et al., 2009; Dikili et al., 2008). The interest in these problems is provoked by the many practical applications and the challenge they provide to researchers. At first glance, they are simple to formulate, but at the same time, they are computationally difficult to solve. The cutting and packing problems (Mukhacheva & Mukhacheva, 2006) belong to the class of NP-hard problems. The solution to these problems extensively relies on mathematical programming and combinatorial methods. There exist many real-life problems that are computationally hard and can be formalized only as NP-hard problems. On the other hand, the continuous growth of the prices of the materials and of the energy requires minimization of the production expenses for every element.

The classical 1D cutting-stock problem addresses the problem of cutting-stock materials of length in order to satisfy the demand of smaller pieces while minimizing the overall trim loss. Kantorovich first formulates 1D cutting-stock problem (Kantorovich, 1960; Kantorovich & Zalgaller, 1971), and Gilmore and Gomory (Gilmore & Gomory, 1961; Gilmore & Gomory, 1963) propose the first solution methodology for the cutting-stock problems.

Table 4.1 Elements length and demand

Element	Length, mm	Demand
1	650	4
2	730	8
3	1400	4
4	1700	4
5	2000	4
6	2100	4
7	2200	8

4.2.1 Combinatorial Optimization Model for 1D Cutting-Stock Problem

The blanks usually are supplied from the factory with a fixed length and are used to cut out different elements. These elements differ in size and number that are specific to each particular project. The goal is to determine the optimal length of blanks (which are usually ordered with equal size in large quantities) in order to satisfy the demand for all elements. It is also necessary to find the optimal cutting patterns for each blank to minimize waste. In Korsemov et al. (2013) in-depth research of joinery modules used in a wide range of buildings is performed and a certain number of unified modules are identified. To solve such a particular task, the following 1D cutting-stock problem with several modules is considered:

- Module 1 with dimensions 2200 mm × 730 mm
- Module 2 with dimensions 2000 mm × 650 mm
- Module 3 with dimensions 1400 mm × 1400 mm
- Module 4 with dimensions 1700 mm × 2100 mm

The investigated cutting-stock problem can be narrowed down to definition of optimal length of blanks and optimal cutting patterns for modules. The problem can be described as follows: a factory has to fulfill order of blanks with certain length needed to assemble a given number of modules, consisting of elements with known length and number. For the sake of simplicity of the presentation, only casement elements for the modules in the example above are summarized as a manufacturing order shown in Table 4.1.

In practice, all profiles come with a fixed length usually of 6 *m*. However, this is not a mandatory requirement, and it is possible to order blanks with different lengths. There are no obstacles to ordering the manufacturing company to produce a number of blanks with different lengths than the standard 6 meters—for example, any length between 5 *m* and 7 *m*. When the optimal length of blanks is determined, the next step is to define the optimal cutting patterns of elements for each blank.

The problem of optimal manufacturing can be investigated as 1D cutting-stock problem by means of proper mathematical modeling. The described one-dimensional cutting-stock problem is formalized via combinatorial optimization model. In contrast to other similar models, it allows determining optimal length of blanks and optimal cutting patterns minimizing the trim loss, according to given demands of joinery elements. This type of functionality of the model requires introducing inequalities for each of blanks. That means there is a necessity of knowing in advance the number N of the blanks. Number N can be calculated as overall demand of joinery elements divided by the length L of the blanks. On the other hand, the length L of the blanks is to be determined after solution of the optimization task. This "recursive" type of problem can be overcome taking into account that length L will have some value close to the standard length of 6 m. Having this in mind, the number of blanks N can be calculated as overall demand of joinery elements divided by the length of 6 m, rounded to integer value. Then this value of N can be used to formulate the optimization task as (Mustakerov & Borissova, 2014):

$$min\left\{\sum_{i}^{N} L - L_i\right\} \tag{4.1}$$

subject to

$$\forall i : L_i = \sum_{j}^{J} x_{ij} l_j \tag{4.2}$$

$$\forall i : L_i \leq L \tag{4.3}$$

$$\forall j : \sum_{i}^{N} x_{ij} = k_{ij} \tag{4.4}$$

$$(6 - \Delta_{min}) \leq L \leq (6 + \Delta_{min}) \tag{4.5}$$

$$\forall j : x_{ij} = \begin{cases} \text{binary integer } 0 \text{ } or \text{ } 1, \text{if } N \leq k_{ij} \\ \text{integer, otherwise} \end{cases} \tag{4.6}$$

where N is the number of blanks, L is length of blanks, L_i is the utilized length of each blank, l_j is length of joinery elements, x_{ij} are decision variables assigned to each element for particular blank, and k_{ij} represents the demand of each element.

The objective function (4.1) minimizes the sum of trim loss for each blank. The optimal cutting pattern for each of the blanks is defined by decision variables x_{ij} in (4.2). Depending on the given particular joinery project, the decision variables (4.6) could be binary integer variables or integer variables. For example, if the number of the blanks is less than the maximum demand of some element, then the decision variables x_{ij} are to be considered as integers. This statement allows the model to allocate more than 1 element within cutting pattern in the blank to satisfy the elements demand. This elements demand is satisfied by (4.4). Deviation from the standard length of 6 m is represented by $\Delta_{\min}$ and $\Delta_{\max}$ both approximately in the range of 1 m.

4.2.2 Numerical Results

The need for elements for the described problem, which have to be ordered according to Table 4.1, is illustrated in Fig. 4.2.

Using the input data from Table 4.1, the following steps are performed (Mustakerov & Borissova, 2014):

- Determination of total length of all elements considering their demand $L_{sum} = 54840\ mm$;
- Determination number of blanks N as rounded to integer result of the total elements length 54840 mm divided by 6000 mm as 54840/6000 = 9.14 = > $N = 9$ and setting of deviations $\Delta_{min} = \Delta_{\max} = 1000\ mm$.
- Formulation of optimization task:

$$\min\left\{(L\text{–}L_1) + (L\text{–}L_2) + (L\text{–}L_3) + (L\text{–}L_4) + (L\text{–}L_5) + (L\text{–}L_6) + (L\text{–}L_7) + (L\text{–}L_8) + (L\text{–}L_9)\right\} \tag{4.7}$$

subject to

$$x_{11}l_1 + x_{12}l_2 + x_{13}l_3 + x_{14}l_4 + x_{15}l_5 + x_{16}l_6 + x_{17}l_7 = L_1 \tag{4.8a}$$

$$x_{21}l_1 + x_{22}l_2 + x_{23}l_3 + x_{24}l_4 + x_{25}l_5 + x_{26}l_6 + x_{27}l_7 = L_2 \tag{4.8b}$$

$$x_{31}l_1 + x_{32}l_2 + x_{33}l_3 + x_{34}l_4 + x_{35}l_5 + x_{36}l_6 + x_{37}l_7 = L_3 \tag{4.8c}$$

$$x_{41}l_1 + x_{42}l_2 + x_{43}l_3 + x_{44}l_4 + x_{45}l_5 + x_{46}l_6 + x_{47}l_7 = L_4 \tag{4.8d}$$

$$x_{51}l_1 + x_{52}l_2 + x_{53}l_3 + x_{54}l_4 + x_{55}l_5 + x_{56}l_6 + x_{57}l_7 = L_5 \tag{4.8e}$$

$$x_{61}l_1 + x_{62}l_2 + x_{63}l_3 + x_{64}l_4 + x_{65}l_5 + x_{66}l_6 + x_{67}l_7 = L_6 \tag{4.8f}$$

$$x_{71}l_1 + x_{72}l_2 + x_{73}l_3 + x_{74}l_4 + x_{75}l_5 + x_{76}l_6 + x_{77}l_7 = L_7 \tag{4.8g}$$

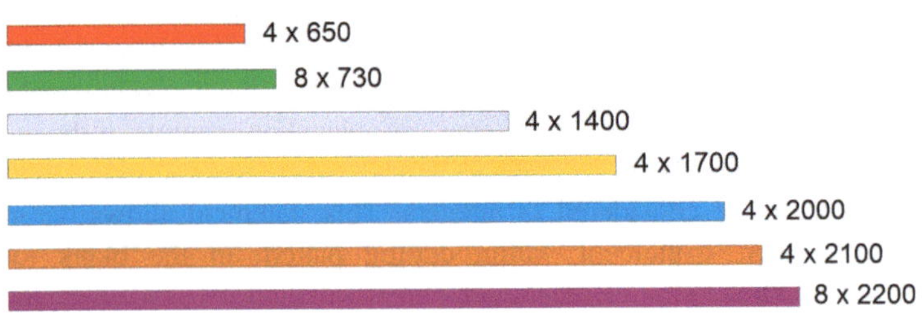

Fig. 4.2 Required type elements

$$x_{81}l_1 + x_{82}l_2 + x_{83}l_3 + x_{84}l_4 + x_{85}l_5 + x_{86}l_6 + x_{87}l_7 = L_8 \quad (4.8h)$$

$$x_{91}l_1 + x_{92}l_2 + x_{93}l_3 + x_{94}l_4 + x_{95}l_5 + x_{96}l_6 + x_{97}l_7 = L_9 \quad (4.8i)$$

$$L_1 \leq L \quad (4.9a)$$

$$L_2 \leq L \quad (4.9b)$$

$$L_3 \leq L \quad (4.9c)$$

$$L_4 \leq L \quad (4.9d)$$

$$L_5 \leq L \quad (4.9e)$$

$$L_6 \leq L \quad (4.9f)$$

$$L_7 \leq L \quad (4.9g)$$

$$L_8 \leq L \quad (4.9h)$$

$$L_9 \leq L \quad (4.9i)$$

$$x_{11} + x_{21} + x_{31} + x_{41} + x_{51} + x_{61} + x_{71} + x_{81} + x_{91} = 4 \quad (4.10a)$$

$$x_{12} + x_{22} + x_{32} + x_{42} + x_{52} + x_{62} + x_{72} + x_{82} + x_{92} = 8 \quad (4.10b)$$

$$x_{13} + x_{23} + x_{33} + x_{43} + x_{53} + x_{63} + x_{73} + x_{83} + x_{93} = 4 \quad (4.10c)$$

$$x_{14} + x_{24} + x_{34} + x_{44} + x_{54} + x_{64} + x_{74} + x_{84} + x_{94} = 4 \quad (4.10d)$$

$$x_{15} + x_{25} + x_{35} + x_{45} + x_{55} + x_{65} + x_{75} + x_{85} + x_{95} = 4 \quad (4.10e)$$

$$x_{16} + x_{26} + x_{36} + x_{46} + x_{56} + x_{66} + x_{76} + x_{86} + x_{96} = 4 \quad (4.10f)$$

$$x_{17} + x_{27} + x_{37} + x_{47} + x_{57} + x_{67} + x_{77} + x_{87} + x_{97} = 8 \quad (4.10g)$$

$$5 \leq L \leq 7 \quad (4.11)$$

$$x_{ij} - \text{binary integer } 0 \text{ } or \text{ } 1 \quad (4.12)$$

The relations (4.8a)–(4.8i) in combination with inequalities (4.9a)-(4.9i) define optimal cutting patterns for each particular blank. The optimal cutting patterns are defined not to exceed the length of the blanks and to satisfy the requested demand of elements expressed by (4.10a)–(4.10g). The objective function (4.7) seeks a solution that minimizes the waste of all blanks. The optimal length of blanks is to be defined within interval of 5–7 m (4.11). In this example, the decision variables for optimal cutting patterns are binary integer variables (4.12).

The solution of the formulated mixed-integer optimization task (4.7)–(4.12) determines the optimal length of blanks, total waste, waste for each blank, and used length of each blank, as shown in Table 4.2.

The optimal cutting patterns defined by the values of the binary integer variables for each blank are shown in Table 4.3.

Table 4.2 Optimal solution results

Optimal length of blanks L, mm	Used length of each blank, mm	Waste for each blank, mm	Total waste for order, mm
6550	L1 = 6330	220	4110
	L2 = 6330	220	
	L3 = 6030	520	
	L4 = 6030	520	
	L5 = 5680	870	
	L6 = 5680	870	
	L7 = 5680	870	
	L8 = 6530	20	
	L9 = 6550	0	

Table 4.3 Optimal cutting patterns for each blank

Blank #	Element1	Element2	Element3	Element4	Element5	Element6	Element7
L1	0	1	1	0	1	0	1
L2	0	1	1	0	1	0	1
L3	0	1	1	1	0	0	1
L4	0	1	1	1	0	0	1
L5	1	1	0	0	0	1	1
L6	1	1	0	0	0	1	1
L7	1	1	0	0	0	1	1
L8	0	1	0	1	1	1	0
L9	1	0	0	1	1	0	1

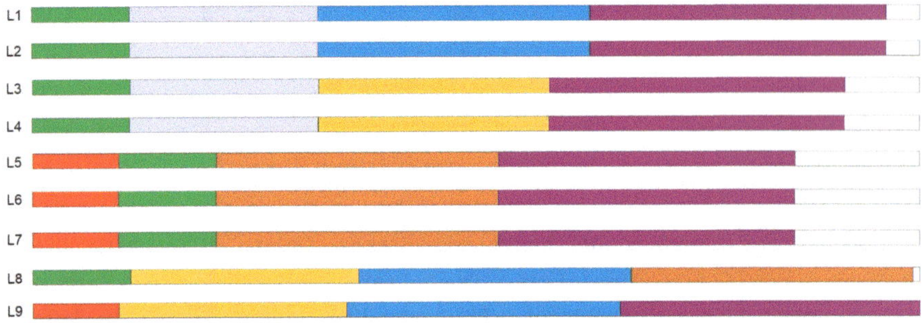

Fig. 4.3 Optimal cutting patterns for blanks ($L = 6550$ *mm*, waste = 4110 *mm*)

The defined optimal length of blanks to fulfill the order is 6550 *mm* and the overall minimum waste is 4110 *mm*. The graphical illustration of optimal cutting patterns for each of the blanks is shown in Fig. 4.3.

It is compared with cutting patterns combinations defined by experienced practitioners for standard length of blanks equal to 6000 *mm*. The comparison shows that without optimization the trim loss is bigger as shown in Fig. 4.4.

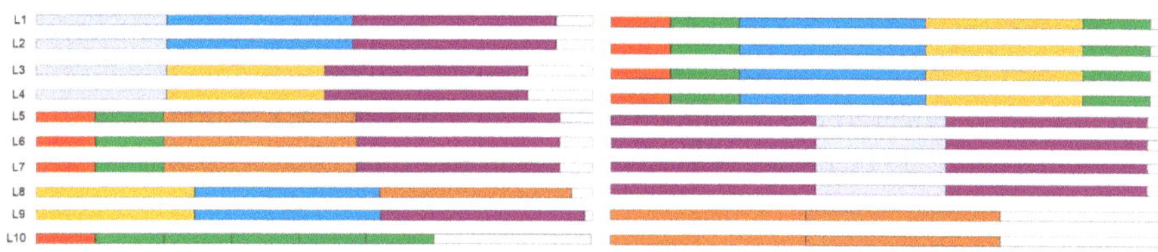

Fig. 4.4 Cutting patterns for standard blank length $L = 6000$ *mm*

<u>*Combinatorial optimization model for one-dimensional cutting stock problem*</u>
l1, l2, ..., l7: length of the elements that are to be cut
ΔLi: waste for each blank = optimal length of blank - utilized length of each blank
Owerall waste = sum of waste of all blanks
Lmin, Lmax: lower and upper limits of optimal length of blank

Lmin =	5000	*l1*	*l2*	*l3*	*l4*	*l5*	*l6*	*l7*	Optimal lenght of blank, *L*
Lmax =	7000	650	730	1400	1700	2000	2100	2200	6550
Demand of elements		4	8	4	4	4	4	8	Waste of each blank, *ΔLi*
Cutting patterns for each blank		0	1	1	0	1	0	1	220
		0	1	1	0	1	0	1	220
		0	1	1	1	0	0	1	520
		0	1	1	1	0	0	1	520
		1	1	0	0	0	1	1	870
		1	1	0	0	0	1	1	870
		1	1	0	0	0	1	1	870
		0	1	0	1	1	1	0	20
		1	0	0	1	1	0	1	0
								Minimal waste:	4110
Elements # in solution:		4	8	4	4	4	4	8	

Fig. 4.5 Optimal cutting-stock solution

The proposed optimization approach determines the optimal length of blanks that is increased toward standard length with 550 *mm*. This reduces the number of needed blanks to fulfill the requested order and waste and costs as compared to the case of standard length using. Using standard length of 6 *m* not only increases the trim loss but also increases the number of required blanks to execute the order. That is important for large joinery work projects in increasing transportation costs.

Due to NP-hard nature of considered problems, the computational time increases essentially with the increasing number of decision variables. The formulated mixed-integer linear optimization task (4.7)–(4.12) is solved by LINGO solver. The solution time for the described example with 64 integer variables amounts to 1 h, 23 min and 50 sec on PC with 2.93 GHz Intel i3 CPU and 4 GB RAM.

By combining a graphical user interface with algebraic modeling language and optimizers implementing different algorithms for linear, nonlinear, and integer problem-solving, the Microsoft Excel Solver can be a good choice for many users (Borissova, 2008). To illustrate this, the described approach to the one-dimensional cutting-stock problem is also implemented as a spreadsheet optimization model in an MS Excel environment as shown in Fig. 4.5 (Mustakerov & Borissova, 2015).

The defined optimal length of blanks to fulfill the order is 6550 *mm* and the overall minimum waste is 4110 *mm*. The optimal cutting patterns visualized in Fig. 4.5 are identical to the obtained results shown in Table 4.3. All of these mean that MS Excel could be used to solve combinatorial optimization problems in a well-known and widely used environment.

4.3 Job Shop Scheduling on Multiple Machines

Once the blanks are cut to the specified length, it is necessary to go through an established sequence of the operation to obtain the specific element for the assembly of the wind turbine. Scheduling is critical for any industry's manufacturing and productivity, including wind turbine production. Effective manufacturing scheduling can improve on-time delivery, reduce inventory, cut lead times, and improve the utilization of bottleneck resources (Chen et al., 2003).

Minimizing the total makespan is one of the basic objectives studied in the scheduling literature. The shortest processing time dispatching rule will give an optimal schedule in the single machine case if the tool life is considered infinitely long (Akturk et al., 2003). The scheduling with sequence-dependent setups is recognized as being difficult and existing results in the literature focus on either a single machine or several identical machines (Kim & Bobrowski, 1994; Ovacik & Uzsoy, 1994; Young et al., 1997). The real-life scheduling problems usually have to consider multiple no identical machines.

Most of the processing machines needed to process the jobs are available in the manufacturer's own factory and are of fixed (finite) number. Sometimes, certain details must be ordered to third-party companies to complete very specific processing as molding for example. In cases like that, the processing schedules are to be agreed for delivery times from the third-party processing. That means generating a schedule to process all jobs, so as to minimize the total cost, including the satisfaction of the due dates of the jobs (Wang et al., 1997). Different manufacturing environments induce different scheduling constraints, some of which may be very specific to the problem under consideration (Baptiste & Pape, 1996).

One of the most studied combinatorial optimization problems is the job shop scheduling problem (Kirilov et al., 2016). Nevertheless, it still remains a very challenging problem to solve optimally. From a complexity point of view, the problem is NP hard; i.e., it can be solved in nondeterministic polynomial time (Garey et al., 1976; Sotskov & Shakhlevich, 1995). The classical job shop scheduling problem is one of the most typical and complicated problems formulated as follows:

- A job shop consists of a set of different machines that perform operations of jobs.
- Each job is composed of a set of operations and the operation order on machines is prescribed.
- Each operation is characterized by the required machine and the processing time.

In the last two decades, numerous techniques were developed on deterministic classical job shop scheduling, such as analytical techniques, rule-based approaches, and meta-heuristic algorithms and algorithms using dynamic programming (Pinedo, 1983; Weber et al., 1986; Golenko-Ginzburg & Gonik, 2002; Lawrence & Sewell, 1997; Gromicho et al., 2012).

The proposed scheduling approach concerns a problem of scheduling for multiple details with fixed processing time and predetermined order of processing operations over different machines. An essential specific of the investigated job shop scheduling problem is the following features: (1) the processing of some details depends on processing of other details, i.e., a group of details have predetermined order of processing, and (2) the processing of the other parts is independent of each other. This variant of job shop scheduling is considered a semi-constrained job shop scheduling problem.

There is a group of details that need to be processed on multiple machines. Some of these details are connected with each other through given order of processing while others can be processed in any order. All details have predetermined sequence of operations on different machines. The details processing times on machines are deterministic and are known in advance. The problem is to determine the minimum makespan for all details processing according to requirements.

4.3.1 Combinatorial Optimization Model for Job Shop Scheduling on Multiple Machines

For clarity of presentation, the investigated job shop problem will be explained by a real-life example for a set of six details (jobs) with given sequences of operations that should be processed on four different machines with known processing time on each machine. All available data are summarized in Table 4.4 where operations' designation O_{ij} means processing of detail i on machine j and processing times are given in hours.

The sequence of operations for each detail is given as $D1$ $\{O_{11}, O_{12}, O_{14}\}$, $D2$ $\{O_{21}, O_{22}, O_{24}\}$, $D3$ $\{O_{31}, O_{33}, O_{32}\}$, $D4\{O_{41}, O_{42}, O_{43}\}$, $D5$ $\{O_{11}, O_{12}, O_{13}\}$, and $D6$ $\{O_{61}, O_{63}\}$. Due to their post-processing specifics, the details $D4$, $D5$, and $D6$ should be processed in a sequential order. All jobs cannot overlap on the machines and one job cannot be processed simultaneously by two or more machines. Each operation needs to be processed during an uninterrupted period of a given length on a given machine.

The goal of the investigated scheduling problem is to determine a schedule that minimizes the total makespan.

The described problem can be represented as machine-oriented Gantt chart visualizing the sequence of details processing as shown in Fig. 4.6.

Most variants of job shop scheduling problem are NP hard in the strong sense and thus defy ordinary solution methods. That is why new techniques are required to

Table 4.4 Input data for details processing

Details (Jobs)	Operations	Processing time on M1	Processing time on M2	Processing time on M3	Processing time on M4
D1	O_{11}	8			
	O_{12}		6		
	O_{14}				6
D2	O_{21}	8			
	O_{22}		9		
	O_{24}				6
D3	O_{31}	8			
	O_{33}			8	
	O_{32}		8		
D4	O_{41}	4			
	O_{42}		2		
	O_{43}			2	
D5	O_{51}	4			
	O_{52}		9		
	O_{53}			5	
D6	O_{61}	6			
	O_{63}			4	

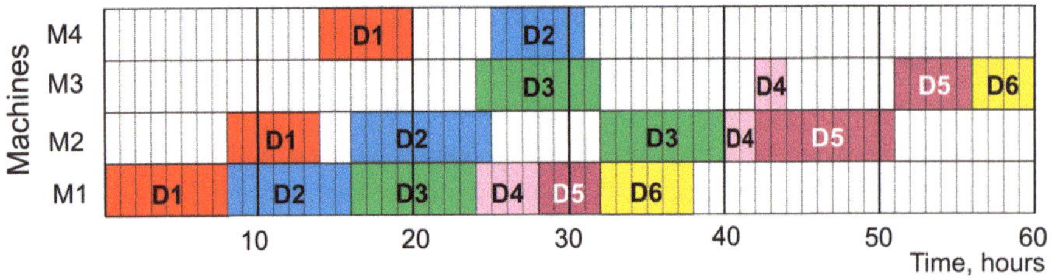

Fig. 4.6 Gantt chart for a schedule

overcome difficulties and to be applied to particular manufacturing job shop scheduling problems. The generalized goal of most of optimal scheduling problems is to minimize the overall costs. Although many costs could be considered for optimization, the minimizing of details processing time duration is one of most frequently used. It provides the effective machines utilization and serves the optimization of details delivering and storage. The overall details processing time duration (makespan) can be defined as difference between end processing moment of the last detail and start processing moment of the first detail and if the processing starts at zero moment then the objective can be minimization of the end processing moment of the last detail. Using those considerations, an optimization model can be formulated following the notations:

- Number of details indexed by $i \in \{1, 2, \ldots, N\}$.
- Number of machines for detail processing, indexed by $j \in \{1, 2, \ldots, M\}$.

- Job processing times $T_{i,\,j}$ of each detail i on machine j are known constants.
- $x_{i,\,j}$ is the moment of time for starting processing of detail i on machine j.

The scheduling problem is formalized via linear programming formulation that minimizes the makespan as (Borissova & Mustakerov, 2014):

$$\min\left\{\sum_{i}^{N} x_{i,end}\right\} \tag{4.13}$$

subject to

$$x_{i,j+1} - x_{i,j} \geq T_{i,j}, \forall i = 1, 2, \ldots, N \tag{4.14}$$

$$x_{i+1,j} - x_{i,j} \geq T_{i,j}, \forall j = 1, 2, \ldots, M \tag{4.15}$$

$$x_{i,j} \geq 0 \tag{4.16}$$

The objective function (4.13) minimizes the processing end time of all details. The relation (4.14) expresses the restriction for operation sequence for each detail, while relation (4.15) illustrates the restriction for the details processing order (if any). For dependent details processing, there exists a certain order of processing, but in case of independent details the processing order is not fixed. This way formulated model can be used to determine the optimal makespan for a given sequence of details processing. To find the optimal makespan among all of the possible sequences of independent details processing, a parallel algorithm can be applied.

To find the optimal scheduling that is minimal in the sense of shortest overall makespan, a parallel algorithm for optimal job shop scheduling of semi-constrained details processing on multiple machines is developed as shown in Fig. 4.7.

In the first step of the algorithm, the independent jobs are to be defined and designated using the name of the corresponding detail. If there exist dependent jobs (as for details D4, D5, and D6) they are considered as one independent job named after the first detail of the processing sequence (D4). Then the overall number of independent details is determined. In the second step, all possible orderings (permutations) for the processing of independent details are defined. In the next step, each of the details processing ordering is formalized by proper optimization tasks following the proposed model (4.13)–(4.16). It is important to stress here that all optimization tasks formulated in this way are independent of each other. The solution to any of them does not depend on data or the solution of other tasks. This makes them perfect candidates for using parallel threads for their solving in step 4. Then, task solution results (makespan values) are compared and ranked. In the last step, the schedule corresponding to the task with the best solution with minimal makespan value is chosen as the optimal job shop schedule.

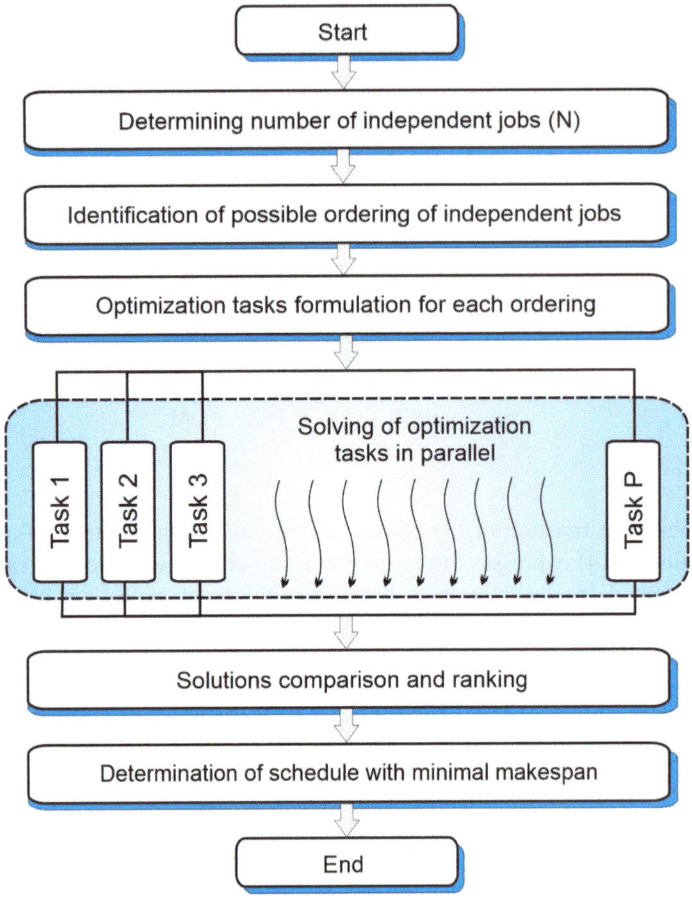

Fig. 4.7 Parallel algorithm for job shop scheduling

4.3.2 Numerical Results

In deterministic job shop scheduling problem, it is assumed that all processing times are fixed and known in advance, so using the input data from Table 4.4, the optimization model (4.13)–(4.16) can be expressed as (Borissova & Mustakerov, 2014):

$$\min \left\{x_{1,end} + x_{2,end} + x_{3,end} + x_{4,end} + x_{5,end} + x_{6,end}\right\} \tag{4.17}$$

subject to

$$x_{1,2} - x_{1,1} \geq 8 \tag{4.18}$$

$$x_{1,4} - x_{1,2} \geq 6 \tag{4.19}$$

$$x_{1,end} - x_{1,4} \geq 6 \tag{4.20}$$

$$x_{2,2} - x_{2,1} \geq 8 \tag{4.21}$$

$$x_{2,4} - x_{2,2} \geq 9 \tag{4.22}$$

$$x_{2,end} - x_{2,4} \geq 6 \tag{4.23}$$

$$x_{3,3} - x_{3,1} \geq 8 \tag{4.24}$$

$$x_{3,2} - x_{3,3} \geq 8 \tag{4.25}$$

$$x_{3,end} - x_{3,2} \geq 8 \tag{4.26}$$

$$x_{4,2} - x_{4,1} \geq 4 \tag{4.27}$$

$$x_{4,3} - x_{4,2} \geq 2 \tag{4.28}$$

$$x_{4,end} - x_{4,3} \geq 2 \tag{4.29}$$

$$x_{5,2} - x_{5,1} \geq 4 \tag{4.30}$$

$$x_{5,3} - x_{5,2} \geq 9 \tag{4.31}$$

$$x_{5,5} - x_{5,3} \geq 4 \tag{4.32}$$

$$x_{5,end} - x_{5,5} \geq 5 \tag{4.33}$$

$$x_{6,3} - x_{6,1} \geq 6 \tag{4.34}$$

$$x_{6,end} - x_{6,3} \geq 4 \tag{4.35}$$

to restrictions for the details priority processing:

$$x_{2,1} - x_{1,1} \geq 8 \tag{4.36}$$

$$x_{3,1} - x_{2,1} \geq 8 \tag{4.37}$$

$$x_{4,1} - x_{3,1} \geq 8 \tag{4.38}$$

$$x_{5,1} - x_{4,1} \geq 4 \tag{4.39}$$

$$x_{6,1} - x_{5,1} \geq 4 \tag{4.40}$$

$$x_{2,2} - x_{1,2} \geq 6 \tag{4.41}$$

$$x_{3,2} - x_{2,2} \geq 9 \tag{4.42}$$

$$x_{4,2} - x_{3,1} \geq 8 \tag{4.43}$$

$$x_{5,2} - x_{4,2} \geq 2 \tag{4.44}$$

$$x_{4,3} - x_{3,3} \geq 8 \quad (4.45)$$

$$x_{5,3} - x_{4,3} \geq 2 \quad (4.46)$$

$$x_{6,3} - x_{5,3} \geq 5 \quad (4.47)$$

$$x_{2,4} - x_{1,4} \geq 6 \quad (4.48)$$

$$x_{i,j} \geq 0 \quad (4.49)$$

The formulated task (4.17)–(4.49) takes into account the details processing sequence $D1 \rightarrow D2 \rightarrow D3 \rightarrow D4 \rightarrow D5 \rightarrow D6$. To define the minimum makespan for other processing sequences, this task should be reformulated.

The group of restrictions for details priority processing (4.36) to (4.48) has to be changed to correspond to other possible details processing sequences. There are 3 details ($D1$, $D2$, and $D3$) that can be processed in any order.

The group of dependent details $D4$, $D5$, and $D6$ can be considered as one independent detail and the number of all possible processing sequences can be calculated as a number of permutations of 4; i.e., the number of different processing sequences that have to be evaluated is equal to $4! = 24$.

For example, if details processing sequence is $D1 \rightarrow D3 \rightarrow D2 \rightarrow D4 \rightarrow D5 \rightarrow D6$ the restrictions (4.36)–(4.38) should be reformulated as:

$$x_{3,1} - x_{1,1} \geq 8 \quad (4.36a)$$

$$x_{2,1} - x_{3,1} \geq 8 \quad (4.37a)$$

$$x_{4,1} - x_{2,1} \geq 8 \quad (4.38a)$$

The objective function (4.17) and the rest of the restrictions remain the same.

If the details processing sequence is $D2 \rightarrow D1 \rightarrow D3 \rightarrow D4 \rightarrow D5 \rightarrow D6$, the restrictions (4.36)–(4.38) have to be changed as follows:

$$x_{1,1} - x_{2,1} \geq 8 \quad (4.36b)$$

$$x_{3,1} - x_{1,1} \geq 8 \quad (4.37b)$$

$$x_{4,1} - x_{3,1} \geq 8 \quad (4.38b)$$

Here again, the objective function and the rest of the restrictions remain the same.

If the group of details $D4$, $D5$, and $D6$ is to be processed in the first place, i.e., details processing order is $D4 \rightarrow D5 \rightarrow D6 \rightarrow D1 \rightarrow D2 \rightarrow D3$, the restrictions (4.36)–(4.48) are transformed to:

$$x_{5,1} - x_{4,1} \geq 4 \quad (4.36c)$$

$$x_{6,1} - x_{5,1} \geq 4 \quad (4.37c)$$

$$x_{1,1} - x_{6,1} \geq 6 \quad (4.38c)$$

$$x_{2,1} - x_{1,1} \geq 8 \quad (4.39c)$$

$$x_{3,1} - x_{2,1} \geq 8 \quad (4.40c)$$

$$x_{5,2} - x_{4,2} \geq 2 \quad (4.41c)$$

$$x_{1,2} - x_{5,2} \geq 9 \quad (4.42c)$$

$$x_{2,2} - x_{1,2} \geq 6 \quad (4.43c)$$

$$x_{3,2} - x_{2,2} \geq 9 \quad (4.44c)$$

$$x_{5,3} - x_{4,3} \geq 2 \quad (4.45c)$$

$$x_{6,3} - x_{5,3} \geq 5 \quad (4.46c)$$

$$x_{3,3} - x_{6,3} \geq 4 \quad (4.47c)$$

$$x_{2,4} - x_{1,4} \geq 6 \quad (4.48c)$$

The objective function is the same as the other constraints.

In a similar way, all possible combinations of detail processing sequences can be reflected in 24 different modifications of basic optimization tasks using the formulations (4.17)–(4.49). The solving of all of the tasks can be done in parallel because all of the tasks are entirely independent of each other. The result of the solutions is 24 job shop schedules corresponding to different details processing sequences.

The solutions of all optimization tasks corresponding to all possible combinations for details processing sequences along with their total makespan are shown in Table 4.5.

The makespan values in task solutions vary within intervals of 65 to 52 h for different details processing sequences. Among them the optimal one with minimal makespan equal to 52 h is for details processing sequence: $D2 \rightarrow D1 \rightarrow D3 \rightarrow D4 \rightarrow D5 \rightarrow D6$.

The corresponding schedules for each processing sequence are illustrated in Fig. 4.8.

For the described example, 11 different makespans have been distinguished that could not be determined by intuitive considerations. Increasing the number of independent details will increase the number of processing sequences and therefore the number of tasks that must be solved but because of the parallel algorithm for solution of each task, this will not affect the computational complexity. Despite the fact that integer problems are difficult to solve (in general they are NP hard), the formulated optimization problems and numerical testing show quite acceptable solution times of a few seconds by means of LINGO solver.

All real-life job shop scheduling problems have their own specifics. When analyzing the resulting schedules, it can be seen that the relationship between the processing details sequence and machines occupation has a significant impact on overall manufacturing process performance. For the described example, it turned out

Table 4.5 Solutions results

Sequence of the details processing	Total makespan, hours
D1, D2, D3, D4, D5, D6	60
D1, D3, D2, D4, D5, D6	61
D2, D1, D3, D4, D5, D6	52
D2, D3, D1, D4, D5, D6	58
D3, D1, D2, D4, D5, D6	59
D3, D2, D1, D4, D5, D6	59
D4, D5, D6, D1, D2, D3	54
D4, D5, D6, D1, D3, D2	61
D4, D5, D6, D2, D1, D3	54
D4, D5, D6, D2, D3, D1	58
D4, D5, D6, D3, D1, D2	60
D4, D5, D6, D3, D2, D1	63
D1, D2, D4, D5, D6, D3	61
D2, D1, D4, D5, D6, D3	59
D3, D2, D4, D5, D6, D1	55
D2, D3, D4, D5, D6, D1	53
D1, D3, D4, D5, D6, D2	58
D3, D1, D4, D5, D6, D2	56
D3, D4, D5, D6, D1, D2	56
D3, D4, D5, D6, D2, D1	56
D1, D4, D5, D6, D3, D2	54
D1, D4, D5, D6, D2, D3	65
D2, D4, D5, D6, D1, D3	54
D2, D4, D5, D6, D3, D1	65
Minimal makespan: 52 h	

that Machine 1 is the busiest machine among others. One possible approach to shorten the overall makespan is to consider more than one machine of type 1 and to estimate the influence of the machine's number on the total makespan.

The proposed approach based on a parallel solving of a set of optimization tasks can be used for other similar problems concerning optimal job shop scheduling including optimal planning of details processing for wind turbine manufacturing.

4.4 Configuration Design Via Group Decision-Making and Combinatorial Optimization

The design process is iterative and often requires testing new ideas. In such situations, these new concepts must implement an element with new parameters. In different industries, the trend shows the use of product family design with different parameters. This trend has also been carried over to wind turbine manufacturing, where products from a single platform with different parameters such as Vestas

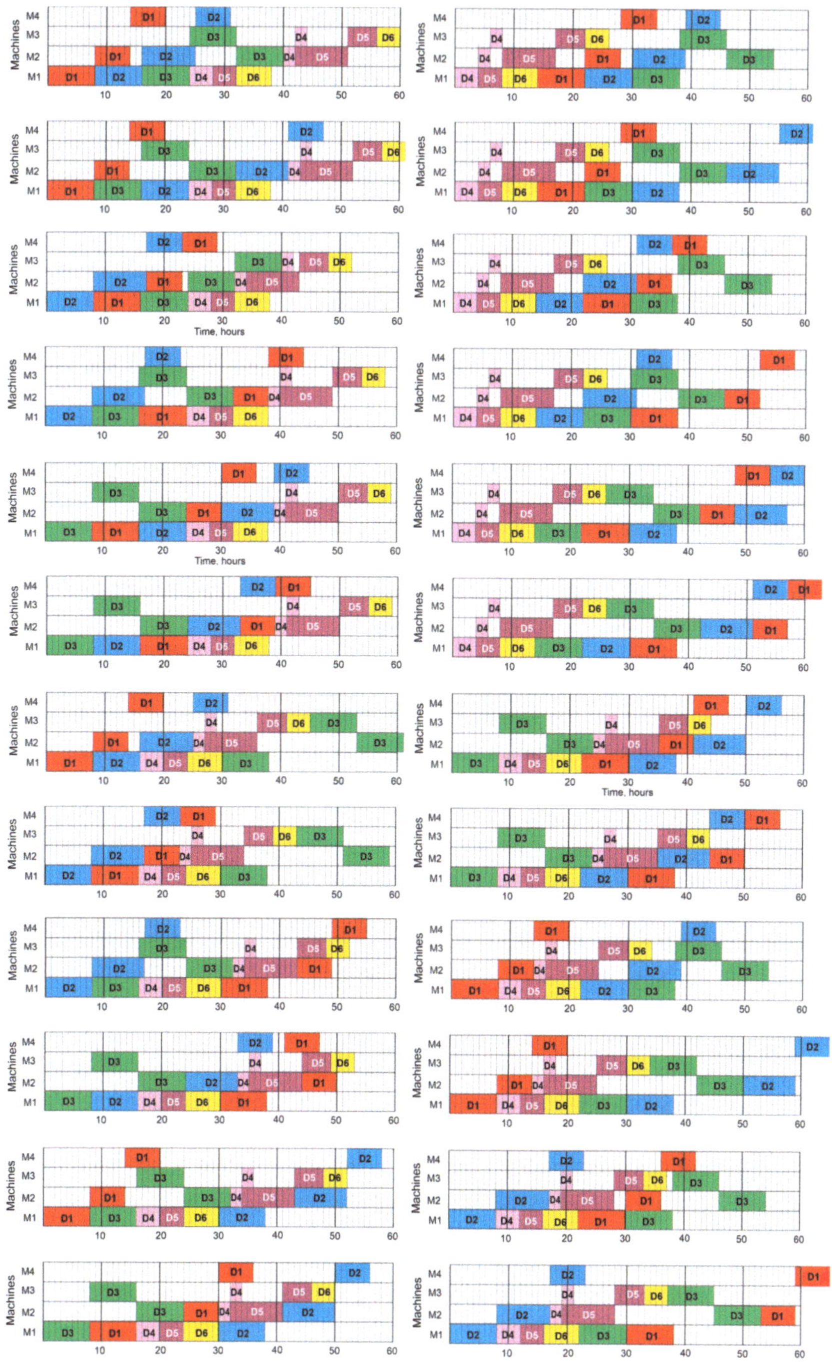

Fig. 4.8 Schedules for different details processing sequences

Platform[2] and ENERCON Platform[3] are available. Therefore, this section presents a group decision-making model for evaluating and selecting compatible modules to design a product configuration that meets expectations.

Nowadays, manufacturing companies are pressured to develop production that is not only low-cost and high-quality but also has to respond to rapid market and customer needs changes. This is due to the frequent market changes caused by global competition. Thus, the modularity in the design systems becomes a promising research direction, able quickly to reflect market changes (Lameche et al., 2017). The product platform design and product family design are based on the concept of modular design that relies on some kind of modules or standardization components. Therefore, modular design has been employed in different areas such as the building industry (Peltokorpi et al., 2018) and night vision devices (Borissova, 2015b).

The modular design proposes an alternative to satisfy consumers' customization needs without sacrificing the scale of economy in production procedures. These new methods are required to support the production system design activity satisfying requirements of modularity, connectivity, and intelligence. On the other hand, the modularity of products is considered a suitable approach in innovative design development, because it contributes to the improvement of quality, cost, and schedule performance and allows flexibility in building.

To be competitive, companies must continuously adapt their capacity investments and sales strategies. The launch of a new generation of products in turn attracts new customers, but before that, the production capacities must be adapted, which leads to a reduction in the available capacities. It is found that companies should increase their production capacity especially when introducing new products and when potential users are significantly more, so that the market is most profitable (Hartl et al., 2020).

The modular design is widely used and different approaches have been proposed. A new GA-based method is proposed for searching a proper balance between the commonality of the baseline product and the performance of the product family derived from the baseline product (Chen & Martinez, 2012). To realize the design of a modular product, multiple criteria for evaluating a modular product design can be used together with a fuzzy group genetic algorithm for grouping components of a design (Mutingi et al., 2017; Peneva & Popchev, 2009). All of these show that the product configuration provides an effective tool to satisfy requirements making product development a cost-efficient way (Zheng et al., 2017).

The design of complex products requires involving different participants that are not equally qualified. This requires using weights for opinions of DMs when aggregating individual preferences (Borissova, 2018) or by using some business intelligence (Borissova et al., 2016). That is why, an approach based on estimations from a group of decision-makers regarding the given evaluation criteria is proposed.

---

[2] https://www.vestas.com/en

[3] https://www.enercon.de/en

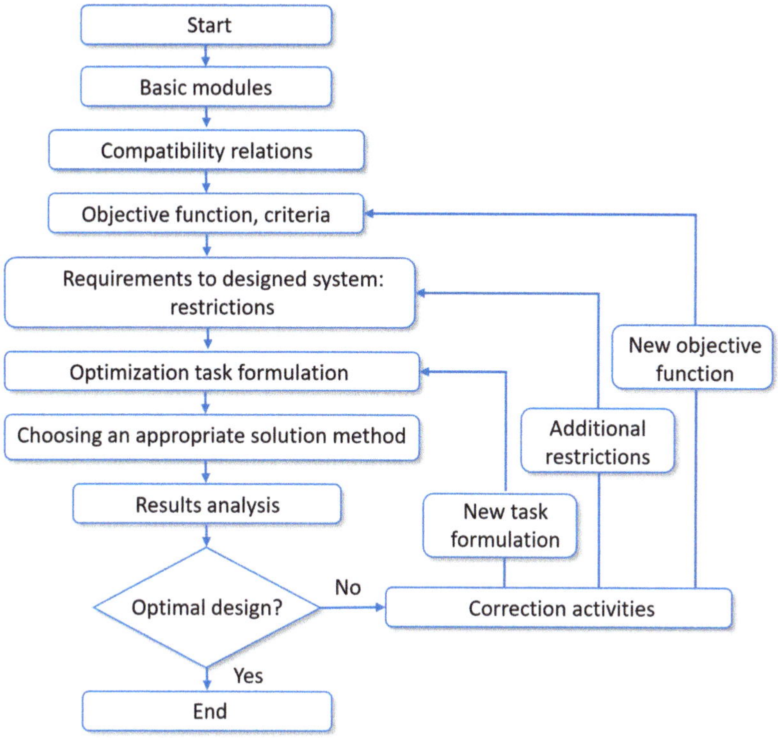

Fig. 4.9 Flowchart of optimal product configuration design

A new utility function is formulated to aggregate the estimations of individual decision-makers, weights for relative importance between evaluation criteria, and weights on decision-makers in forming the final group decision. This utility function is used as the objective function in formulated optimization model with binary decision variables.

4.4.1 *Algorithm for Optimal Product Configuration*

Determining the optimal design has different perspectives depending on the set criteria and constraints. Therefore, as a general approach, the following algorithm can be applied, which aims to find the optimal product configuration or Pareto-optimal combinations of modules for engineering system design, as a consequence of solving single- or multicriteria optimization tasks as shown in Fig. 4.9.

The algorithm begins with a procedure to determine a set of modules that can be used in the design process. It is then necessary to determine the compatible subsets between the individual modules. At the next stage, the criteria to which the designed configuration of the system must meet are defined. On the basis of the requirements set in this way to the parameters and the determined compatibilities between the modules, a corresponding optimization task can be formulated. Since the algorithm is aimed at assisting the DM in making a decision regarding the parameters of the designed configuration of the system, the resulting decision needs to be analyzed from DM. If the DM is not satisfied with the selected modules, it is necessary to make corresponding corrections. The algorithm includes 3 types of corrective actions: (1) reformulation of criteria or addition of new criteria; (2) redefining the set user requirements for the design; and (3) a new formulation of the optimization task. Through these actions, the DM has the opportunity to reflect more accurately his views regarding the designed system.

Optimizing the design of any system is done with respect to certain optimization criteria, often requiring simultaneous consideration of more than one criterion. This means that it is necessary to look for a multicriteria compromise between existing functional and user requirements and limitations. All of them must be properly defined and included in the mathematical model. User requirements are essential to the acceptance of the designed system. The formalization of the optimization problem is based on a mathematical model of the designed system. The design process can be divided into two problems: generating alternative solutions and making an appropriate decision.

4.4.2 Group Decision-Making with Combinatorial Optimization for Product Configuration Design

The design of a module-based product family shares a common platform that could be upgraded with different additional modules. The design of a common platform relies on compatibility between components that can be assembled and conform to some predefined requirements and restrictions. In group decision-making the experts play an essential role. The differences between the experts' knowledge and experiences should be taken into account to express the group decision. All of these considerations are involved in the proposed generalized model for group decision-making in the design of modular product configuration as follows (Borissova & Keremedchiev, 2019):

$$\max\left(\sum_m M_m x_a^m \sum_k \sum_a \lambda^k E_{m,a}^k\right) \tag{4.50}$$

subject to

$$\forall m : \sum_a x_a^m = \alpha^m, x_a^m \in \{0, 1\} \quad (4.51)$$

$$\sum_k \lambda^k = 1, \lambda^k \in [0, 1] \quad (4.52)$$

$$\forall m, \forall k : E_{m,a}^k = \sum_g \left(w_{m,a}^k\right)^g \left(e_{m,a}^k\right)^g \quad (4.53)$$

$$\sum_g \left(w_{m,a}^k\right)^g = 1 \quad (4.54)$$

where M is the number of modules, x_a^m re binary integer variables to select the required number of each module, λ^k is the weight for opinion importance of k-th DM, $E_{m,a}^k$ is the overall performance of each alternative from the module sets according to the point of view of k-th DM, $\left(w_{m,a}^k\right)^g$ express the relative weight for g-th evaluation criterion (modules' parameters) of a-th alternative for m-th module according to k-th DM, and $\left(e_{m,a}^k\right)^g$ is the particular evaluation of a-th alternative for m-th module from k-th DM toward g-th evaluation criterion.

The evaluations $\left(e_{m,a}^k\right)^g$ expressed by score a_i^k,j are limited within the range of (0, 1) where the bigger value means the better performance. Another evaluation scale could be to use also for example the range between 0 and 10 or 0 and 100, but using different scale requires normalization to obtain the comparable measure between coefficients $\left(w_{m,a}^k\right)^g$, evaluation score $\left(e_{m,a}^k\right)^g$, and weight λ^k for DMs.

The Eq. (4.51) provides the required number of elements from each module to be selected. For particular design, the number of elements from different modules could vary and using (4.51) allows selecting the needed number of elements. The feasible interval for coefficients λ^k in relation to (4.52) makes it possible to transform group decision-making into individual decision-making if λ^k takes value equal to 1.

The modular design does not involve the development of new component types but requires knowledge of the modules' characteristics and their compatibility with other modules.

4.4.3 Numerical Results

The described model can be applied to the selection of the main components of gearboxes, where it is necessary to select a certain set of gears, planet wheels, and bearings to fulfill a certain gear ratio. To illustrate this idea, let us consider four sets of different compatible modules, from which it is necessary to determine a suitable combination. Each element is evaluated toward 3 criteria (parameters). The given element is evaluated from a group of 3 DMs composed as shown in Tables 4.6, 4.7, 4.8, and 4.9.

Table 4.6 Elements of type A, scores, and weights for criteria importance

	DM-1			DM-2			DM-3		
	C1	C2	C3	C1	C2	C3	C1	C2	C3
Element of type A	0.41	0.26	0.33	0.22	0.42	0.36	0.33	0.34	0.33
A1	0.8	0.77	0.82	0.84	0.82	0.81	0.79	0.88	0.86
A2	0.75	0.72	0.88	0.83	0.79	0.84	0.86	0.82	0.79
A3	0.71	0.74	0.86	0.91	0.89	0.79	0.83	0.77	0.83
A4	0.78	0.79	0.83	0.88	0.84	0.68	0.95	0.83	0.86
A5	0.82	0.75	0.81	0.87	0.86	0.83	0.8	0.79	0.76
A6	0.84	0.81	0.78	0.85	0.8	0.72	0.86	0.85	0.79
A7	0.8	0.79	0.82	0.8	0.81	0.78	0.89	0.86	0.84
A8	0.82	0.72	0.81	0.75	0.79	0.8	0.81	0.81	0.82
A9	0.79	0.83	0.77	0.76	0.78	0.76	0.82	0.88	0.83
A10	0.74	0.78	0.85	0.8	0.82	0.86	0.81	0.78	0.86
A11	0.76	0.75	0.74	0.86	0.8	0.81	0.75	0.81	0.87
A12	0.84	0.83	0.79	0.93	0.79	0.82	0.81	0.78	0.85

Table 4.7 Elements of type B, scores, and weights for criteria importance

	DM-1			DM-2			DM-3		
	C1	C2	C3	C1	C2	C3	C1	C2	C3
Element of type B	0.4	0.35	0.25	0.3	0.35	0.35	0.2	0.35	0.45
B1	0.85	0.84	0.68	0.95	0.93	0.88	0.84	0.68	0.82
B2	0.86	0.86	0.83	0.8	0.84	0.68	0.95	0.81	0.86
B3	0.84	0.8	0.72	0.9	0.86	0.83	0.8	0.79	0.76
B4	0.8	0.81	0.78	0.89	0.8	0.72	0.88	0.85	0.79
B5	0.81	0.81	0.78	0.89	0.8	0.76	0.82	0.84	0.78
B6	0.82	0.79	0.8	0.81	0.73	0.72	0.87	0.83	0.82
B7	0.84	0.78	0.76	0.82	0.71	0.74	0.86	0.93	0.81
B8	0.78	0.84	0.86	0.82	0.78	0.79	0.89	0.88	0.87
B9	0.88	0.84	0.68	0.82	0.82	0.79	0.8	0.81	0.72
B10	0.82	0.83	0.8	0.78	0.78	0.84	0.68	0.75	0.78

Numerical testing is realized by using input data from Tables 4.6, 4.7, 4.8, and 4.9 and formulation of corresponding optimization tasks based on the proposed model (4.50)–(4.54).

Two different scenarios are studied for:

1. Choosing one element from each module
2. Choosing more than one element for one of the modules

These two tasks are solved by using three different cases regarding the weights for the DMs scores as follows:

Table 4.8 Elements from Module-C, scores, and weights for criteria importance

	DM-1			DM-2			DM-3		
	C1	C2	C3	C1	C2	C3	C1	C2	C3
Element of type C	0.38	0.27	0.35	0.25	0.35	0.4	0.35	0.4	0.25
C1	0.81	0.9	0.85	0.79	0.72	0.88	0.81	0.9	0.85
C2	0.79	0.89	0.9	0.84	0.76	0.9	0.79	0.82	0.9
C3	0.83	0.81	0.81	0.82	0.7	0.77	0.83	0.89	0.81
C4	0.86	0.82	0.88	0.83	0.72	0.81	0.86	0.82	0.88
C5	0.76	0.78	0.79	0.89	0.69	0.79	0.76	0.78	0.79
C6	0.83	0.72	0.81	0.82	0.79	0.89	0.88	0.79	0.82
C7	0.79	0.83	0.77	0.77	0.83	0.81	0.83	0.72	0.81
C8	0.87	0.85	0.89	0.81	0.86	0.82	0.79	0.83	0.77
C9	0.72	0.81	0.86	0.82	0.86	0.79	0.89	0.9	0.84
C10	0.69	0.79	0.76	0.78	0.79	0.83	0.81	0.81	0.82

Table 4.9 Elements of type D, scores, and weights for criteria importance

	DM-1			DM-2			DM-3		
	C1	C2	C3	C1	C2	C3	C1	C2	C3
Element of type D	0.36	0.24	0.4	0.28	0.33	0.39	0.34	0.4	0.26
D1	0.82	0.7	0.77	0.78	0.79	0.76	0.78	0.79	0.81
D2	0.8	0.72	0.81	0.79	0.82	0.82	0.75	0.8	0.88
D3	0.79	0.69	0.79	0.72	0.88	0.88	0.72	0.81	0.79
D4	0.81	0.82	0.79	0.74	0.86	0.89	0.69	0.79	0.77
D5	0.76	0.77	0.83	0.79	0.83	0.8	0.76	0.82	0.83

a) Case-1: $\lambda_1 = 0.33$ (DM-1), $\lambda_2 = 0.33$ (DM-2), $\lambda_3 = 0.34$ (DM-3);
b) Case-2: $\lambda_1 = 0.54$ (DM-1); $\lambda_2 = 0.23$ (DM-2); $\lambda_3 = 0.23$ (DM-3) and
c) Case-3: $\lambda_1 = 0.13$ (DM-1); $\lambda_2 = 0.31$ (DM-2); $\lambda_3 = 0.56$ (DM-3).

The resulting combinations of modules for the designed block according to the first scenario, where one element from each module is selected, are shown in Fig. 4.10.

The results of a single choice of elements for each type in Case-1, where the opinion of DMs with equal importance, determine the value of the objective function equal to 3.2961 and the value for the decision variables for the four types of elements as follows: $x_4 = 1$ (for type A), $y_1 = 1$ (for type B), $z_2 = 1$ (for type C), and $u_2 = 1$ (for type D).

That means that the design of configuration for the block should integrate the fourth element from type *A* set, namely *A*4, the first element from the set about type *B* that is *B*1, the second element from the set about type *C* that is *C*2, and also the second element from set about type *D* that is *D*2 as shown in Fig. 4.10a.

Case-2 expresses the situation where the opinion of DM-1 is the most important, while the opinions of DM-2 and DM-3 are considered with equal importance. For this case, the result of the corresponding optimization task determines objective

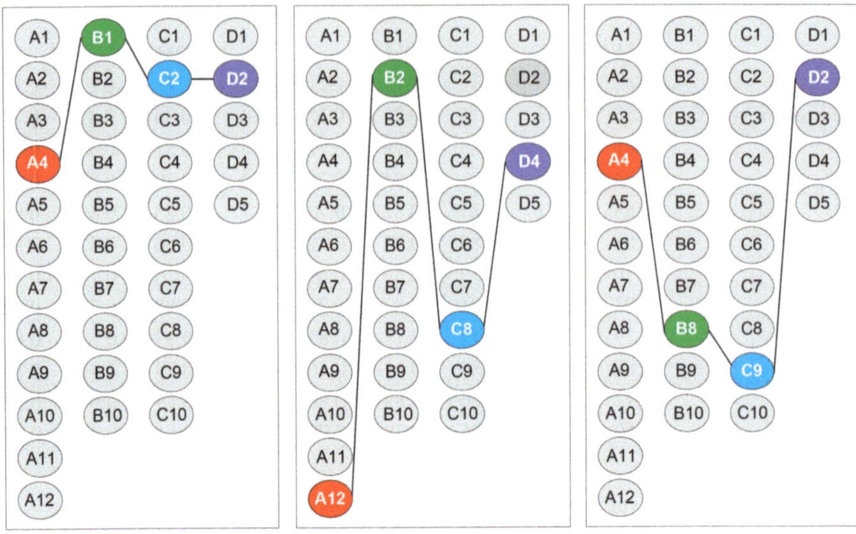

Fig. 4.10 Design configurations of the single-choice block for each of the modules A, B, C, and D: (**a**) case-1; (**b**) case-2; (**c**) case-3

function value equal to 3.3033 and the following values for binary decision variables: $x_{12} = 1$, $y_2 = 1$, $z_8 = 1$, and $u_4 = 1$. Thus, the block configuration should be composed of the following elements $A12$, $B2$, $C8$, and $D4$ as shown in Fig. 4.10b.

Case-3 illustrates the situation where the opinion of DM-3 is dominated and is considered the most important followed by the opinion of DM-2 and the less important is the opinion of DM-1. The obtained results for the selected elements of each type determine the objective function value equal to 3.3417 and values for decision variables for the elements of type A, type B, type C, and type D as follows: $x_4 = 1$, $y_8 = 1$, $z_9 = 1$, and $u_2 = 1$. That means that the block configuration should be composed of the following elements $A4$, $B8$, $C9$, and $D2$ considering DMs' point of view together with their importance as shown in Fig. 4.10c.

To demonstrate the flexibility of the proposed model, a corresponding optimization task is formulated to select two number of elements of type C for the block configuration. The resulting combinations of elements for this second scenario using the same weights for the DMs' scores are shown in Fig. 4.11.

The proposed model (4.50)–(4.54) also allows selecting more than one item for different element types by using the relation (4.51). This could be imposed by the nature of the considered configured block system. The corresponding optimization task is formulated and solved to get the corresponding configurations with two numbers for the element of type C. The obtained solutions determine the following values for the decision variables under different cases as follows:

d) For Case-1: $x_4 = 1$, $y_1 = 1$, $z_1 = 1$, and $z_2 = 1$, $u_2 = 1$.
e) For Case-2: $x_{12} = 1$, $y_2 = 1$, $z_2 = 1$ and $z_8 = 1$, $u_4 = 1$.
f) For Case-3: $x_4 = 1$, $y_8 = 1$, $z_1 = 1$ and $z_9 = 1$, $u_2 = 1$.

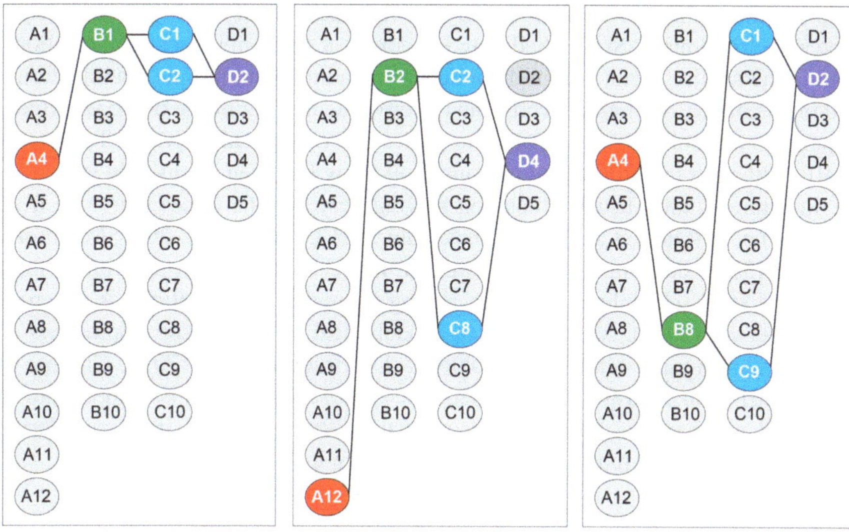

Fig. 4.11 Design configurations with two number of element of type C: (**a**) Case-1; (**b**) Case-2; (**c**) Case-3

The above-described results demonstrate the practical applicability of the proposed model for group decision-making for the block configuration design. Beyond the basic functionality elements, further different elements could be added to get a full block configuration.

The optimization model could be easily extended with additional restrictions to express compatibilities relations when input sets of elements from different types are not compatible with each other. The proposed model is useful to manage the design of product platform configuration using a group of experts with different knowledge areas. In addition, the described model could be realized as a software application where experts can express their preferences.

4.5 Staff Scheduling

Many management decisions involve the efficient use of the organization's resources. This is also the case in the production of elements for wind turbines. Once the blanks are cut and the sequence of operations to obtain the particular wind turbine element is clear, it is necessary to designate a specific team to service the machine for particular operations. Personnel scheduling is the process of constructing work timetables for its staff so that an organization can satisfy the demand for its goods or services (Ernst et al., 2004). Personnel scheduling is usually nontrivial due to the many sources of variability inherent in real service systems (Defraeye & Van Nieuwenhuyse, 2016). Optimizing staffing does not mean reducing the number of staff but finding the right number of employees at the right time in the right place. Staff scheduling is a common problem in most organizations, both in the service sector and in the industry including wind turbine manufacturing and maintenance. Staffing concerns the issues of the maximum use of the workforce to perform a set of activities in order to achieve a certain goal (Miwa & Takakuwa, 2010). Although the problem of personnel planning has been intensively studied in the literature, studies usually focus on solving very specific issues that arise from practical needs (Quan et al., 2007; Bai & Tang, 2013; Aksin et al., 2007; Mayrhofer et al., 2013). A description of mathematical models and solution algorithms for a family of staff scheduling problems arising in real-life applications is given in Caprara et al. (2003). Most proposed models deal with heuristic approaches to optimize staffing and scheduling problems (De Bruecker et al., 2015; Gunther & Nissen, 2014). In contrast to known heuristic approaches, the aim is to formulate an integrated approach to determine an optimal or Pareto-optimal staff considering given restrictions and requirements for minimizing the staff idle.

4.5.1 Combinatorial Optimization Model for Minimal Staff Idle in Open Shop Environment

The investigated staffing problem under an open shop environment addresses a single job consisting of a given number of operations that have to be processed on a number of machines. Each operation is assigned to a particular machine with a known processing time. Each machine is served by a known number of operators. The operations of the job can be processed in any order in sequence or in parallel. The processing times are independent of the processing sequence. There is only one of each type of machine and the machine operators have competence to serve all of the machines. The corresponding problem input data (operations, machines, processing times, and operators) are shown in Table 4.10.

Let us consider a specific problem for the determination of an optimal number of operators required to complete all work operations aiming to minimize the overall staff idle. The particular problem for staff scheduling is visualized as a Gantt-type chart in Fig. 4.12.

Each operation assigned to a particular machine is illustrated in Fig. 4.12 as a rectangle with length (processing time) and width (number of operators serving that

Table 4.10 Input data

Input data	Operation 1 on Machine 1	Operation 2 on Machine 2	...	Operation *m* on Machine *m*
Task processing time	h_1	h_2	...	h_m
Number of operators serving the machine	w_1	w_2	...	w_m

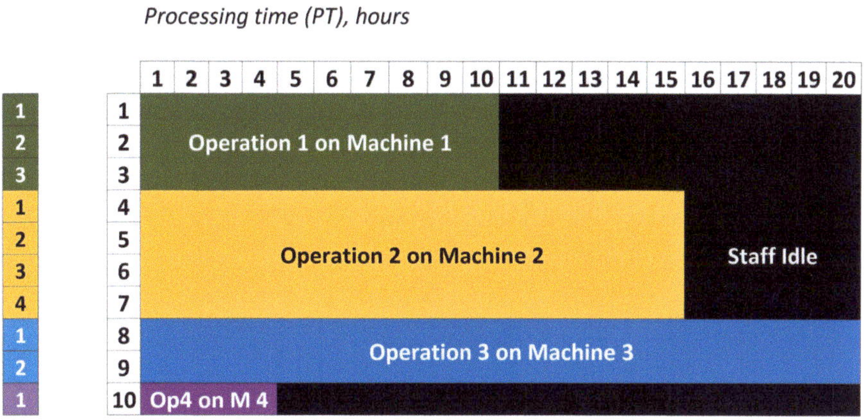

Fig. 4.12 Gantt-type chart of investigated open shop staffing problem

machine). It can be seen from Fig. 4.12 that the graphical representation of the investigated problem resembles a two-dimensional cutting-stock problem with fixed orientation of rectangles where the waste (staff idle) and overall length (processing time) are to be minimized. Along with this there exists a difference from the classical cutting-stock problem that consists in variable width of the strip (operators' number). The number of all operators serving machines should be optimized to get minimal staff idle. The innovative contribution of the chapter is that it models the open shop staffing problem by means of cutting-stock problems approach. Based on this model, mixed-integer linear programming task is formulated and used in an algorithm for optimal staffing in an open shop environment.

Using a mathematical model, it is possible to evaluate various staffing policies subject to personnel and machines constraints. A mixed-integer programming model to address this aspect of the problem is developed based on cutting-stock modeling. The corresponding notations are as follows:

- PT overall job processing time (makespan)
- $PT_{\min}$ minimal job processing time
- $PT_{\max}$ maximal job processing time
- M number of machines
- W number of operators to serve the machines
- $SA = W * PT$ strip area in Gantt diagram
- $J = \{1, 2, \ldots, M\}$ set of operations' indexes
- $h_j, j \in J,$ operation's processing time on *j-th* machine
- $w_j, j \in J$ number of operators serving *j-th* machine
- $W_{\min} = \max\{w_j\}$ minimal number of operators to serve the machines (for sequential processing scenario)
- $W_{max} = \sum w_j$ maximal number of operators to serve the machines (for parallel processing scenario)
- $OA = \sum w_j h_j$ all operations area in Fig. 4.12

- $SI = (SA - OA)$ staff idle
- $l_{jm} = 0/1$ variable; equal to 1 if *j-th* rectangle is located to the left of the *m*-th rectangle or 0 otherwise
- $b_{jm} = 0/1$ b variable; equal to 1 if *j-th* rectangle is located below the *m*-th rectangle or 0 otherwise,
- x_j, y_j coordinates of left upper corner of the *j-th* rectangle representing *j-th* operation on *j-th* machine in Fig. 4.12

The formulation of the mixed-integer optimization task is stated as (Borissova & Mustakerov, 2013):

$$min\ \{PT\} \tag{4.55}$$

subject to

$$PT \geq PT_{\min} \tag{4.56}$$

$$PT \leq PT_{\max} \tag{4.57}$$

$$x_j \geq 0, \text{integer}, j \in J \tag{4.58}$$

$$y_j \geq 0, \text{integer}, j \in J \tag{4.59}$$

$$x_j + h_j \leq PT \tag{4.60}$$

$$y_j + w_j \leq W \tag{4.61}$$

$$l_{jm} + l_{mj} + b_{jm} + b_{mj} = 1 \tag{4.62}$$

$$x_j + h_j \leq x_m + \left(1 - l_{jm}\right) PT_{\max} \tag{4.63}$$

$$y_j + w_j \leq y_m + \left(1 - b_{jm}\right) W \tag{4.64}$$

$$SA = W * PT \tag{4.65}$$

$$SI = (SA - OA) \tag{4.66}$$

where

$$PT_{\min} = \max\{h_j\} \tag{4.67}$$

$$PT_{\max} = \max\left\{\sum_{j=1}^{J} h_j\right\} \tag{4.68}$$

$$OA = \sum_{j=1}^{J} w_j h_j \tag{4.69}$$

In the current modeling approach, each operation is represented as a rectangle in a Gantt-type chart (Fig. 4.12) with processing time in hours as the x axis and machines' operators number as the y axis. The objective function (4.55) minimizes the overall job processing time PT (makespan). The variable PT is limited by its minimal and maximal values in (4.56) and (4.57), which can be defined as PT_{min} equal to the largest operation's processing time (for parallel processing scenario) and PT_{max} equal to the sum of all operations processing times (for sequential processing scenario). The coordinate x_j, y_j of left upper corner of the *j-th* rectangle represents the start of *j-th* operation on *j-th* machine. Equations (4.60) and (4.61) ensure that all rectangles are in the strip area, while (4.62) expresses the fact that a particular rectangle can only be in one position to another rectangle—left, right, top, or bottom. Equations (4.63) and (4.64) mean that no rectangles overlap each other. The solution of the task (4.55)–(4.69) defines minimal job processing time PT for the given W. Since the total operations area OA (4.66) is constant and the total strip area SA can be calculated by (4.65), the staff idle SI for that number of operators W can be evaluated by (4.66).

This model definition assumes that operators number W has a preliminary fixed constant value. To define the optimal number of operators W in regard to minimal staff idle, the defined optimization task (4.55)–(4.69) should be solved for all feasible operators' number combinations. Then the combination corresponding to minimal staff idle can be determined. For the goal an iterative algorithm based on multiple solving of optimization task is proposed.

The flowchart of the optimal staffing algorithm in an open shop environment is presented in Fig. 4.13.

Applying this algorithm the optimal number of operators in regard to minimal staff idle is defined as follows:

Step 1: Define

- Set of indexes J of operations and corresponding machines
- Processing times h_j of each operation j on j-th machine
- Number of needed operators w_j to process operation j on j-th machine

Step 2: Calculate

- Summary area of all operations in Gantt-type chart (Fig. 4.12) as $OA = \sum w_j h_j$,
- Minimal feasible job processing time $PT_{min} = max \{PT_j\}$ as the largest processing time of the various operations (for parallel processing)
- Maximal feasible job processing time $PT_{max} = \sum PT_j$ as sum of all operations processing times (for sequential processing)
- Minimal operators number $W_{min} = \max \{w_j\}$, as largest operators' number of the various operations (for sequential processing)
- Maximal operators number $W_{max} = \sum w_j$ as sum of operators for all operations (for parallel processing).

The calculations of minimal and maximal values of PT and W on Step 2 are based on two boundary scenarios for job completion—execution of all operations in parallel or sequentially.

Step 3: Mathematical modeling of two-dimensional cutting-stock problem and definition of optimization task formulation (4.55)–(4.69).

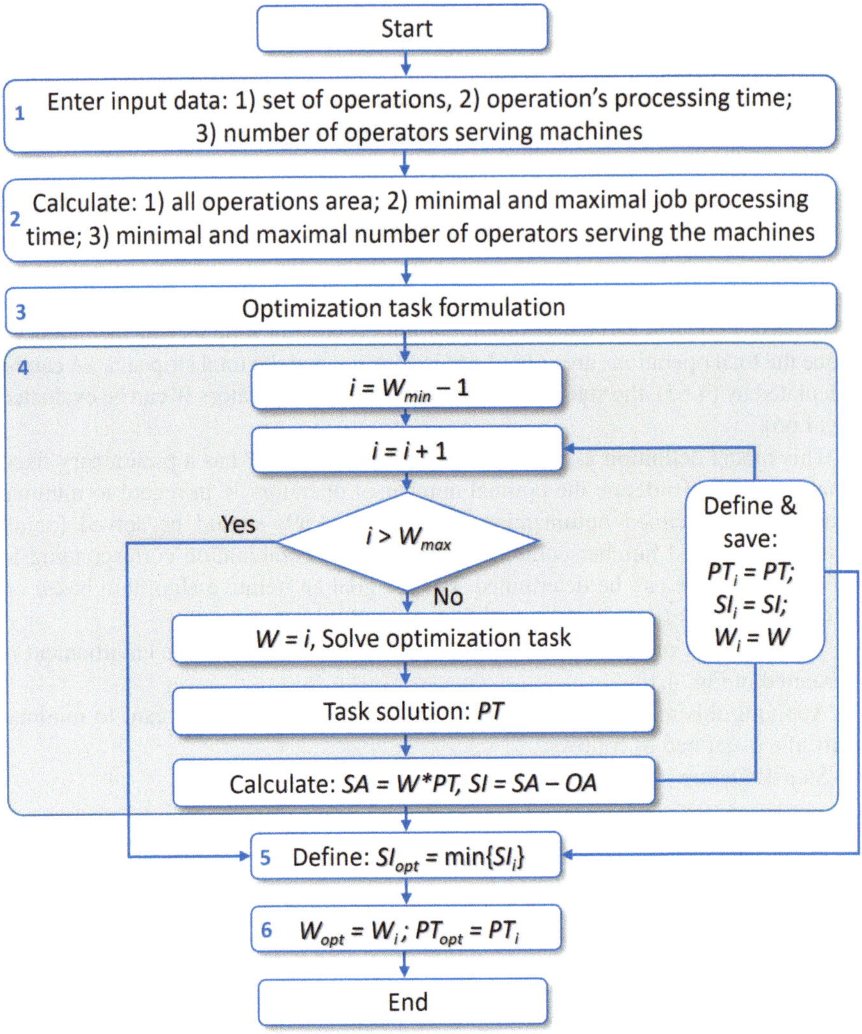

Fig. 4.13 Flowchart of optimal staffing algorithm

Step 4: The formulated optimization task is solved repeatedly with different numbers of operators W. The process starts with $W_{\min}$, then W increments by 1, and repeats until $W_{\max}$ is reached. After each task solution SA and SI are calculated and together with the optimal PT value are saved as SA_i, SI_i, PT_i, where $i \in \{W_{\min}, W_{\min} + 1, \ldots, W_{\max}\}$ is index of each particular optimization task.

Step 5: Define

- $$SI_{\text{opt}} = \min\{SI_i\},$$
- Index i corresponding to SI_{opt}

Step 6: Define

$$W_{\text{opt}} = W_i$$

$$PT_{\text{opt}} = PT_i$$

The final result of algorithm execution is optimal number of operators W_{opt} for the investigated problem and optimal job processing time PT_{opt} for this number of operators.

4.5.2 *Numerical Results*

The proposed optimal staffing approach is numerically tested for a real manufacturing problem with job processing data shown in Table 4.11.

In accordance with Step 4 of the described iterative algorithm, the optimization tasks (4.55)–(4.69) are to be solved sequentially for all feasible numbers of machines' operators, i.e., for $W = W_{\min}$, $W = W_{\min} + 1$, $W = W_{\min} + 2$, etc., to $W = W_{\max}$, where $W_{\min} = 4$ and $W_{\max} = 10$.

The summarized numerical results of task solutions are shown in Table 4.12.

Table 4.11 Real manufacturing problem data

	Operation 1 on Machine 1	Operation 2 on Machine 2	Operation 2 on Machine 2	Operation 4 on Machine 4
Task processing time	10	15	20	4
Number of operators serving the machine	3	4	2	1
Number of machines	1	1	1	1

Table 4.12 Numerical results after algorithm completion

Task #	Operators (W), number	Job processing time (PT), hours	Staff idle (SI), man-hours
1	4	45	46
2	5	35	41
3	6	25	16
4	7	25	41
5	8	25	66
6	9	20	46
7	10	20	66

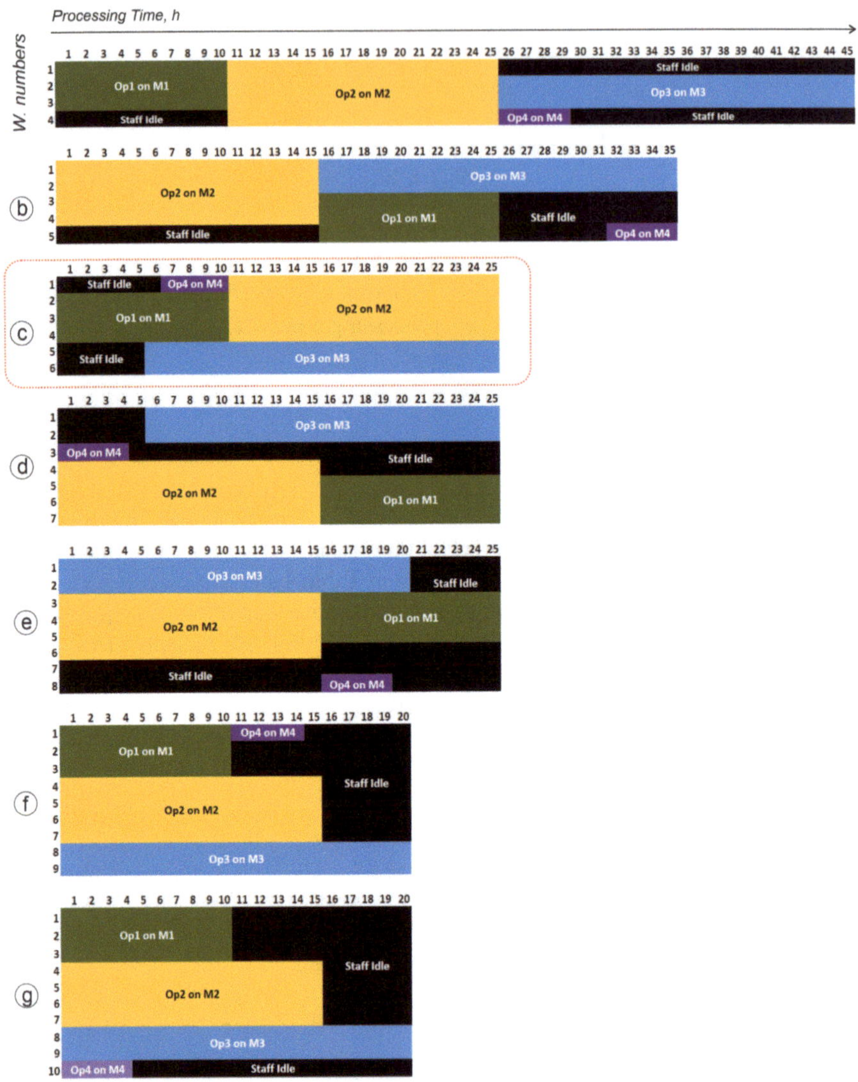

Fig. 4.14 Graphical representation of results after algorithm completion

As can be seen, the optimal solution is achieved with 6 operators, with overall jobs processing time equal to 25 h, while the staff idle is 16 h. The corresponding numerical results of the problem solutions are illustrated in Fig. 4.14.

In Fig. 4.14, case (a) represents scenario of sequential operations execution with minimal number of operators $W = W_{\min}$ and maximal job processing time $PT = PT_{\max}$. Case (g) conforms to scenario of parallel operations processing with maximal number of operators $W = W_{max}$ and minimal job processing time

$PT = PT_{min}$. These are the boundary cases of the investigated optimal staffing problem.

As it is seen from Table III and Fig. 4.14 (c) the optimal number of operators is $W = 6$ corresponding to minimal staff idle $SI = 16$. Although the job processing time PT is not minimal compared to case (g), it is optimal for this particular number of operators W.

The computational difficulty of most open shop problems is known as being NP hard. The computational difficulty of special cases like this is to be investigated for any particular real-life problem. However, the powerful computational capacity of modern computers and availability of efficient mathematical programming solvers makes attractive using the proposed algorithm. The numerical example used to illustrate the proposed algorithm was provoked by a real industry staffing problem. For this problem, it is easy to determine the optimal staff number just by looking at the results in Table 4.11 or in Fig. 4.14. For more complex problems, steps 5 and 6 have to be automated by proper software modules.

The optimization task solutions on Step 4 of algorithm have solution times in order of seconds by means of Lingo solver on desktop PC with Intel® Celeron® 2.80 GHz CPU and 1.96 GB of RAM. Other real-life problems would be more complex and will need many more iterations in Step 4. The proposed algorithm can be implemented in a decision support system, by coding all of the algorithm steps in software modules.

The used modeling approach leads to degenerate integer programming in the sense that several combinations of placement of rectangles have identical objective values. In Fig. 4.14, the rectangle corresponding to Operation 4 on Machine 4 can have other coordinates (in staff idle areas) without changing the objective function value. In some practical cases, this could be an advantage giving managers the freedom to choose other schedule parameters for such operations on the basis of other subjective criteria.

There are many details that need to be processed and assembled to obtain a working wind turbine, and the proposed optimization model for minimal staff idle in an open shop environment would be the appropriate tool to determine the required personnel.

4.5.3 Combinatorial Optimization Model for Staff Scheduling in Open Shop Environment with Different Criteria

The problems for optimal staffing and scheduling could be solved considering different criteria or strategies for (1) the determination of minimum staff idle, (2) the determination of minimum processing time, or (3) the simultaneous determination of minimum staff idle and minimum staff number. The relevant optimization tasks to solve these problems are:

Task 1: Determination of operators' number to achieve minimum staff idle:

$$\min\left(WT - \sum_{j}^{J} w_j h_j\right), j \in J \tag{4.70}$$

subject to:

$$x_j \geq 0, \text{integer}, j \in J \tag{4.71}$$

$$y_j \geq 0, \text{integer}, j \in J \tag{4.72}$$

$$x_j + h_j \leq T, j \in J \tag{4.73}$$

$$y_j + w_j \leq W, j \in J \tag{4.74}$$

$$\max\{h_j\} \leq T \leq \sum_{j}^{J} h_j \tag{4.75}$$

$$\max\{w_j\} \leq W \leq \sum_{j}^{J} w_j \tag{4.76}$$

$$x_j + h_j \leq x_m + (1 - l_{im})T_{\max}, j, m \in J \tag{4.77}$$

$$y_j + wh_j \leq y_m + (1 - b_{im})W_{\max}, j, m \in J \tag{4.78}$$

$$l_{jm} + l_{mj} + b_{jm} + b_{mj} = 1, j \in J \tag{4.79}$$

$$I = WT - \sum_{j}^{J} w_j h_j, j \in J \tag{4.80}$$

where I expresses staff idle, W is the total number of operators serving the machines, $W_{\min} \leq W \leq W_{\max}$, $W_{\min} = \max\{w_j\}$ is the minimal number of operators serving the machines (for sequential processing scenario), $W_{\max} = \sum_{j}^{J} w_j$ is the maximal number of operators to serve machines for parallel processing scenario, T is the overall job processing time, w_j is the number of operators serving j-th machine, and h_j expresses the processing time on j-th machine. The set $J = \{1, 2, \ldots, n\}$ expresses the set of operations; w_j and h_j express the needed operators and duration of j-th operation; x_j and y_j are coordinates of the upper left corner of the rectangle representing the j-th operation; l_{jm} is equal to 1, if j-th operation is located on left to the m-th operation and 0 otherwise (Kallrath et al., 2007); and b_{jm} is equal to 1, if j-th operation is below the m-th operation and 0 otherwise.

Task 2: Determination of operators' number providing minimum processing time (Borissova, 2015a):

$$\min\{T_i\} \tag{4.81}$$

subject to restrictions (4.71)–(4.80).

Task 3: Simultaneously determination of minimum staff idle and minimum staff number:

$$\begin{cases} \min \{T_i\} \\ \min \{W_i\} \end{cases} \tag{4.82}$$

subject to the same restrictions (4.71)–(4.80).

These optimizations tasks demonstrate the flexibility of the proposed model to adjust to different real-life open shop scheduling problems. Using multicriteria optimization leads to more informed and better decisions when conflicting objectives exist. In multicriteria optimization methods, the DM plays a major role in providing information for his preferences that influence the final solution (Miettinen, 2008). The most widely used methods are based on a priori expressing of DM preferences (Marler & Arora, 2004).

4.5.4 *Numerical Results*

The proposed modeling approach for optimal staffing in an open shop environment is numerically tested for a real manufacturing problem with input data shown in Table 4.13.

Task 1 and Task 2 are solved sequentially for each value of *i*. The index *i* takes values from 4 (for sequential processing scenario) to 10 (for parallel processing scenario). Formulated single-objective Task 1 and Task 2 are solved by means of branch-and-bound algorithm implemented in Lingo solver. Multicriteria Task 3 is solved by lexicographic solution method. The solution results of formulated Task 1, Task 2, and Task 3 are summarized in Table 4.14.

Table 4.13 Input data for optimal staff manufacturing problem

	Operation 1 on Machine 1	Operation 2 on Machine 2	Operation 3 on Machine 3	Operation 4 on Machine 4
Task processing time	10	15	20	4
Number of operators serving the machine	3	4	2	1
Number of machines	1	1	1	1

Table 4.14 Solution results for Task 1, Task 2, and Task 3

	Operators, number	Staff idle, man-hours	Processing time, hours
Task 1	6	16	25
Task 2	9	46	20
	10	66	20
Task 3	5	35	41

The results of solutions for Task 1 show that the minimal staff idle is equal to 16 h which corresponds to 6 operators. The solutions for Task 2 show that the minimal processing time of 20 h is reached for 9 operators and for 10 operators. Obviously, the better alternative in this case is using 9 operators. Multicriteria Task 3 is solved by the lexicographic method. This method is based on priority aggregation of preference information by arranging objective functions in order of importance. Then a sequence of single-objective optimization problems is solved one at a time in order of their importance. For Task 3, the used lexicographic ordering is defining processing time as more important than defining staff number. In the first step of lexicographic method, a single-objective optimization task for minimization of overall processing time is solved and the solution defines minimum processing time of 20 h. This value is used to define an upper limit for the processing time that is used at the second stage of lexicographic method. The lexicographic method implementation of Task 3 uses a value of 40 h. The final Pareto-optimal solution determines the overall processing time equal to 35 h with staff number equal to 5 corresponding to 41 h staff idle.

The advantage of the proposed approach is the ability to use different formulations of optimization tasks based on different objective functions depending on the company's strategy for wind turbine element manufacturing. This modeling approach and the formulated optimization tasks are not heuristic, and therefore they maintain a provable upper and lower bound on the (globally) optimal objective value.

It is worth mentioning that similar problems also arise when building a wind farm. For example, it is needed to determine the right number of people able to install the wind turbine considering the specifics of a particular wind farm site—onshore or offshore—and knowing the needed time for each operation during the installation.

References

Aksin, O. Z., Armony, M., & Mehrotra, V. (2007). The modern call centers: A multi-disciplinary perspective on operations management research. *Production and Operations Management, 16*(6), 665–688. https://doi.org/10.1111/j.1937-5956.2007.tb00288.x

Akturk, M. S., Ghosh, J. B., & Gunes, E. D. (2003). Scheduling with tool changes to minimize total completion time: A study of heuristics and their performance. *Naval Research Logistics, 50*(1), 15–30. https://doi.org/10.1002/nav.10045

Alsaleh, A., & Sattler, M. (2019). Comprehensive life cycle assessment of large wind turbines in the US. *Clean Technologies and Environmental Policy, 21*, 887–903. https://doi.org/10.1007/s10098-019-01678-0

Bai, D., & Tang, L. (2013). Open shop scheduling problem to minimize makespan with release dates. *Applied Mathematical Modelling, 37*(4), 2008–2015. https://doi.org/10.1016/j.apm.2012.04.037

Baptiste, P., & Pape, C. L. (1996). Disjunctive constraints for manufacturing scheduling: Principles and extensions. *International Journal of Computer Integrated Manufacturing, 9*(4), 306–310. https://doi.org/10.1080/095119296131616

Borissova, D., & Keremedchiev, D. (2019). Group decision making in evaluation and ranking of students by extended simple multi-attribute rating technique. *Cybernetics and Information Technologies, 18*(3), 45–56. https://doi.org/10.2478/cait-2019-0025

Borissova, D., Mustakerov, I., & Korsemov, D. (2016). Business intelligence system via group decision making. *Cybernetics and Information Technologies, 16*(3), 219–229. https://doi.org/10.1515/cait-2016-0045

Borissova, D., & Mustakerov, I. (2014). A parallel algorithm for optimal job shop scheduling of semi-constrained details processing on multiple machines. In *Circuits, systems, communications and computers* (Vol. 1). https://doi.org/10.13140/2.1.4406.0805

Borissova, D., & Mustakerov, I. (2013). An algorithm for an optimal staffing problem in open shop environment. *WASET International Scholarly and Scientific Research & Innovation, 7*(4). https://doi.org/10.5281/zenodo.1329761

Borissova, D. (2018). A group decision making model considering experts competency: An application in personnel selections. *Comptes Rendus de L'Academie Bulgare des Sciences, 71*(11), 1520–1527.

Borissova, D. (2015a). An optimal staffing and scheduling approach in open shop environment. *Comptes Rendus de L'Academie Bulgare des Sciences, 68*(10), 1295–1300.

Borissova, D. (2015b). *Night vision devices–Modeling and optimal design*. Prof. Marin Drinov Academic Publ.

Borissova, D. (2008). Optimal scheduling for dependent details processing using MS excel solver. *Cybernetics and Information Technologies, 8*(2), 102–111.

Caprara, A., Monaci, M., & Toth, P. (2003). Models and algorithms for a staff scheduling problem. *Mathematical Programming, Series B, 98*, 445–476. https://doi.org/10.1007/s10107-003-0413-7

Chen, A. L., & Martinez, D. H. (2012). A heuristic method based on genetic algorithm for the baseline-product design. *Expert Systems with Applications, 39*(5), 5829–5837. https://doi.org/10.1016/j.eswa.2011.11.084

Chen, D., Luh, P. B., Thakur, L. S., & Moreno, J., Jr. (2003). Optimization-based manufacturing scheduling with multiple resources, setup requirements, and transfer lots. *IIE Transactions, 35*, 973–985. https://doi.org/10.1080/07408170309342349

Cherri, C., Arenales, M. N., & Yanasse, H. H. (2009). The one-dimensional cutting stock problem with usable leftover–A heuristic approach. *European Journal of Operational Research, 196*, 897–908. https://doi.org/10.1016/j.ejor.2008.04.039

Cui, Y., & Lu, Y. (2009). Heuristic algorithm for a cutting stock problem in the steel bridge construction. *Computers & Operations Research, 36*, 612–622. https://doi.org/10.1016/j.cor.2007.10.019

De Bruecker, P., Van den Bergh, J., Belien, J., & Demeulemeester, E. (2015). A model enhancement heuristic for building robust air craft maintenance personnel rosters with stochastic constraints. *European Journal of Operational Research, 246*(2), 661–673. https://doi.org/10.1016/j.ejor.2015.05.008

Defraeye, M., & Van Nieuwenhuyse, I. (2016). Staffing and scheduling under nonstationary demand for service: A literature review. *Omega, 58*, 4–25. https://doi.org/10.1016/j.omega.2015.04.002

Dikili, A. C., Sarioz, E., & Pek, N. A. (2007). A successive elimination method for one-dimensional stock cutting problems in ship production. *Ocean Engineering, 34*, 1841–1849. https://doi.org/10.1016/j.oceaneng.2006.11.008

Dikili, A. C., Takinaci, A. C., & Pek, N. A. (2008). A new heuristic approach to one-dimensional stock-cutting problems. *Ocean Engineering, 35*(7), 637–645. https://doi.org/10.1016/j.oceaneng.2008.01.003

Ernst, A. T., Jiang, H., Krishnamoorthy, M., & Sier, D. (2004). Staff scheduling and rostering: A review of applications, methods and models. *European Journal of Operational Research, 153*(1), 3–27. https://doi.org/10.1016/S0377-2217(03)00095-X

Garey, M. R., Johnson, D. S., & Sethi, R. (1976). The complexity of flowshop and job shop scheduling. *Mathematics of Operations Research, 1*(2), 117–129. https://www.jstor.org/stable/3689278

Gilmore, P., & Gomory, R. (1961). A linear programming approach to the cutting stock problem. *Operations Research, 9*(6), 848–859. http://www.jstor.org/stable/167051

Gilmore, P., & Gomory, R. (1963). A linear programming approach to the cutting stock problem, part II. *Operations Research, 11*, 863–888. http://www.jstor.org/stable/167827

Golenko-Ginzburg, D., & Gonik, A. (2002). Optimal job-shop scheduling with random operations and cost objectives. *International Journal of Production Economics, 76*(2), 147–157. https://doi.org/10.1016/S0925-5273(01)00140-2

Gromicho, J. A. S., van Hoorn, J. J., Saldanha-da-Gama, F., & Timmer, G. T. (2012). Solving the job-shop scheduling problem optimally by dynamic programming. *Computers & Operations Research, 39*, 2968–2977. https://doi.org/10.1016/j.cor.2012.02.024

Gunther, M., & Nissen, V. (2014). A comparison of three heuristics on a practical case of sub-daily staff scheduling. *Annals of Operations Research, 218*, 201–219. https://doi.org/10.1007/s10479-012-1259-2

Hartl, R. F., Kort, P. M., & Seidl, A. (2020). Decisions on pricing, capacity investment, and introduction timing of new product generations in a durable-good monopoly. *Central European Journal of Operations Research, 28*, 497–519. https://doi.org/10.1007/s10100-019-00659-4

Kallrath, J., Kochetov, Y., & Rudnev, A. (2007). Strip packing problem for circles and rectangles. In *4th ESICUP Meeting* (pp. 1–20). University of Tokyo.

Kantorovich, L. V., & Zalgaller, V. A. (1971). *Rational cutting of stock.* [in Russian]. Nauka.

Kantorovich, L. V. (1960). Mathematical methods of organizing and planning production. *Management Science, 6*, 366–422.

Kim, S. C., & Bobrowski, P. M. (1994). Impact of sequence-dependent setup time on job shop scheduling performance. *International Journal of Production Research, 32*(7), 1503–1520. https://doi.org/10.1080/00207549408957019

Kirilov, L., Guliashki, V., Genova, K., & Angelova, V. (2016). An overview of multiple objective job shop scheduling techniques. *Jökull Journal, 66*(2), 172–206.

Korsemov, C., Toshev, H., Mustakerov, I., Borissova, D., & Grigorova, V. (2013). An optimal approach to design of joinery for renovation of panel buildings. *International Journal of Science and Engineering Investigations, 2*(18), 123–128.

Lameche, K., Najid, N. M., Castagna, P., & Kouiss, K. (2017). Modularity in the design of reconfigurable manufacturing systems. *IFAC-PapersOnLine, 50*(1), 3511–3516. https://doi.org/10.1016/j.ifacol.2017.08.939

Lawrence, S. R., & Sewell, E. C. (1997). Heuristic optimal static and dynamic schedules when processing times are uncertain. *Journal of Operations Managementm, 15*(1), 71–82. https://doi.org/10.1016/S0272-6963(96)00090-3

Marler, R. T., & Arora, J. S. (2004). Survey of multi-objective optimization methods for engineering. *Structural and Multidisciplinary Optimization, 26*, 369–395. https://doi.org/10.1007/s00158-003-0368-6

Mayrhofer, W., Marz, L., & Sihn, W. (2013). Simulation-based optimization of personnel assignment planning in sequenced commercial vehicle assembly: A software tool for iterative improvement. *Journal of Manufacturing Systems, 32*(3), 423–428. https://doi.org/10.1016/j.jmsy.2013.04.009

Miettinen, K. (2008). Introduction to multiobjective optimization: Noninteractive approaches. In J. Branke, K. Deb, K. Miettinen, & R. Słowiński (Eds.), *Multiobjective optimization* (Vol. 5252, pp. 1–26). Lecture Notes in Computer Science. https://doi.org/10.1007/978-3-540-88908-3_1

Miwa, K., & Takakuwa, S. (2010). Optimization and analysis of staffing problems at a retail store. In *Proc. of the Winter Simulation Conference (WSC '10). Winter Simulation Conference* (pp. 1911–1923). https://doi.org/10.1109/WSC.2010.5678879

Mobasher, A., & Ekici, A. (2013). Solution approaches for the cutting stock problem with setup cost. *Computers & Operations Research, 40*, 225–235. https://doi.org/10.1016/j.cor.2012.06.007

Mukhacheva, E. A., & Mukhacheva, A. S. (2006). L. V. Kantorovich and cutting-packing problems: New approaches to combinatorial problems of linear cutting and rectangular packing. *Journal of Mathematical Sciences, 133*(4), 1504–1151. https://doi.org/10.1007/s10958-006-0065-2

Mustakerov, I., & Borissova, D. (2015). Combinatorial optimization modeling approach for one-dimensional cutting stock problems. *International Journal of Systems Applications, Engineering & Development, 9*, 13–18.

Mustakerov, I., & Borissova, D. (2014). One-dimensional cutting stock model for joinery manufacturing. In *Circuits, systems, communications and computers, Santorini Island, Greece* (Vol. 1). https://doi.org/10.13140/2.1.2845.5049

Mutingi, M., Dube, P., & Mbohwa, C. (2017). A modular product design approach for sustainable manufacturing in a fuzzy environment. *Procedia Manufacturing, 8*, 471–478. https://doi.org/10.1016/j.promfg.2017.02.060

Ovacik, I. M., & Uzsoy, R. (1994). Rolling horizon algorithm for a single-machine dynamic scheduling problem with sequence-dependent setup times. *International Journal of Production Research, 32*(6), 1243–1263. https://doi.org/10.1080/00207549408956998

Peltokorpi, A., Olivieri, H., Granja, A. D., & Seppanen, O. (2018). Categorizing modularization strategies to achieve various objectives of building investments. *Construction Management and Economics, 36*(1), 32–48. https://doi.org/10.1080/01446193.2017.1353119

Peneva, V., & Popchev, I. (2009). Models for fuzzy multicriteria decision making based on fuzzy relations. *Comptes Rendus de L'Academie Bulgare des Sciences, 62*(5), 551–558.

Pinedo, M. (1983). Stochastic scheduling with release dates and due dates. *Operations Research, 31*(3), 559–572. https://www.jstor.org/stable/170623

Quan, G., Greenwood, G. W., Liu, D., & Hu, S. (2007). Searching for multiobjective preventive maintenance schedules: Combining preferences with evolutionary algorithms. *European Journal of Operational Research, 177*(3), 1969–1984. https://doi.org/10.1016/j.ejor.2005.12.015

Sotskov, Y. N., & Shakhlevich, N. V. (1995). NP-hardness of shop-scheduling problems with three jobs. *Discrete Applied Mathematics, 59*, 237–266. https://doi.org/10.1016/0166-218X(95)80004-N

Wang, J., Luh, P. B., Zhao, X., & Wang, J. (1997). An optimization-based algorithm for job shop scheduling. *Sadhana, 22*, 241–256. https://doi.org/10.1007/BF02744491

Weber, R. R., Varaiya, P., & Walrand, J. (1986). Scheduling jobs with stochastically ordered processing times on parallel machines to minimize expected flowtime. *Journal of Applied Probability, 23*(3), 841–847. https://www.jstor.org/stable/3214023

Young, H. L., Bhaskaran, K., & Pinedo, M. (1997). A heuristic to minimize the total weighted tardiness with sequence-dependent setups. *IIE Transactions, 29*(1), 45–52. https://doi.org/10.1080/07408179708966311

Zheng, P., Xu, X., Yu, S., & Liu, C. (2017). Personalized product configuration framework in an adaptable open architecture product platform. *Journal of Manufacturing Systems, 43*, 422–435. https://doi.org/10.1016/j.jmsy.2017.03.010

Chapter 5
Decision-Making in Structural Health Monitoring and Predictive Maintenance of Wind Turbines

Abstract The fourth stage of the life cycle of a wind farm refers to operation and maintenance activities. This stage includes routine maintenance and repairs that are necessary to achieve the design life of the wind turbine. Structural health monitoring for wind turbines is focused on the condition monitoring of the foundation, tower, and rotor blades for damage detection at an early stage, life cycle prognosis, and the optimization of wind farm operations. Regardless of location, a sustainable reduction of operating costs with a simultaneous increase in yields is gaining more and more relevance. Therefore, it is more important than ever to detect structural damage and icing at an early stage in order to avoid severe damage. Structural health monitoring of the foundation, tower, and rotor blade and conventional condition monitoring on the drive train are becoming increasingly important in this context. Based on a detailed system condition analysis, it is possible to plan ahead, optimize the operation of wind turbines, and extend their lifetime; thus, the operation of wind farms becomes more efficient and economical. Once the wind farm is ready for operation, the next stage involves regular health monitoring and predictive maintenance. Due to the construction of wind turbine blades and many blade-breaking accidents caused by blade vibration, proper models for monitoring predictive maintenance could be helpful (Zhou et al., 2023; Liu et al., 2023; Rokicki et al., 2023).

Predictive maintenance helps to determine the condition of in-service equipment in order to predict when proper maintenance should be performed. Implementation of condition monitoring and fault detection system entails initial investment, but these costs are being offset by the benefits of continuous production, minimum downtimes, and early planning for the replacement of defective parts.

Therefore, in this chapter, various combinatorial models could be used for monitoring and predictive maintenance of wind turbines/farms is presented. A strategy for predictive maintenance of the machine as a whole and of its individual components is presented, which is based on optimization models determining whether repair or replacement is necessary. A concept for an intelligent decision-making system for e-support that integrates different stratagems capable of expressing different uncertainties when forecasting the deterioration of the systems is proposed. A group decision-making approach for the choice of specialized

D. Borissova, *Decision-Making in Design, Maintenance, Planning, and Investment of Wind Energy*, International Series in Operations Research & Management Science 355, https://doi.org/10.1007/978-3-031-52219-2_5

software considering different strategies of uncertainty based on cost–benefit analysis is described too.

Keywords Decision-making in structural health monitoring · Predictive maintenance strategies · e-maintenance in uncertain conditions · Group decision-making under uncertainty

5.1 Structural Health Monitoring

Many condition monitoring techniques are applied to monitor and inspect the components in a wind turbine. One of the most important is the vibration analysis of different components such as the nacelle, tower, blade, bearings, shaft, gearbox, and generator (Ren et al., 2021). The placement of sensors for structural health monitoring plays a key role in structural control and damage detection. Predictive maintenance monitoring is essential for any successful system application. Maintenance monitoring is based on dynamic response information (mode shape curves) to identify the possibility of system faults. In structural vibration measurements, the locations of sensors are essential for the determination of structure response to dynamic loads. The structure could be simplified as a system with more degrees of freedom or as a lumped mass system. The goal is to have more nodal points for detailed data on structural responses and a part of these nodes could be used as sensor locations. In general, the more sensors are used; the more detailed information about the structure can be obtained. However, the more sensors are used, the more instruments and workloads are required, and in practice, a fixed number of sensors should be located on the structure positions that best characterize the structure response to dynamic loading.

As a result of dynamic analysis, the natural frequencies and mode shapes of the structure with dumping neglected can be defined. The natural frequencies of the structure are the frequencies at which structure vibrates if it is subjected to a disturbance. The deformed shape of structure at a specific frequency of vibration is known as mode shape. The optimization of damage identification could be realized through methods of optimal positioning of sensors for the acquisition of the data that must be evaluated so that conclusions can be made (Gomes et al., 2018).

Let us consider a vibrating cantilever beam that is subject to a time-varying distributed load. By the method of dynamic equilibrium are determined the corresponding mode shapes shown in Fig. 5.1 (Tejada, 2009).

If more modal points for data structural responses are used as sensors locations, the more detailed information of the structure status can be obtained. In practice, some fixed number of sensors is to be located on the optimal positions. The optimal location of sensors is determined by past experience of knowledge of structure or by finite elements analysis. In the chapter, for the sake of clarity we assume 10 sensors evenly distributed on the cantilever beam (Fig. 5.1).

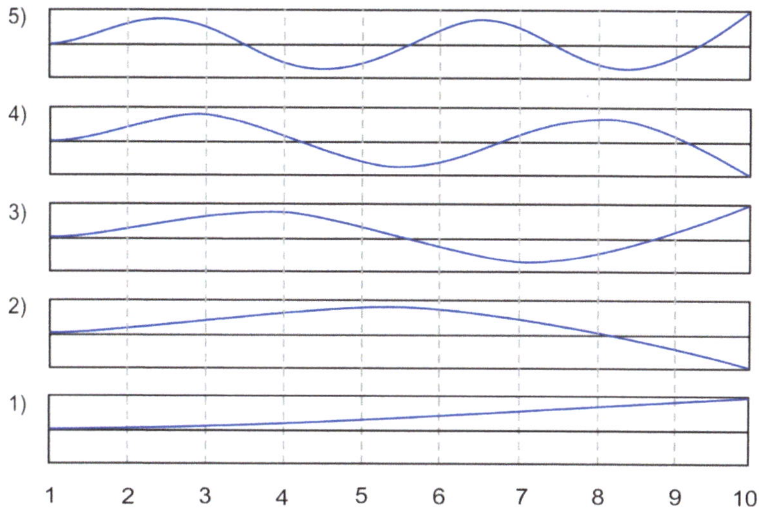

Fig. 5.1 Cantilever beam vibration mode shape functions

5.1.1 Single-Objective Combinatorial Optimization Model for Optimal Sensors Locations for Structural Health Monitoring

The problem of determining the optimum sensors locations has received considerable attention recently. In structural vibration measurements, the locations of sensors can be determined by past experience and knowledge of a structure or by finite element analysis. The mode shapes of a structure have influences on the measurements. Usually, the structural dynamic characteristics are analyzed first before making the measurements of the structural dynamic response. At the analysis phase, the structure could be simplified as a system with more degrees of freedom or as a lumped mass system. The goal is to have more nodal points for detailed data of structural responses and a part of these nodes could be used as sensors locations. In general, the more sensors are used; the more detailed information of the structure can be obtained. However, the more number of sensors are used, the more instruments and workloads are required and in practice a fixed number of sensors should be located on the best structure positions.

Assume the mode shape matrix of a structure for p modes is (Li et al., 2004):

$$\Phi = \left[\phi_1, \phi_2, \phi_3, \ldots, \phi_p,\right] \tag{5.1}$$

A widely used approach to a best sensor location problem is the degree of similarity of the real mode shape and the measured mode shape. That is usually done by defining proper fitness functions.

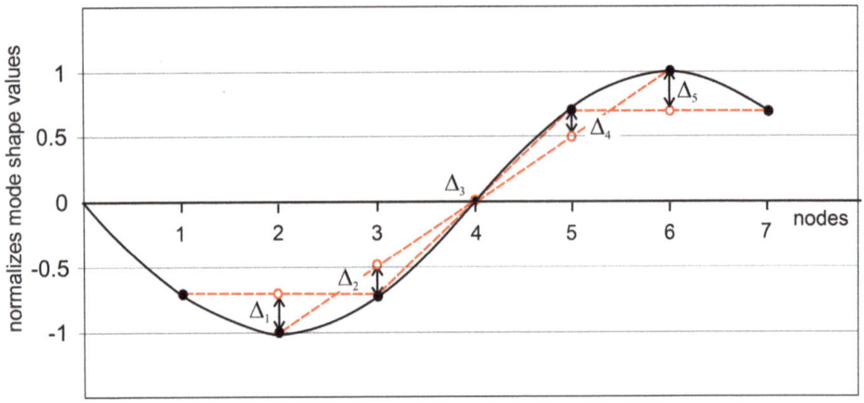

Fig. 5.2 Deviation of the measurement accuracy

A new type of fitness function could be used to formulate combinatorial optimization task for determination of optimal locations of a given number of sensors. Let us assume that the structure has n degrees of freedom or nodes where the sensors can be located and the goal is to choose the best m of them as sensors locations. The data for $(n - m)$ locations (without sensors placement) could be obtained by linear interpolation of the two neighboring points data. If the i-th node has no sensor, its data could be calculated taking into account the structural dynamic response curve from the vibration analysis. In this case, the deviation of the measurement accuracy Δ_i is shown in Fig. 5.2.

The basic idea is to drop out the $(n - m)$ sensors with smallest information accuracy loss, i.e., the sensors locations with smallest absolute value of Δ_i. That idea is realized in a fitness function used as objective function in a combinatorial optimization problem as:

$$\max\left\{\sum_{i}^{p}\sum_{j=2}^{n-1} x_j\left|\phi_j^i - \frac{\phi_{j-1}^i - \phi_{j+1}^i}{2}\right|\right\} \tag{5.2}$$

where $x_j \in \{0, 1\}$, $i = 1, 2, \ldots, n$ are binary integer variables assigned for all sensors locations and ϕ_j^i is the value of the i-th mode shape of the j-th node.

The requirement for finding of m best sensors locations is introduced by the constraint:

$$\sum_{j=1}^{n} x_j = m \tag{5.3}$$

In that approach, the first and last sensors locations nodes are always present, i.e.:

$$x_1 = 1, x_{10} = 1 \tag{5.4}$$

Dropping out the sensors with smallest Δ_i means that the function of structural dynamic response will be the closest one to the function with all sensors present. It is obvious that when the selection of sensor locations is optimum, then (5.2) should be maximized and the selection of the sensors locations is equivalent to solving a combinatorial optimization problem (5.2) subject to (5.3) and (5.4). The solution of the formulated optimization task defines m optimal sensors locations. It should be pointed out that the measure (5.2) is applied simultaneously for all p mode shapes, thus defining a compromise choice of sensors to be dropped out, optimal for all p mode shapes.

5.1.1.1 Numerical Results

A numerical example based on published data for vibration displacement of a cantilever structure is used. The first three mode shapes calculated from free vibration analysis are shown in Table 5.1 (Li et al., 2004) and Fig. 5.3. Every mode shape is normalized to have maximum value of each mode shape equal to 1.

Every mode shape is normalized to have maximum value of each mode shape equal to 1. The structure is divided into 10 equal sections with node number 10 located at the free end of the cantilever.

The mixed-integer linear programming formulation problem with objective function (5.2) subject to (5.3) and (5.4) is used for 4 practical examples. The goal of these 4 cases is to select 9, 8, 7, and 6 optimal sensors locations ($m = 9, 8, 7, 6$) among the 10 node locations. The results of the optimization tasks solutions are illustrated in Fig. 5.4.

The results from numerical experiments demonstrate that it is possible to use the proposed approach for optimal placements of fixed number of sensors. The optimal solutions define sensors to be removed as best compromise for all mode shapes simultaneously. The task with 9 sensors locations (Fig. 5.4a) is better approximation

Table 5.1 The first three mode shapes of structure

Node	Mode 1	Mode 2	Mode 3
1	0.15643	0.45400	−0.70706
2	0.30902	0.80901	−0.99993
3	0.45400	0.98770	−0.70706
4	0.58779	0.95107	0.00004
5	0.70711	0.70711	0.70701
6	0.80901	0.30902	1.00000
7	0.89103	−0.15643	0.70704
8	0.95107	−0.58779	0.00005
9	0.98770	−0.89103	−0.70713
10	1.00000	−1.00000	−0.99986

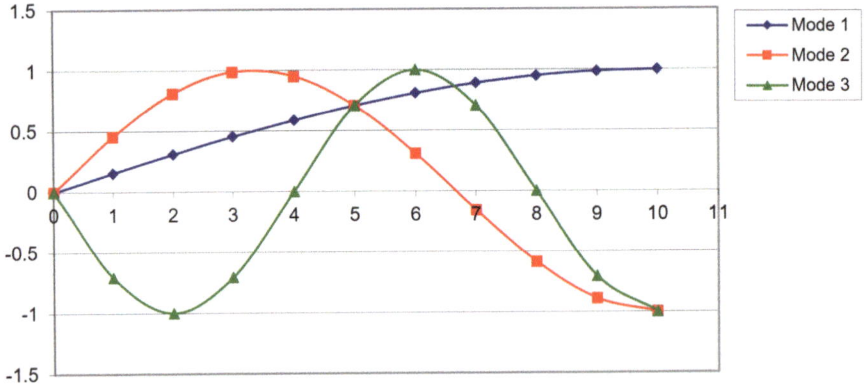

Fig. 5.3 Dynamic response functions from Table 5.1 (10 sensors)

to the original response function curves than the task with 8 sensors locations (Fig. 5.4b).

The difference in accuracy as a result of removing the sensors is shown in Table 5.2.

The summarized difference of the measurement accuracy Δ_i (the last 3 columns in Table 5.2) as a result of sensor removing is increased. The last task shows that the choice of sensors to be removed is a compromise between mode shapes; for some mode shapes, the curve approximation is not too good (Mode 3). The signal processing results are to be used for determining the number of the sensors to be removed in such a way that the quality measurements and accurate vibration data for identifying faults are still adequate.

5.1.2 *Multi-objective Combinatorial Optimization Model for Determining the Optimal Number and Placement of Sensors*

Let us assume that the structure has n degrees of freedom or nodes where the sensors can be located. The goal is to define optimal number of sensors and their locations considering all p mode shapes of a structure simultaneously. To define optimal number of sensors, some of them are to be dropped without considerable loss of information. If the j-th sensor for the i-th mode shape is missing, its data can be calculated as linear interpolation of data of its neighbors as $\phi_j^{i*} = \frac{\phi_{j-1}^i + \phi_{j+1}^i}{2}$. The deviation of data if the j-th sensor for the i-th mode shape is missing Δ_j^i is calculated as $\Delta_j^i = \phi_j^i - \phi_j^{i*}$. The selection of sensors to be present or to be dropped out is done by assigning binary integer variables x_j to each of sensors.

Using all of these assumptions an optimization multi-objective and combinatorial problem is formulated as follows (Mustakerov & Borissova, 2014):

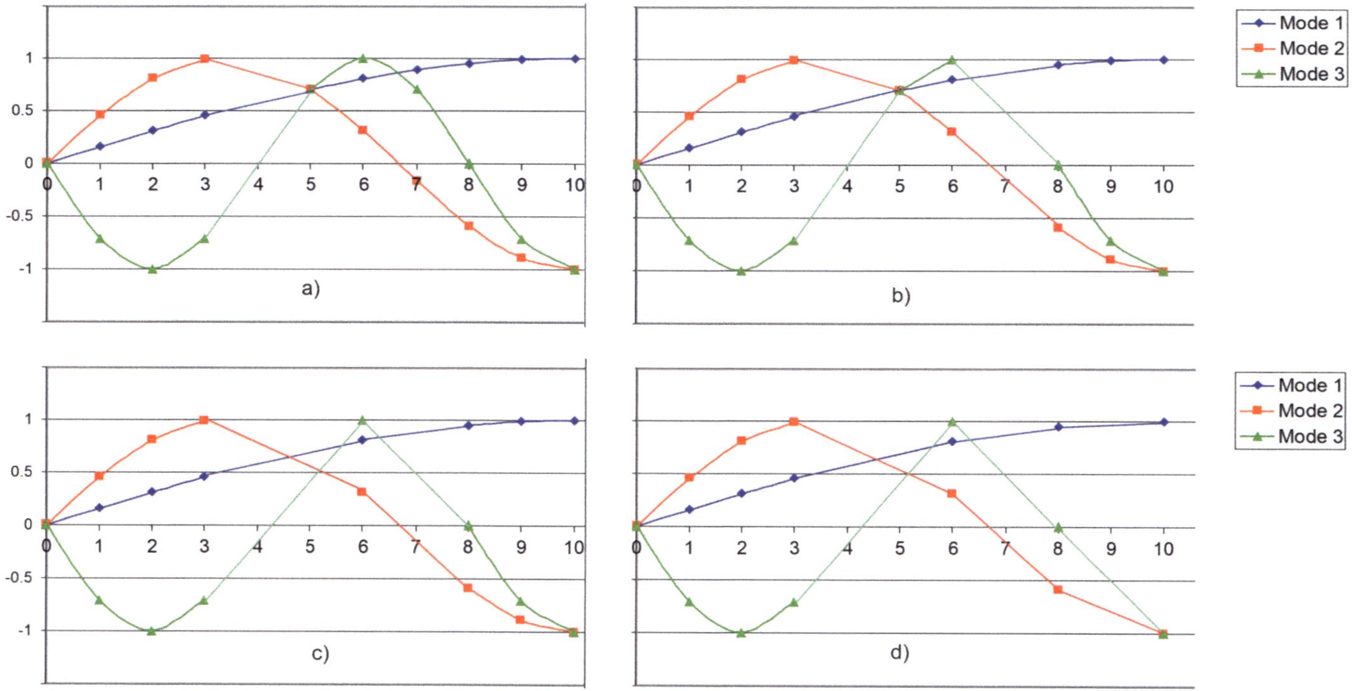

Fig. 5.4 Dynamic response functions with **(a)** 9 sensors, **(b)** 8 sensors, **(c)** 7 sensors, **(d)** 6 sensors

Table 5.2 Tasks solution results

Task	Sensors to be locate, m	Removed sensors	Mode 1 $\sum\boldsymbol{\Delta}_i$	Mode 2 $\sum\boldsymbol{\Delta}_i$	Mode 3 $\sum\boldsymbol{\Delta}_i$
1	9	4	0.007235	0.10366	0.000065
2	8	4 and 7	0.018225	0.12071	0.207080
3	7	4, 5 and 7	0.026935	0.197775	0.414070
4	6	4, 5, 7 and 9	0.039100	0.294910	0.621295

$$\max \sum_i^p \sum_j^n \left| x_j \phi_{i,j} + y_j \phi_{i,j}^* \right| \tag{5.5}$$

$$\min \sum_j^n x_j \tag{5.6}$$

subject to

$$\Phi_{i,j}^* = \left| \frac{\left(\phi_{i,j-1} + \phi_{i,j+1}\right)}{2} \right| \tag{5.7}$$

$$x_j + y_j = 1 \tag{5.8}$$

$$x_j \in \{0, 1\} \tag{5.9}$$

$$y_j \in \{0, 1\} \tag{5.10}$$

where x_j are binary integer variables assigned to sensors; $\phi_{i,j}^*$ is linear interpolation value if sensor j is missing; y_j are binary integer variables assigned to $\phi_{i,j}^*$; p is the number of mode shapes; and n is the number of sensors. The relation (5.8) represents the fact that if sensor j is present there is no interpolation of data and vice versa.

There exist different approaches for solving multicriteria problems. Multicriteria decision-making approach is characterized by the use of mathematical programming techniques and some decision-making method. In most multicriteria decision-making methods, the DM plays a major role in providing information for his preferences that influence the final solution (Miettinen, 2008). In general, the multicriteria problems can be handled in different ways depending on when the decision-maker expresses his preference for the different objectives: never, before, during, or after the actual optimization procedure. Depending on this, different multicriteria solution methods can be used. The most widely used methods are based on a priori expressing of the DM preferences. The specificity of the investigated problem requires minimizing the number of sensors while keeping the structure health monitoring information as close to the maximal (when all sensors are available) as it is possible. For the goal, the lexicographic method is chosen as most appropriate.

5.1.2.1 Numerical Results

In order to verify the developments in the chapter, the example of vibrating cantilever beam structure is used (Tejada, 2009). The adapted for the example data for first 5 mode shapes of structure are shown in Table 5.3.

Every mode shape is normalized to have maximum value of each mode shape equal to 1.

The implementation of lexicographic method based on the data from Table 5.3 and the formulated combinatorial optimization model (5.2)–(5.7) requires solving at the first place single-criterion optimization task toward the most preferable criterion for maximizing structure health monitoring information as follows:

$$\max\left(\sum_{i}^{5}\sum_{j}^{10}\left|x_j\phi_{i,j}+y_j\phi^*_{i,j}\right|\right) \tag{5.11}$$

subject to

$$\phi^*_{i,j}=\left|\frac{(\phi_{i,j-1}+\phi_{i,j+1})}{2}\right| \tag{5.12}$$

$$x_j+y_j=1, j=1,2,\ldots,10 \tag{5.13}$$

$$x_j\in\{0,1\} \tag{5.14}$$

$$y_j\in\{0,1\} \tag{5.15}$$

$$x_1=1, x_{10}=1 \tag{5.16}$$

It is assumed that the first and last sensors have to be present always that is expressed by (5.16). The solution of this task represents the ideal case where all 10 sensors are present, thus providing the maximal value of objective function (5.11) that can be used in the next step.

In the second step of the lexicographic method implementation, a single-criterion task for the second important criterion of multicriteria problem is solved. The

Table 5.3 Data for first 5 mode shapes of structure

Node #	Mode 1	Mode 2	Mode 3	Mode 4	Mode 5
1	0.102	0.294	0.370	0.576	0.610
2	0.202	0.581	0.766	0.820	0.722
3	0.335	0.770	0.872	0.370	−0.090
4	0.467	0.870	0.604	−0.384	−0.724
5	0.590	0.807	−0.020	−0.810	−0.020
6	0.720	0.579	−0.670	−0.430	0.726
7	0.810	0.269	−0.854	0.410	0.050
8	0.890	−0.187	−0.465	0.820	−0.735
9	0.950	−0.603	0.251	0.230	−0.037
10	1.000	−1.000	1.000	−1.000	1.000

objective function of the task solved in the first step is set as restriction in second single-criterion task with some acceptable tolerance toward its optimal value. The corresponding combinatorial optimization task solved in the second step is:

$$\min\left(\sum_{j}^{10} x_j\right), \forall x_j \in \{0, 1\} \tag{5.17}$$

subject to

$$\sum_{i}^{5}\sum_{j}^{10}\left|x_j\phi_{i,j} + y_j\phi_{i,j}^{*}\right| \geq \alpha\left(\sum_{i}^{5}\sum_{j}^{10}\left|x_j\phi_{i,j} + y_j\phi_{i,j}^{*}\right|\right)^{\text{optimum}} \tag{5.18}$$

$$\phi_{i,j}^{*} = \left|\frac{\left(\phi_{i,j-1} + \phi_{i,j+1}\right)}{2}\right| \tag{5.19}$$

$$x_j + y_j = 1 \tag{5.20}$$

$$x_j \in \{0, 1\} \tag{5.21}$$

$$y_j \in \{0, 1\} \tag{5.22}$$

$$x_1 = 1, x_{10} = 1 \tag{5.23}$$

where $\left(\sum_{i}^{5}\sum_{j}^{10}\left|x_j\phi_{i,j} + y_j\phi_{i,j}^{*}\right|\right)^{\text{optimum}}$ is the optimal value of (5.11) defined in the first step and α is a coefficient that defines the required data accuracy when some sensors are to be dropped out.

The results of applying a lexicographic method to determine the optimal number and placement of sensors are illustrated in Fig. 5.5.

Figure 5.5a represents the case when all 10 sensors are present. In this case, all of the information for health structure monitoring is available. Figure 5.5b, c, d represents the solutions for α set to be equal to 98%, 97%, and 95%, respectively. Setting $\alpha = 0.98\%$ results in dropping out sensor #9. In Fig. 5.5b, the corresponding mode shapes curves are plotted with dotted line for the missing sensor. Analogically, for $\alpha = 0.97\%$ sensors #3 and #9 are dropped out (Fig. 5.5c) and for $\alpha = 0.95\%$ sensors #3, # 5, and #9 are dropped out (Fig. 5.5d).

The main problem of sensors' number and locations optimization for structure health monitoring is to provide as minimum loss of information as possible when some of sensors are dropped out. This problem is mathematically formalized by the proposed combinatorial optimization model (5.5)–(5.10). This model takes into account both different sensors locations and mode shapes corresponding to a time-varying distributed load.

An essential advantage of using the lexicographic method is its easiness and intuitive understanding from engineers who are not familiar with multicriteria optimization. Along with this, the implementation of a lexicographic method for described problem allows the setting of acceptable accuracy for sensors' information as a required deviation from the ideal case when all sensors are present. The

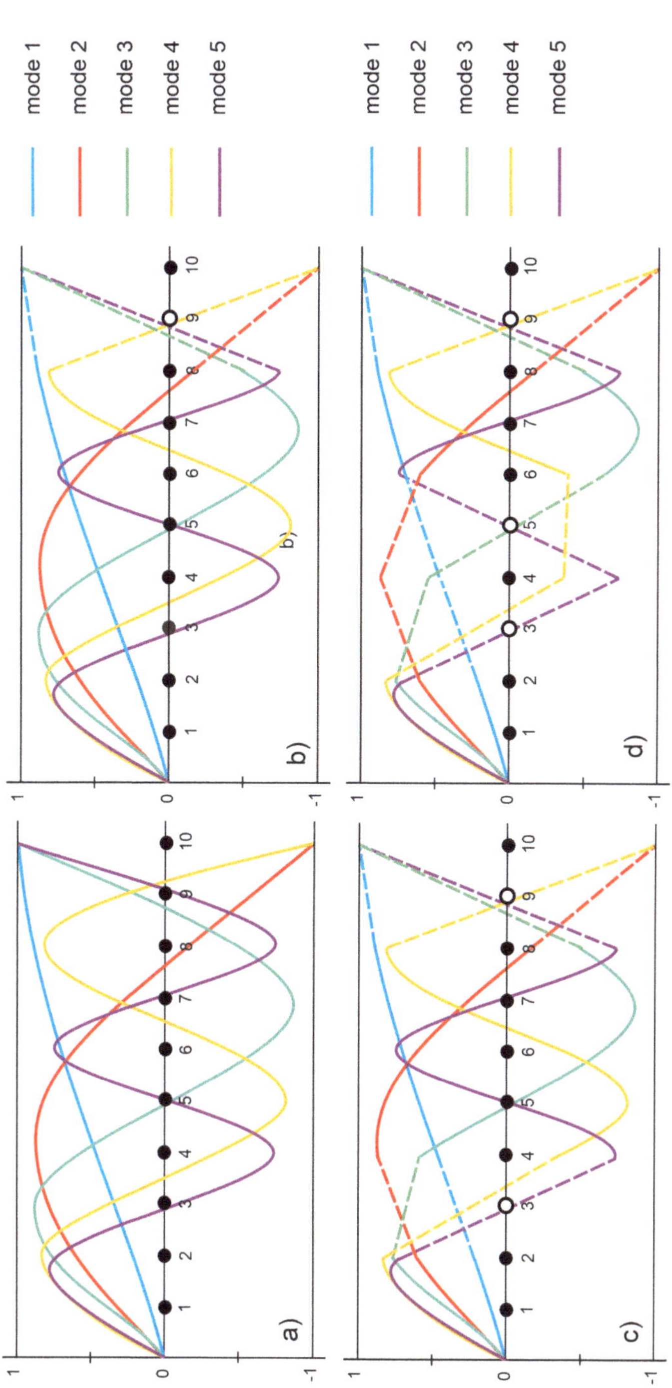

Fig. 5.5 Results of lexicographic method implementation: **(a)** ideal case for 10 sensors; **(b)** 9 sensors and $\alpha = 0.98\%$; **(c)** 8 sensors and $\alpha = 0.97\%$; **(d)** 7 sensors and $\alpha = 0.95\%$

multicriteria task solution based on real data shows that sensors corresponding to minimal data loss are dropped out. The described approach defines sensors on particular locations that can be dropped out in order to provide the best fit to the given acceptable accuracy. In this way, both the number of sensors and their locations are defined considering all mode shapes simultaneously. Regardless of the various mode shape curves, the results of the solution provide an accuracy of 98%, 97%, and 95% when dropping out one, two, or three sensors. The results of the proposed approach approbation prove its practical applicability.

5.2 Predictive Maintenance Strategy to Repair or Replace the Machine as a Whole

After the investment in the feasibility study and the acquisition and installation of a wind turbine, the main costs incurred during the useful life of a wind power generation project are those corresponding to operation and maintenance (Merizalde et al., 2019). Wind turbine technologies are rapidly evolving in terms of complexity and size and need cost-effective operation and maintenance strategies. Component reliability, including maintenance strategies, has a bigger effect on the planning of wind farms. That is why many different approaches are proposed to cope with such problems (Tavner et al., 2007; Faulstich et al., 2011; Carroll et al., 2016). In this regard, this section proposes a generalized approach for maintenance decisions to pick an optimum maintenance strategy depending on failure analysis.

Technological development resulted in increased complexity in both industrial machinery and production systems. The modern industry is constantly demanding high reliability, low environmental risks, and human safety while operating their processes at maximum yield. Therefore, the prevention of failures and early detection of incipient problems increase the useful operating life of plant machinery. Fault detection and diagnosis in the early stages of damage is necessary to prevent their malfunctioning and failure during operation. The premise of condition-based maintenance is that regular monitoring of the actual mechanical condition of equipment and operating efficiency of process systems will ensure the maximum interval between repairs, minimize the number and cost of unscheduled outages created by machines failures, and improve the overall availability of operating plants (Mobley, 2010). One of the most cost-effective maintenance techniques is condition-based maintenance.

Condition-based maintenance in a plant management program provides the ability to optimize the availability of process machinery and greatly reduce the cost of maintenance. Major improvements can be achieved in maintenance costs, unscheduled machine failures, repair downtime, spare parts inventory, and both direct and in-direct overtime premiums. This data provides the means to determine the most cost-effective time to replace machinery rather than continue to absorb high maintenance costs.

Many efforts have been made to develop methods and tools to diagnose failures for predictive maintenance goals (Nielsen et al., 2012; Mahadevan et al., 2010; Ferreira et al., 2009; Bouvard et al., 2011). The essence of prognostics is the estimation of remaining life in meaningful terms that would lead to a profound and intelligent maintenance decision process. This in turn would lead to proactive maintenance processes minimizing downtime of machinery and production and increasing of manufacturing efficiency (Swanson, 2001). Prognostics are viewed as add-on capabilities to diagnosis. They assess the current health of a system and predict its remaining life, based on features that capture the gradual degradation in the operational capabilities of a system (Luo et al., 2003). Prognostics are critical to improve safety, plan successful missions, schedule maintenance, and reduce maintenance cost and downtime (Dragomir et al., 2009).

Condition-based maintenance techniques provide an assessment of the system's condition, based on data collected from the system by continuous monitoring. The goal is to determine the required maintenance plan prior to any predicted failure. Therefore, the maintenance strategies aim to minimize the cost by improvement of operational safety and reduce the severity and number of in-service system failures. According to the ISO 13381-1:2004 standard, the activities start with monitoring, followed by diagnostic, prediction, and posterior actions (Borissova et al., 2011).

Various decision-making approaches have been proposed for maintenance strategy selection such as analytic hierarchy process, fuzzy set theory, genetic algorithm, mathematical programming, factor analysis, simple multi-attribute rating technique, and multicriteria optimization (Gandhare & Akarte, 2012; Akhshabi, 2011; Cavalcante et al., 2010; Ginevicius, 2011; Rigopoulos et al., 2010).

5.2.1 *Predictive Maintenance Methodology with Two Types of Cost–Benefit Estimations*

The management of particular system can be realized by collecting information about the system, defining the objectives of its management, formulating an appropriate task for decision-making, solving this problem, implementation of the decision, assessment of its effectiveness, and interactive influence of the decision-maker during various stages. The corresponding methodology is proposed in Fig. 5.6.

Condition monitoring is an essential part of predictive maintenance where the condition of specific equipment is monitored. This can be done automatically with the use of proper sensors or manually. During stage 1, phase 1a of the methodology implementation, quantitative information about the investigated system is collected by means of sensors. For example, the frequency of the vibrations can be mapped since certain frequencies will only be present when conditions that indicate an impending defect are present. Equipment can be monitored using sophisticated instrumentation such as vibration analysis equipment or the expert senses. Therefore,

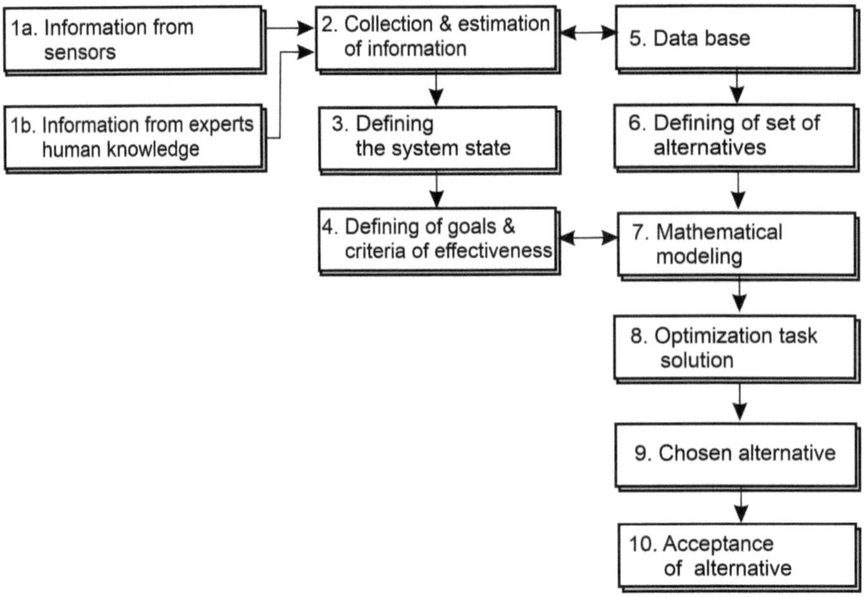

Fig. 5.6 Generalized methodology for predictive maintenance

phase 1b collects information from professionals serving the system and experts to ensure additional knowledge from human information.

The goal of condition monitoring of an item is to collect condition data to make it possible to detect incipient failure so that maintenance tasks can be planned at a proper time. Another goal is to increase the knowledge about failure cause and effect and deterioration pattern.

Stage 2 is carrying out by processing and summarizing the obtained information for the goals of the implementation of stage 3 and 4 aiming to establishing a system description and definition of objectives and criteria effectiveness. Aggregated and processed data from stage 2 is used in stage 5 to establish a database of investigated system. In stage 6 through interactive engagement of experts a set of alternatives is determined. The results of stage 4, 5, and 6 are used to define a mathematical model of the investigated system and to formulate an appropriate optimization task (stage 7). In stage 8, the optimal solution is determined. In stage 9 the alternative defined by the solution is applied. In stage 10 the alternative application is verified against the goals. If the results differ from the expected, the corresponding correction is setting in the system' mathematical model, the optimization task for decision-making is redefined on step 7, and steps 8, 9, and 10 are repeated. In some cases, because of incompleteness of a priori information about the investigated system, a redefinition of the objectives and criteria in stage 4 is required to achieve the desired objectives for management of the system. The implementation of stages 6, 7, 8, 9, and 10 is visualized by structural diagram shown in Fig. 5.7.

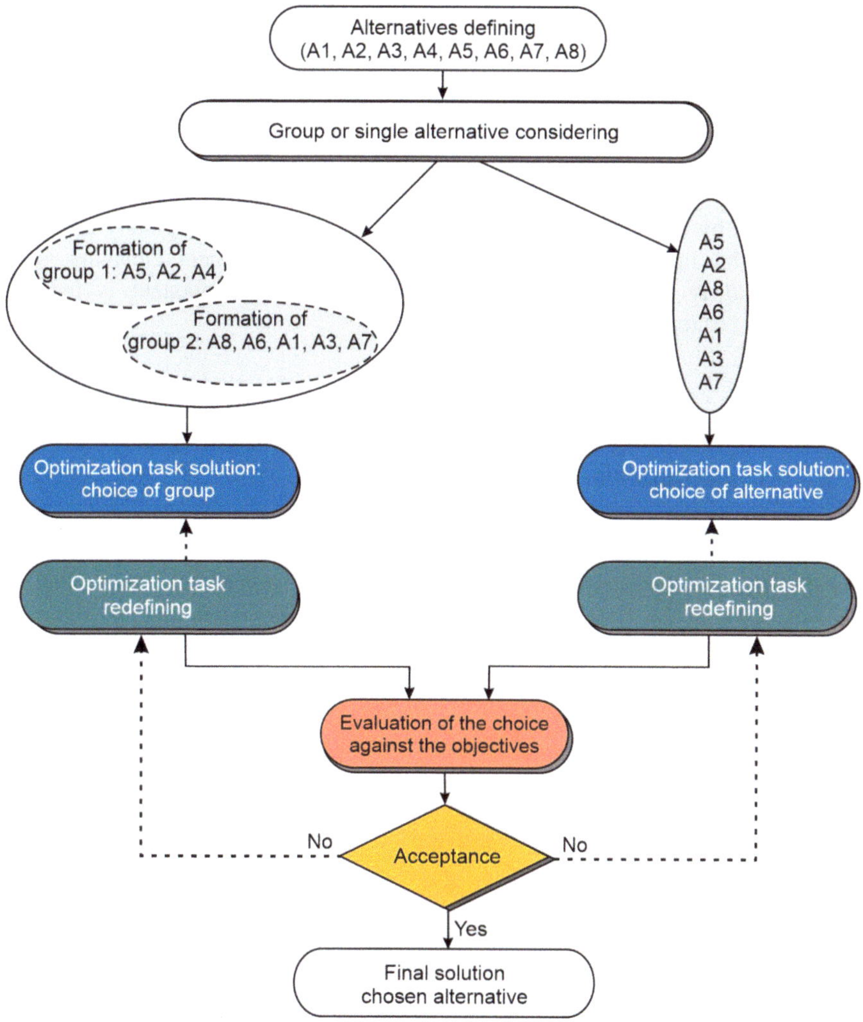

Fig. 5.7 A generalized structural diagram for decision-making

As a result of optimization tasks solutions a group or single alternative is chosen and evaluated against all the objectives. Finally, the selected alternative should be approved by the expert.

Since the predictive maintenance is a well-proven strategy that can significantly reduce maintenance costs, the cost–benefit estimation can be used to define the most appropriate decision-making strategy. Two types of cost–benefit estimations are proposed (Mustakerov & Borissova, 2013):

$$\mathrm{CBE}^{\mathrm{repair}} = \frac{\alpha C_{\mathrm{repair\_component}}}{\beta P_{\mathrm{remaining}}} \tag{5.24}$$

$$\mathrm{CBE}^{\mathrm{new}} = \frac{C_{\mathrm{new\_component}}}{P_{\mathrm{remaining}}} \tag{5.25}$$

where $C_{\mathrm{repair\_component}}$ is the cost for component repair, $C_{\mathrm{new\_component}}$ is the cost of replacement by new component, α is coefficient indicating how many times the component has failed, $P_{\mathrm{remaining}}$ is the profit during the component remaining useful time, and β is coefficient that indicates whether the component is repaired ($0 < \beta < 1$) or replaced with a new one ($\beta = 1$).

5.2.2 *Determining a Predictive Maintenance Strategy Using Combinatorial Optimization Models*

The question that arises when an alert alarm from the condition monitoring equipment appears is what decision should be taken. The decision process can be realized in two steps. In the first step two alternative exist—*to repair or to replace the machine as a whole*. If the decision in the first step is to replace the machine by new one the decision process ends. If the decision in the first step is to repair machine, the next question is—*what to do with any particular machine component—to repair or to replace*? The algorithm of this decision-making process is shown in Fig. 5.8.

The mathematical modeling of optimal maintenance strategy (replace or repair) is based on optimization tasks formulations taking into account (5.24) and (5.25). These formulations are based on maximization of the cost–benefit estimations. This is because maximal cost–benefit values define the worst-case scenario requiring proper maintenance reaction.

Step 1: The answer of the question *to repair or to replace the machine as a whole* is the result of solving the following optimization task:

$$\max\left(\sum_{i}^{n} x\mathrm{CBE}_i^{\mathrm{repair}} + \sum_{i}^{n} y\mathrm{CBE}_i^{\mathrm{new}}\right) \tag{5.26}$$

subject to

$$x + y = 1 \tag{5.27}$$

$$x, y \in \{0, 1\} \tag{5.28}$$

where $\mathrm{CBE}_i^{\mathrm{repair}}$ represents the cost–benefit evaluations for repairing the i-th component, $\mathrm{CBE}_i^{\mathrm{new}}$ represents the cost–benefit evaluations for replacing the i-th component, and x, y are binary integer variables assigned for each alternative.

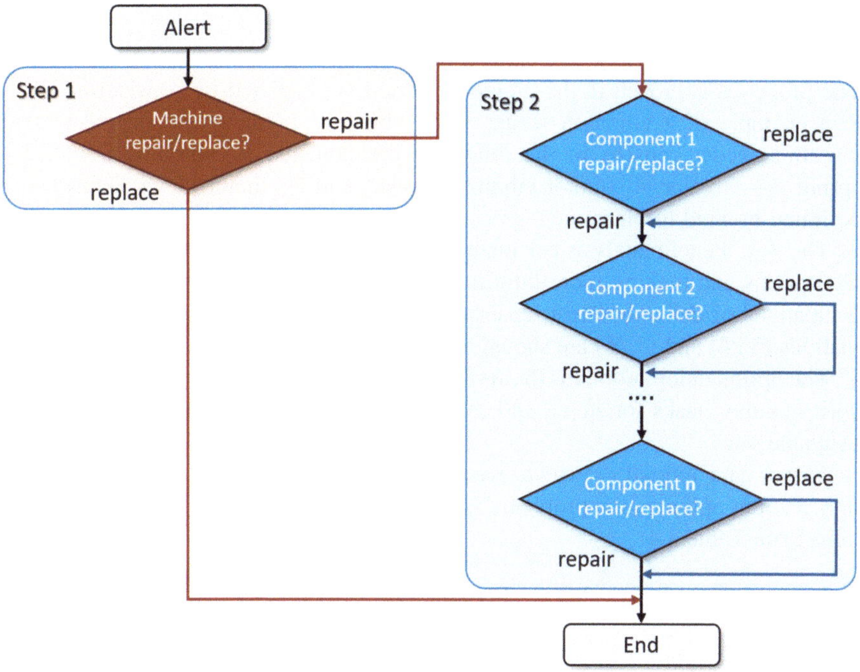

Fig. 5.8 Predictive maintenance decision-making algorithm

Step 2: The repair/replace strategy for each particular component is defined by a single run of the following optimization problem:

$$\max\left(\sum_{i}^{n} x_i \mathrm{CBE}_i^{\mathrm{repair}} + \sum_{i}^{n} y_i \mathrm{CBE}_i^{\mathrm{new}}\right) \tag{5.29}$$

subject to

$$\forall i \in \{1, 2, \ldots, n\} : \sum_{i}^{n} x_i + y_i = 1 \tag{5.30}$$

$$x_i, y_i \in \{0, 1\} \tag{5.31}$$

where $\mathrm{CBE}_i^{\mathrm{repair}}$ represents the cost–benefit evaluation for repairing the i-th component, $\mathrm{CBE}_i^{\mathrm{new}}$ represents the cost–benefit evaluation for replacing the i-th component, and x_i, y_i are binary integer variables assigned for each machine component alternative.

In this step, the maintenance strategy considers each component as independent variable in contrast to the previous step where the machine maintenance is considered as a whole.

5.2.3 Numerical Results

The proposed approach in the previous section is illustrated by considering a real data example of a vibration feeder as an object of predictive maintenance. The vibrating feeder consists of the following components: 1—vibrating frame, 2—spring, 3—vibrator, 4—motor vibrating device, and 5—motor. Such elements can be found in wind turbines.

The cost–benefit analysis calculations are useful, because they allow examining the options and making more informed choices. The vibration feeder cost–benefit evaluations of the component repair/replace alternatives corresponding to the estimations (5.24) and (5.25) are shown in Table 5.4.

The optimization problems (5.26)–(5.28) and (5.29)–(5.31) are used to formulate corresponding tasks for step 1 and step 2 to define the "best" alternative from the available ones.

Step 1: The optimal strategy *to repair or to replace the machine as a whole* takes into account optimization problem (5.26)–(5.28) involving the cost–benefit evaluations from Table 5.4:

$$\max\left\{\begin{array}{c} x(0.85+0.70+0.65+0.90+0.80+)+ \\ +y(0.75+0.80+0.75+0.85+0.70) \end{array}\right\} \tag{5.32}$$

subject to

$$x + y = 1 \tag{5.33}$$

$$x, y \in \{0, 1\} \tag{5.34}$$

The solution results of combinatorial optimization task (5.32)–(5.34) are shown in Table 5.5.

According to solutions results data in Table 5.2, the optimal maintenance strategy is to repair the machine. This means step II should be executed.

Step 2: The optimization problem (5.29)–(5.31) is transformed into the following optimization task:

Table 5.4 Matrix of evaluations of alternatives

	Cost–benefit estimations				
Alternatives	Component 1	Component 2	Component 3	Component 4	Component 5
Repair	0.85	0.70	0.65	0.90	0.80
Replace	0.75	0.80	0.75	0.85	0.70

Table 5.5 Solution results for step 1

Objective function value	x (repair)	y (replace)	Decision
3.90	1 (yes)	0 (no)	Repair

Table 5.6 Optimal component maintenance strategy

Alternatives	Component 1	Component 2	Component 3	Component 4	Component 5
x_i (repair)	1 (yes)	0 (no)	0 (no)	1 (yes)	1 (yes)
y_i (replace)	0 (no)	1 (yes)	1 (yes)	0 (no)	0 (no)

$$\max\{ \quad (0.85x_1+0.75y_1)+(0.7x_2+0.8y_2+)++(0.65x_3+0.75y_3)+(0.9x_4+0.85y_4)++(0.8x_5+0.7y_5) \tag{5.35}$$

subject to

$$\forall i \in \{1,2,3,4,5\} : \sum_{i}^{5} x_i + y_i = 1 \tag{5.36}$$

$$x_i, y_i \in \{0, 1\} \tag{5.37}$$

The solution results of optimization task (5.35)–(5.37) defining the particular strategies for each component are shown in Table 5.6.

The results of the first step execution of the proposed algorithm show that the optimal maintenance decision is to repair the machine (Table 5.5). According to the algorithm of predictive maintenance decision-making, step 2 should be executed too. This is accomplished by a single run of formulated optimization task. Its solution results in Table 5.6 suggest that component 1, 4, and 5 should be repaired, while components 2 and 3 should be replaced by new ones.

The example of vibration feeder is used just for illustration of the proposed approach. In practice, defining an optimal predictive maintenance decision strategy for other machines consisting of much more components will be a complex mathematical optimization problem. In such cases, the proposed intelligent approach to optimal predictive maintenance strategy defining could be a better alternative to intuitive decision-making.

The units in cost–benefit analysis are standard monetary units, so costs and benefits can be directly compared. In some cases where it is difficult to express the benefits in money, cost-effectiveness analysis can be used as a cost minimization technique. It should be noted that the model allows the use of other objective functions in the evaluation of the maintenance strategy of machine components.

5.3 A Concept of Intelligent e-Maintenance Decision-Making

The failure of one minor wind turbine component can cause escalated damage to a major component, which can increase repair and or replacement costs. That is why, the reliability-centered maintenance approach could be considered as a promising

Fig. 5.9 Stages of ISO 13381-1: 2004 standard

approach to the management of wind turbine operations since it includes an appreciation of the impact of faults on operations (Singh et al., 2015). The condition-based maintenance enables real-time diagnosis of impending failures and prognosis of future equipment health, where the decision to perform maintenance is reached by observing the "condition" of the system and its components (Peng et al., 2010). Condition-based e-maintenance method is used to reduce the uncertainty of maintenance activities and is carried out according to the need indicated by the equipment condition. Maintenance decisions depend on actual measured abnormalities and incipient faults, and the prediction of the trend of equipment deterioration (Yam et al., 2001). The condition-based maintenance is conceived to detect the onset of a failure, avoiding critical damages of high-cost components before they might happen, thus reducing overall maintenance costs. Possible faults are detected by monitoring representative parameters by signal analysis techniques and comparing signals during normal and abnormal conditions (Velarde-Suarez et al., 2006; Charbonnier et al., 2005). Therefore, e-maintenance avoids failure by detecting early deterioration or potential failures involving condition-based maintenance. The ISO 13381-1: 2004 standard prescribes to start with monitoring, followed by diagnostic, prediction, and posterior actions as shown in Fig. 5.9.

E-maintenance is the synthesis of two major trends in today's society: the growing importance of maintenance as a key technology and the rapid development of information and communication technology (ICT). ICT and in particular web and agent-based technologies for data acquisition, data transfer, and integration in condition monitoring and maintenance of mechanical systems are becoming more and more important (Mathew et al., 2006; Chowdhury & Akram, 2011). E-maintenance has been discussed in one of four ways: (1) a maintenance strategy (i.e., a management method), (2) a maintenance plan, (3) a maintenance type, or (4) a maintenance support (Muller et al., 2008). E-maintenance as a maintenance type describes replacing traditional maintenance with technological maintenance where predictive maintenance is done with the help of artificially intelligent systems that provide only condition monitoring and predictive prognostic functions. Continuous insights into the present and future health of machines and their components, as well as the information flow infrastructure, enable the move to e-maintenance based on intelligent prognostics, where maintenance actions are synchronized with the overall operation of the system as well as with the necessary maintenance resources and spare parts (Lee et al., 2006).

5.3.1 Intelligent e-Maintenance Decision Support Systems

E-maintenance intelligent system aims to recommend maintenance actions is based on information collected through condition monitoring—data acquisition, data processing, and maintenance decision-making. That means to obtain relevant data to equipment's health, to analyze collected data, and to recommend efficient maintenance policies, based on data acquired and processed in previous steps. Collected data via e-maintenance program represent the measurements dealing with the health condition of the physical machine (like vibration data). Condition monitoring is suitable for engineering equipment management since it is able to identify potential failure symptoms and suggest actions before any operational interruptions occur. Conditions of machines can be investigated through various tests and then decision has to be made whether the machine should be repaired or replaced or decide whether further in-depth test is needed. The first step in data processing is cleaning data so that it can be used for efficient analysis and modeling. With respect to condition monitoring data, errors may be caused by the sensor itself, whereupon sensor fault isolation is the best solution. Once data are cleaned, the next step is data analysis. Maintenance decision support is the last step of e-maintenance. Diagnostics and prognostics are the two main techniques used to perform efficient maintenance policies and decision-making. Within diagnostics, several statistical approaches have been developed, as well as artificial intelligence approach—a computational model that imitates the human brain structure, and model-based approaches.

The main objective of e-maintenance system is to assess machine condition and predict any impending failure, through data acquisition and the trending of the results. Potential problems can be identified through monitoring, comparing its actual performance with design capacity. Combining the data and knowledge from different sources allows maximizing the useful content of the information. The integration of condition-based maintenance technologies in e-maintenance system assists the users in automating the maintenance process from point of alert to completion of the maintenance work order.

The concept of an intelligent e-maintenance decision-making system is realized via four basic layers as shown in Fig. 5.10 (Borissova & Mustakerov, 2013).

The first layer represents the condition monitoring module which collects and processes data acquiring and information fusion combined with signal processing. The signal processing of sensor data is one of the most important functions of structural health monitoring systems for the goal of predictive maintenance. The input raw data is adjusted (the data is removed to get rid of noises and is normalized if required) in the preconditioning of data stage. Immediately after, useful information is extracted. The condition monitoring module derives directly from routinely collected condition monitoring signals from the sensors and from historical records to produce prediction outputs directly in terms of condition monitoring data. The results from signal processing are collected in an actual sensor database for additional processing and feature extraction and selection, pattern recognition, and information fusion. The external data sources as the Internet, other networks, and

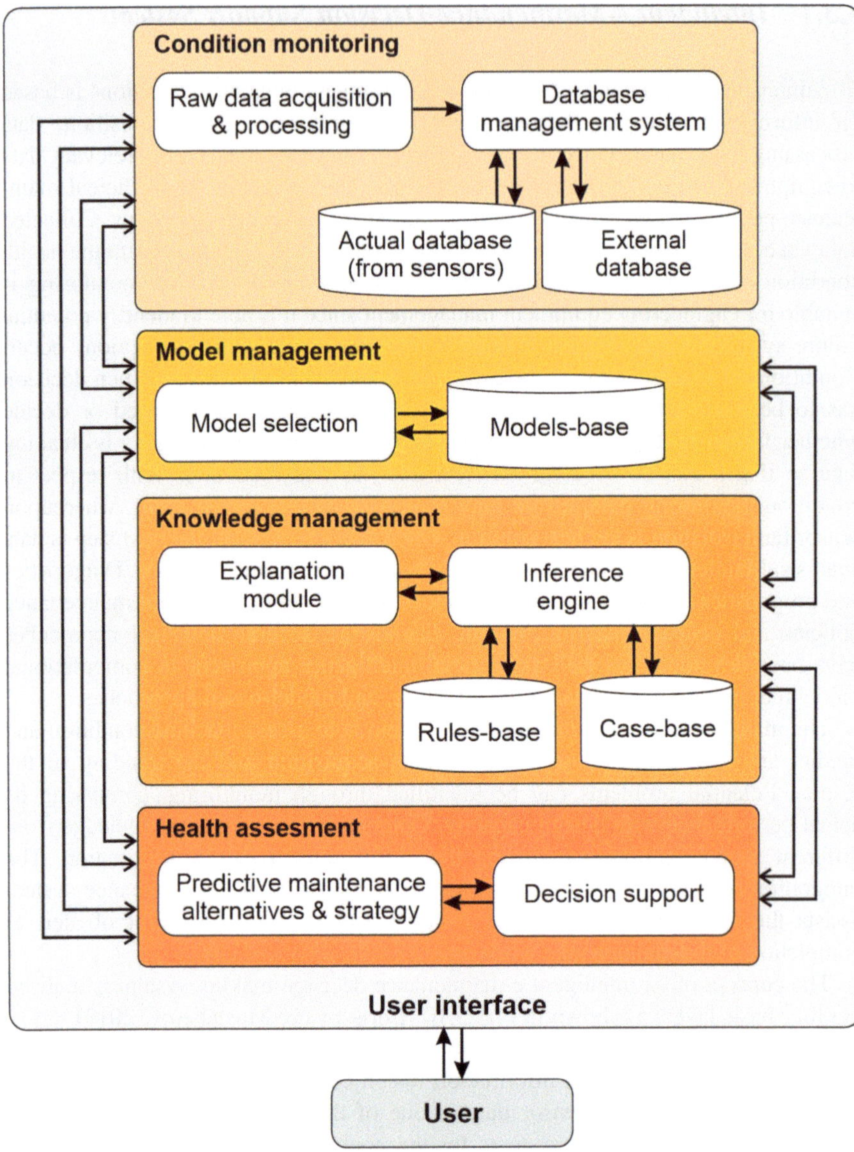

Fig. 5.10 Architecture of an e-maintenance decision support system

information systems are also used. The condition monitoring layer should also be able to generate alerts based on preset operational limits or changes in the trend. The collected data are assessable through database management system to aid in the decision-making process.

The second layer consists of module for model management. Some of models can estimate the time to failure of a machine as well as forecasting the fault condition.

The way in which a prognostic model is developed differs therefore from the method for building an explanatory model. The prognostic models focus on the search for a combination of factors which are as strongly as possible related to the outcome. Accurate prognostic models are based on algorithms that are capable of predicting future component failure rates or performance degradation rates. Development of strategies for assessment prognostic modeling of machinery working life involves various methods including structural system reliability, probabilistic-based life cycle assessment and maintenance, optimization of multiple criteria under uncertainty, and integration of monitoring in life cycle management.

A knowledge management subsystem is proposed as a third layer in intelligent e-maintenance system. The engineering condition-based maintenance is a specific area which is characterized by the availability of past information about the possible problems, their diagnosis, assessment of situations, and effective solutions. That information can be incorporated into an expert system used for structure health management to recommend courses of action and decisions. This layer is composed of knowledge management system, knowledge base (rules and cases base), inference engine, and explanation subsystem. The knowledge management system is responsible for knowledge extraction and storing in a knowledge base in the form of rules or cases. It also interacts with DM to show him the results of incorporated inference engine and how these results are obtained through the explanation subsystem. Knowledge extraction is conducted primarily during interviews with experts, field visits, service order, manuals, technical documentation and actual operations. During the interview process, conversations were recorded in detail and then transformed into acceptable format for diagnosing rules. Usually, rules are expressed in the form: *IF* condition, *THEN* consequence. The outcomes can also be used to test other conditions/rules, or even add a new fact to the knowledge base. These rules can be specific domain rules or heuristic rules and can be chained together using logical operators. The knowledge consists of concepts, objects, relationships, and inference rules. The set of rules constitute the knowledge base. Another part of the knowledge base stores a set of problems and answers (cases) needed for case-based reasoning. Case-based reasoning solves new problems by retrieving relevant prior cases and comparing and adapting them to fit new situations. Case-based reasoning is to find the case that is most closely related to the new problem and present a case's solution as an output, with suitable modifications. Case-based reasoning can be effective even if the knowledge base or domain theory is incomplete. Certain techniques of automated learning, such as explanation-based learning, work well when only a strong domain theory exists, whereas case-based reasoning can use many examples to overcome the gaps in a weak domain theory while still taking advantage of the domain theory. These characteristics of case-based reasoning make it appropriate for diagnosis, prognosis, and predictive maintenance. The inference engine applies the knowledge base to the particular fact under consideration to derive conclusions that can be used by the DM. An important feature is the possibility to justify the conclusions by the explanation subsystem. It should answer the questions of type "how the conclusion is reached" or "why the conclusion is reached" or "trace the conclusion" and could be a great help for decision-maker to take decisions about the maintenance activities. The expert system included in the DSS compensates for the user's limitations about some expertise areas. It makes use of all available expert resources to present the most complete picture of the problem possible.

Health assessment layer is intended to receive data from different condition monitoring sources. The primary focus of the health assessment is to prescribe if the health of the monitored component, subsystem, or system has degraded. Health assessment decision-making relies heavily on the availability of relevant information in the right format and at the right time. On this layer, data from all the prior layers are taken into account. The primary focus is to calculate the future health of equipment (machine), taking into account the future usage prognostics or the so-called remaining useful life.

A similar architecture can be used to build a digital twin of a wind farm where the health assessment layer will give information about the predicted wind energy. The purpose of this digital twin is to predict the generated wind power from the wind power plant in virtual reality considering the same wind conditions in reality. This is needed to preserve the energy balance considering the generated and consumed electricity. The working principle of the prediction system is to collect weather data in real time through sensors and after sorting substitute it directly into a forecasting model (Liu et al., 2023). Through the obtained predicted value of the digital twin, the power grid planning can be realized effectively and the wind energy can be fully utilized.

Forecasting deterioration of the engineering systems characteristics during the time and proper decision-making is associated with considerable uncertainty. The DM in uncertain conditions has an idea about the goals to be achieved, but information about alternatives and future events is incomplete. Usually, there is no sufficient data to assess the risk of each alternative.

In cases like these, the DM preferences can be considered as a function called preference utility (optimization) function $f(x)$, on the set of possible alternatives $x = \{x_1, x_2, \ldots, x_m\}$ under environment states $s = \{s_1, s_2, \ldots, s_n\}$.

The DM can assess the usefulness of alternatives x_i, $i = 1, 2, \ldots, m$ by using an optimization approach as:

$$x_{\text{opt}} = \max f(x) \tag{5.38}$$

subject to

$$x = \{x_1, x_2, \ldots, x_m\} \tag{5.39}$$

$$s = \{s_1, s_2, \ldots, s_n\} \tag{5.40}$$

The following optimization criteria are used according to the decision-maker's point of view in the health assessment layer in Fig. 5.10.

Wald Criterion In this case, the decision-maker selects the strategy associated with the best possible worst outcome regardless of whether probabilities are available or not. For each alternative solution x_i ($i = 1, 2, \ldots, m$), the worst output min E_{ij} ($j = 1, \ldots, n$) is defining. Next, an alternative solution is determined for which the minE_{ij} ($j = 1, \ldots, n$) has maximum magnitude (Harrell, 2001; Pawitan, 2001):

$$x_{\text{opt}} => \max\ \min\ E_{ij}\ (i = 1, 2, \ldots, m), (j = 1, 2, \ldots, n) \tag{5.41}$$

Savage Minimax Regret The Savage minimax regret criterion is looking at a small loss of efficiency due to missed opportunities and is calculated by the relation:

$$R_{ij} = \mid E_{ij} - \max\ E_{ij} \mid \tag{5.42}$$

Based on a "regret matrix" which compares (subtracts) the highest outcomes of each strategy from other outcomes (Linares, 2002). The Wald solution rule (maximin/minimax) is applied to this new matrix to gain the minimax regret solution. Optimum would be the minimum losses value R_{ij} among all alternatives:

$$x_{\text{opt}} => \min\ \max\ R_{ij}\ (i = 1, 2, \ldots, m), (j = 1, 2, \ldots, n) \tag{5.43}$$

Laplace Criterion The core of Laplace principle is based on the fact that if there is no information to determine a condition as more likely than another, then the optimal solution is determined as:

$$x_{\text{opt}} => \max{}_i \left(\sum_{j=1}^{n} \frac{E_{ij}}{n} \right) \tag{5.44}$$

Here, all possible states have equal probability when no other information is available (Cabulea & Aldea, 2004).

Hurwicz Criterion This decision criterion is a simplified version of Laplace principle and involves the identification of the worst and best outcomes for each strategy. Under certain probabilities of particular states, the arithmetic average of the results of the best solutions is taken (Cabulea & Aldea, 2004). The optimal solution is determined taking into account both a minimum and maximum profit:

$$x_{\text{opt}} => \max \left\{ \alpha \max\ E_{ij} + (1 - \alpha) \min\ E_{ij} \right\} \tag{5.45}$$

where α is optimism coefficient ($0 < \alpha < 1$). When $\alpha = 0$, the Hurwicz solution is the same as the pure Wald solution; when $\alpha = 0.5$, it corresponds to the equivalent antagonistic and friendly environment and in case of $\alpha = 1$ corresponds to the maximum favorable environment.

5.3.2 Numerical Results

To illustrate numerically the advantage of the proposed intelligent concept for decision-making, an easy-to-understand example is described. A manufacturing company (such as those that produce components for wind turbines) is equipped with various machines, which have different reliability requirements, safety levels, failure effects, and impacts on the company's gains. Let us consider three maintenance alternatives (to repair, to replace, to do nothing) for each of the machines. The maintenance decision depends on the machines' health assessment (high, average, low). The alternatives' benefits evaluations E_{ij} are shown in Table 5.7.

Using Wald criterion (pessimistic) the minimum value on each table row is chosen and then the maximum of them is selected; i.e., the decision is:

$$x_{\text{opt}} = \max\{30; 20; -50\} = 30$$

The Savage minimax regrets are calculated as:

$$r_{i1} = \max\{60; 40; 80\} - E_{i1}$$

$$r_{i2} = \max\{30; 20; 10\} - E_{i2}$$

$$r_{i3} = \max\{65; 70; -5\} - E_{i3}$$

The calculations' results are shown in Table 5.8.

$$x_{\text{opt}} = \min\{20; 40; 75\} = 20$$

This criterion indicates that alternative A1 has the lowest regret and is the best choice.

Applying Laplace criterion results in calculation of:

Table 5.7 Alternatives' benefits evaluations

Alternatives	Benefits evaluations		
	P1	P2	P3
A1	60	30	65
A2	40	20	70
A3	80	10	–5

Table 5.8 Savage minimax regrets

Regrets			Minimax
20	0	5	20
40	10	0	40
0	20	75	75

$$x_{\text{opt}} = \max\left\{\left(\frac{1}{3}(60+30+65)\right);\left(\frac{1}{3}(40+20+70)\right);\left(\frac{1}{3}(80+10-5)\right)\right\}$$
$$= \max(51.6;43.3;28.2) = 51.6$$

The decision is to choose alternative $A1$ with expected benefit of 51.6.

According to the Hurwicz criterion, an optimism coefficient value of 0.25 is set. It is used to calculate the following values:

$$x_{\text{opt}} = \max\{(0.2*65+0.8*30),(0.2*70+0.8*20),(0.2*80+0.8*(-15))\}$$
$$= \max(37;30;12) = 37$$

Again the decision is to choose alternative $A1$ with expected benefit of 37.

Combining the information of all layers with optimization approaches would assist the decision-maker in his/her decision-making for engineering systems' maintenance under uncertainty or incomplete information conditions.

The described concept is built on four abstraction levels, to serve as a reference for developing intelligent e-maintenance decision support system. It extends the concept of a decision support system with expert system advantages. The advances of expert systems embrace symbolic reasoning and explanation capabilities. This is an efficient addition to making strategic decisions through intelligent interpretations of various alternatives. The essence of the proposed concept of an intelligent e-maintenance decision support system is the ability to integrate actual information from structure health monitoring and structure health modeling modules with knowledge management. The condition-based maintenance decisions rely not only on decision-maker preferences but also on the knowledge of the actual sensor database, model base, and rule base knowledge. The health assessment turbine's layer allows the DM to assess the usefulness of different alternatives by using optimization approaches to forecast the deterioration of the system during the time.

In conclusion, leveraging the information from monitoring technologies and predictive capabilities to determine the system condition and remaining lifetime, it is possible to provide criteria for optimal decision-making with regard to the selection of operation and maintenance activities of wind turbines.

5.4 Integrated Group Decision-Making Approach for Evaluation and Choice Under Uncertainty Conditions

Decision-making and policymaking have become a complex process not only due to their very nature but also due to various conditions of uncertainty. Planning and investing in projects such as wind farms involves making many decisions under various uncertain conditions. The main paradigm of decision-making is the subjective human factor that cannot be avoided as the decisions are made by humans.

Therefore, any attempt to overcome such subjectivism is to be encouraged to reduce bigger and wrong decisions with subsequent accidents (Klapproth, 2008; Moura et al., 2017). The usage of group decision-making aims to involve more experts from different areas to reduce the subjectivism that arises in activities such as planning and investment in wind farm projects.

During the decision-making process, a set of different alternatives with very different likely consequences must be analyzed, i.e., presence of uncertainty in decision-making. In some situations, the proposed models of multicriteria decision-making by fuzzy criteria can be used (Peneva & Popchev, 2006, 2008). When making decisions in conditions of uncertainty, it is assumed that the DM has an idea of the goals to be achieved, but the information about the alternatives and future events is incomplete. In the decision-making process, the preferences of DM can be represented by a utility function *f*(a) over a set of alternatives $A = \{a_1, a_2, \ldots, a_m\}$ in different states $S = \{s_1, s_2, \ldots, s_n\}$.

5.4.1 Algorithm for Evaluation Considering Different Criteria to Cope with Uncertainty Conditions

The proposed decision-making algorithm for evaluation involving different criteria to cope with uncertainty conditions is shown in Fig. 5.11.

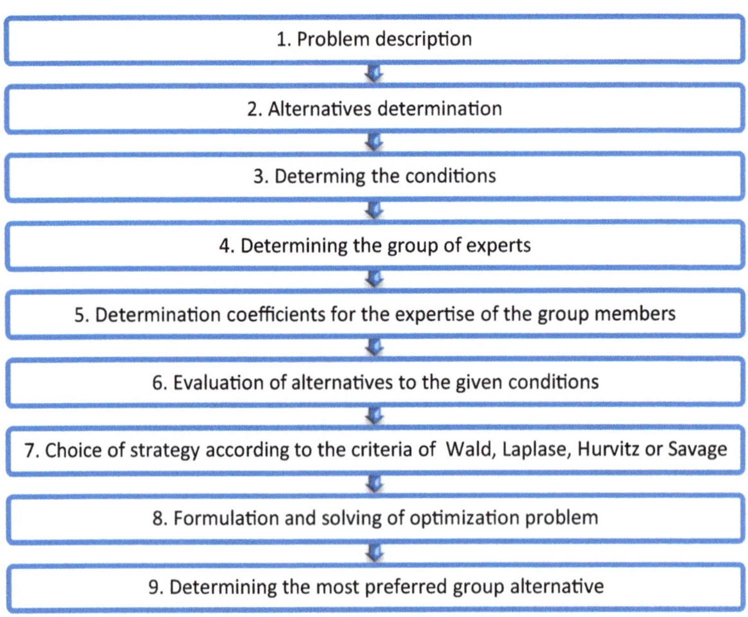

Fig. 5.11 Algorithm for group decision-making under uncertainty conditions

The first stage for the implementation of group decision-making in the conditions of uncertainty is related to the description of the specific problem for choosing an alternative. Stage 2 is related to identifying the acceptable alternatives appropriate to the problem at hand. In stage 3, the possible states of the environment in which the identified alternatives are implemented are determined. Depending on the specifics of the identified problem, a group of experts capable of evaluating the alternatives is selected in stage 4. Once the group of experts are determined, the respective weighting coefficients are to be determined in accordance with the expertise and the degree of importance of each DM in the group. Stage 6 is related to the process of alternatives assessment toward identified conditions and to the point of view of each expert in the group. To cope with uncertainty in stage 7, one of the well-known Wald, Laplace, Hurwitz, or Savage criteria is to be chosen to determine the optimal strategy. Next, a corresponding optimization task for choosing an alternative in the conditions of group decision-making is to be formulated and solved in stage 8. In the last stage (9), the most preferable group alternative is determined.

To realize the group decision-making under uncertainty conditions, it is needed to formulate mathematical models able to aggregate experts' evaluation with respect to the selected criterion to overcome uncertainty.

5.4.2 Mathematical Models for Group Decision-Making Considering Uncertainty When Choosing Specialized Software

As a quantitative analytical tool, revenue and expenditure analysis can be used to support decision-making (Boardman et al., 2010). In this case, a utility function is formulated based on the estimation of expected revenues and expenses. In the general case, the utility function is:

$$\mathrm{CBE} = \left(\frac{\mathrm{Costs}}{\mathrm{Benefits}}\right) \tag{5.46}$$

The overall Costs could be expressed as a sum of different aspects from purchasing to installation and maintenance. The Benefits can be expressed by revenue related to some activities that contribute to the reducing labor costs, automated processes, etc.

Using the generalized utility function (5.46) and the abovementioned criteria to cope with uncertainty, four modified group decision-making models are formulated as follows (Borissova & Dimitrova, 2021):

Group decision-making model based on Wald criterion (strategy):

$$\text{maximin} \sum_{i=1}^{M} \sum_{k=3}^{K} \lambda^k \text{CBE}_{ij}^k \tag{5.47}$$

subject to

$$\sum_{k=1}^{K} \lambda^k = 1 \tag{5.48}$$

where CBE_{ij}^k expresses evaluation of the i-th alternative toward j-state from k-expert and λ^k represents the importance of a particular expert by corresponding weighted coefficient for the expertise.

Wald's principle is based on a pessimistic view and conservative moderation. It consists of the fact that for any of the chosen strategies, the objective circumstances will always represent the most unfavorable situation. For each alternative solution, the worst outcome can be determined.

Group decision-making model based on Laplace's criterion (strategy):

$$\max\left(\sum_{i=1}^{M} \sum_{k=1}^{K} \frac{\lambda^k \text{CBE}_{ij}^k}{M}\right) \tag{5.49}$$

subject to relation (5.48).

According to the Laplace's criterion, when is no other additional information all possible states are considered equally probable.

Group decision-making model based on Hurwitz's criterion (strategy):

$$\max\left\{\begin{array}{c} \alpha \max \sum_{i=1}^{M} \sum_{k=1}^{K} \lambda^k \text{CBE}_{ij}^k + \\ +(1-\alpha) \min \sum_{i=1}^{M} \sum_{k=1}^{K} \lambda^k \text{CBE}_{ij}^k \end{array}\right\} \tag{5.50}$$

subject to relation (5.48) and α is the coefficient of optimism $(0 < \alpha < 1)$.

The value of the coefficient $\alpha = 0$ corresponds to environment considered to be completely antagonistic, $\alpha = 0.5$ corresponds to an equivalent environment (neither antagonistic nor friendly), and at a value of $\alpha = 1$, the medium is the most favorable. In essence, the Hurwitz's criterion is a simplified version of the Laplace principle, namely—with certain probabilities of the individual states, the arithmetic mean of the results of the best decisions is taken.

Group decision-making model based on the Savage criterion (strategy):

$$\text{minimax} \sum_{i=1}^{M} \sum_{k=1}^{K} \lambda^k R_{ij}^k \tag{5.51}$$

subject to relation (5.48) and one more relation

$$\forall i = 1, 2, \ldots M : \left(\forall k = 1, 2, \ldots M : R_{ij}^k = \left| \text{CBE}_{ij}^k - \max \text{CBE}_{ij}^k \right| \right) \tag{5.52}$$

where R_{ij}^k represent the regret as a result of opportunity loss if A_i is chosen and state S_j happens in accordance with k-th expert point of view.

It is possible to use different ranges for the coefficients that express the importance of DMs' opinions according to the particular problem, background, and expertise, for example between 0 and 10 or range from 1 to 100. In this situation, a proper normalization has to be done to get compatibility between dimensionless weighted coefficients (λ^k) and the CBE.

5.4.3 Numerical Testing

The applicability of the proposed models for group decision-making in conditions of uncertainty is made for a specific example of choosing specialized software. Such type of software is an integral part of wind projects and could be used for design and planning of wind farms projects (windPRO[1]), for wind turbine blade design and aerodynamic simulation (QBlade[2]), for real-time diagnosis (DewesoftX[3]), for low-carbon planning, implementation, monitoring, and reporting (The RETScreen® Clean Energy Management Software platform[4]), etc.

Due to the dynamics of the economic environment, the companies' prospects can be generally represented by three possible situations: increasing revenues, reducing revenues, or maintaining the current state. Therefore, the subject of the decisions is in the conditions of uncertainty, as the possible situations are known, but it is not known which of them will come true and there is no information about the probabilities for the realization of these situations. The selection of the most appropriate software system could be made through cost–benefit analysis as a tool to determine whether the investment decision is good taking into account the abovementioned possible situations. For the goal, the following representation of the cost–benefit relation is formulated:

---

[1] https://www.emd-international.com/windpro/

[2] https://qblade.org/

[3] https://dewesoft.com/applications/wind-turbine-monitoring

[4] https://natural-resources.canada.ca/maps-tools-and-publications/tools/modelling-tools/retscreen/7465

Table 5.9 Modified matrix for group decision-making under the conditions of uncertainty

Experts (DMs)	Weighted coefficient for DMs' expertise	Alternatives	Conditions		
			Increase	Reduction	Unchanged
E-1	0.25	A-1	0.52	0.83	0.75
		A-2	0.53	0.86	0.72
		A-3	0.54	0.84	0.71
E-2	0.35	A-1	0.56	0.87	0.78
		A-2	0.55	0.80	0.74
		A-3	0.58	0.81	0.72
E-3	0.40	A-1	0.62	0.85	0.72
		A-2	0.60	0.88	0.76
		A-3	0.61	0.82	0.70

$$\text{CBE} = \frac{\text{Costs}}{\text{Benefit}} = \frac{C_{\text{ac}} + C_{\text{cust}} + C_{\text{inst}} + C_{\text{test}} + C_{\text{staff}} + C_{\text{proc}} + C_{\text{file}} + C_{\text{un}} + C_{\text{lc}}}{R_{\text{lcr}} + R_{\text{stock}} + R_{\text{lc}} + R_{\text{ap}}}$$

The overall Costs are expressed as a sum of the costs of product acquisition (C_{ac}), customization (C_{cust}), installation (C_{inst}), post-installation testing (C_{test}), staff training (C_{staff}), file conversion (C_{file}), uninstallation of the old system (C_{un}), and loyalty policy (C_{lc}). Revenue is related to labor costs reducing (R_{lcr}), reducing the need to maintain the stock (R_{stock}), improved reliability through a new policy for loyal customers (R_{lc}), and the use of more automated processes (R_{ap}).

Thus, the use of cost-benefit analysis makes it possible to determine the value of CBE as a result of the realization of certain situations.

For the purpose of a small company, it is needed to make a decision to implement new software for the design and planning of wind farms project. The higher management together with the chief information officer determined three suitable providers (alternatives) from which the selection should be done. For brevity, these suppliers are considered as three alternatives labeled A-1, A-2, and A-3. In addition, three experts are also determined to conduct the evaluation. These experts are as follows: an expert in the database (E-2), one end-user of software (E-1), and one expert from the IT support team (E-3).

The obtained cost–benefit estimations by taking into account the above relation for the given three providers of software for the design and planning of wind farm projects under the possible three situations for increasing, decreasing, and maintaining the current state of the company are shown in Table 5.9.

The first two columns of Table 5.9 contain the weighted coefficients about the point of view importance of each expert (DM) from the formed group.

Using the data from Table 5.9 and formulated group decision-making models in the previous section based on the Wald, Laplace, and Hurwitz criteria, corresponding tasks are formulated and solved. The Savage criterion requires drawing up a regret matrix. The values of the elements of this matrix represent the losses due to missed opportunities. The calculation is performed according to the point of view of the DMs from the group according to the following formula:

Table 5.10 Regret matrix as a result of opportunity loss

Experts (DMs)	Alternatives	Regret as result of opportunity loss		
E-1	A-1	0.02	0.03	0
	A-2	0.01	0	0.03
	A-3	0	0.02	0.04
E-2	A-1	0.02	0	0
	A-2	0.03	0.07	0.04
	A-3	0	0.06	0.06
E-3	A-1	0	0.03	0.04
	A-2	0.02	0	0
	A-3	0.01	0.06	0.06

Table 5.11 Two cases for the expertise of group members

Cases	Weighted coefficient for the experts' expertise		
	λ^1 for E-1	λ^2 for E-2	λ^3 for E-3
Case-1	0.25	0.35	0.40
Case-2	0.34	0.46	0.20

$$R_{ij}^k = \left| \mathrm{CBE}_{ij}^k - \max \mathrm{CBE}_{ij}^k \right|$$

The values of regret resulting from lost profits are shown in Table 5.10.

The described above problem is numerically tested for two cases that express different combination between DMs' opinions importance as shown in Table 5.11.

The Case-1 illustrates the scenario in which expert E-1 is the most important with a weight of 0.4, closely followed by expert E-2 with a weight of 0.35 and expert E-1 the least important with a weight of 0.25. Case-2 represents the experts' importance in the following order E-2, followed by E-1 and finally E-3.

The formulated four models for group decision-making are based on criteria behind a certain strategy for dealing with uncertainty. Therefore, each of the models can be considered as a model of the respective specific strategy. The results obtained from solving the respective optimization tasks based on the proposed models for group decision-making, taking into account two scenarios for the importance of DM's opinions, are illustrated in Fig. 5.12.

When the Wald criterion is used, the obtained solutions identified the alternative A-3 as the most preferred in both cases of weights for the experts in the group. Choosing this strategy, the worst possible consequences of each strategy are considered, choosing the least bad or the best of them, according to the point of view of the experts in the group. The choice of this strategy determines the worst possible consequences of each condition and choosing the least bad or the best of them, according to the point of view of the experts in the group.

According to the Laplace criterion, it is assumed that all objective conditions have the same probability of occurrence due to a lack of other grounds. The result is obtained by calculating the arithmetic mean for each strategy, according to the views of the experts in the group and the determined weighting factors, choosing the

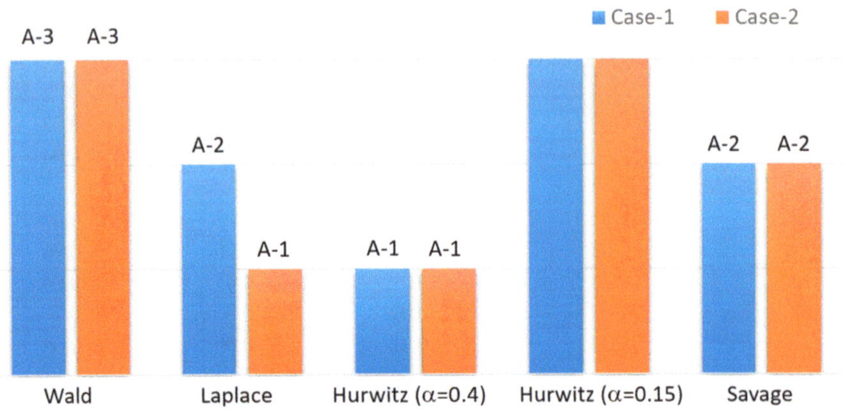

Fig. 5.12 Selected alternative under different criteria to overcome the uncertainty conditions and different weights for DMs' importance

strategy with the highest score. When using the first combination of weights for experts by Case-1 ($\lambda^1 = 0.25$; $\lambda^2 = 0.35$; $\lambda^3 = 0.40$) the solution determines A-2 as the most preferred alternative, while in the second combination of weights noted as Case-2 ($\lambda^1 = 0.34$; $\lambda^2 = 0.46$; $\lambda^3 = 0.20$) the most preferred alternative is A-1.

Unlike Wald and Laplace criteria, the Hurwitz criterion requires the determination of the so-called coefficient of optimism and the coefficient of pessimism, which are applied to each strategy, and the choice is determined by the highest result obtained. Two cases corresponding to two different values for the coefficient of optimism ($\alpha = 0.40$ and $\alpha = 0.15$) are considered. When using both combinations of weights for experts, for the specific example, and a value of the coefficient of optimism $\alpha = 0.40$, the choice for the most preferred is the alternative A-1, and when using $\alpha = 0.15$, the solution identifies alternative A-3 as the most preferred alternative.

The use of Savage's criterion requires the compilation of a regret matrix as a result of lost profits, choosing this alternative in which maximum losses are minimal. In the particular numerical experiment for both use combinations (Case-1 and Case-2) of weights for the opinions of experts, the most preferred alternative is A-2.

According to the Wild and Hurwitz ($\alpha = 0.15$) strategies and particular input data, the final group decision is to choose alternative A-3 (Fig. 5.12) in both cases for the importance of DMs' opinions. If the strategy behind the Savage principle is chosen, the final group decision is to select the alternative A-2 for both cases for DMs' opinions importance, but according to the strategy behind Laplace principle, the importance of DMs' opinions determines alternative A-2 for Case-1 and alternative A-1 for Case-2 (Fig. 5.12). The alternative A-1 is also preferable group decision in accordance with the Hurwitz strategy and coefficient of optimism $\alpha = 0.40$.

As a result of the analysis, it is found that the use of different strategies according to the principles of Wald, Laplace, Hurwitz, and Savage, and in combination with the experts' point of views (Case-1 and Case-2), leads to a different selection of group alternative. It is therefore important to choose in advance the most appropriate decision-making strategy in the face of uncertainty. In summary, it can be concluded that the results of numerical testing of the proposed models for group decision-making in conditions of uncertainty show the applicability of the described modifications based on the optimization criteria of Wald, Laplace, Hurwitz, and Savage.

Using the proposed modified optimization strategies of Wald, Laplace, Hurwitz, and Savage allows us to consider experts' opinions with different importance. These opinions are based on the usage of the calculated value for costs–benefits ratio for each particular alternative. The identified components that make up the costs and benefits are applied in the selection of specialized software for wind energy. It is shown that the final group decision depends not only on the used strategy according to the principles of Wald, Laplace, Hurwitz, and Savage, but is influenced also by the introduced weighed coefficients expressing the experts' opinion importance when aggregating the final group decision.

References

Akhshabi, M. (2011). A new fuzzy multi criteria model for maintenance policy. *Middle-East Journal of Scientific Research, 10*, 33–38.

Boardman, A., Greenberg, D., Vining, A., & Weimer, D. (2010). *Cost benefit analysis: Concepts and practice. The Pearson series in economics* (4th ed.). Pearson.

Borissova, D., & Dimitrova, Z. (2021). An integrated group decision-making approach considering uncertainty conditions. *Business Information Systems, 1*, 307–316. https://doi.org/10.52825/bis.v1i.52

Borissova, D., Mustakerov, I., & Grigorova, V. (2011). Engineering systems maintenance by optimal decision making strategies under uncertainty conditions. *Problems of Engineering Cybernetics & Robotics, 63*, 14–21.

Borissova, D., & Mustakerov, I. (2013). A concept of intelligent e-maintenance decision making system. In *2013 IEEE INISTA*, Albena, Bulgaria (pp. 1–6). https://doi.org/10.1109/INISTA.2013.6577668

Bouvard, K., Artus, S., Berenguer, C., & Cocquempot, V. (2011). Condition-based dynamic maintenance operations planning & grouping. Application to commercial heavy vehicles. *Reliability Engineering and System Safety, 96*(6), 601–610. https://doi.org/10.1016/j.ress.2010.11.009

Cabulea, L., & Aldea, M. (2004). Making a decision when dealing with uncertain conditions. *Acta Universitatis Apulensis – Mathematics - Informatics, 7*, 85–92.

Carroll, J., McDonald, A., & McMillan, D. (2016). Failure rate, repair time and unscheduled O&M cost analysis of offshore wind turbines: Reliability and maintenance of offshore wind turbines. *Wind Energy, 19*(6), 1107–1119. https://doi.org/10.1002/we.1887

Cavalcante, C. A. V., Ferreira, R. J. P., & Teixeira de Almeida, A. (2010). A preventive maintenance decision model based on multicriteria method PROMETHEE II integrated with Bayesian approach. *IMA Journal of Management Mathematics, 21*(4), 333–348. https://doi.org/10.1093/imaman/dpn017

Charbonnier, S., Garcia-Beltan, C., Cadet, C., & Gentil, S. (2005). Trends extraction and analysis for complex system monitoring and decision support. *Engineering Applications of Artificial Intelligence, 18*, 21–36. https://doi.org/10.1016/j.engappai.2004.08.023

Chowdhury, S., & Akram, A. (2011). E-maintenance: Opportunities and challenges. In *Proceedings IRIS* (pp. 68–81).

Dragomir, E., Gouriveau, R., Dragomir, F., Minca, E., & Zerhouni, N. (2009). Review of prognostic problem in condition-based maintenance. In *European Control Conference (ECC)*, Budapest, Hungary, 2009 (pp. 1587–1592). https://doi.org/10.23919/ECC.2009.7074633

Faulstich, S., Hahn, B., & Tavner, P. J. (2011). Wind turbine downtime and its importance for offshore deployment. *Wind Energy, 14*(3), 317–470. https://doi.org/10.1002/we.421

Ferreira, R. J. P., Teixeira de Almeida, A., & Cavalcante, C. A. V. (2009). A multi-criteria decision model to determine inspection intervals of condition monitoring based on delay time analysis. *Reliability Engineering & Systems Safety, 94*, 905–912. https://doi.org/10.1016/j.ress.2008.10.001

Gandhare, B., & Akarte, M. (2012). Maintenance strategy selection. In *Proceedings of Ninth AIMS International Conference on Management* (pp. 1330–1336).

Ginevicius, R. (2011). A new determining method for the criteria weights in multicriteria evaluation. *International Journal of Information Technology and Decision Making, 10*(6), 1067–1095. https://doi.org/10.1142/S0219622011004713

Gomes, G. F., Mendéz, Y. A. D., da Silva Lopes Alexandrino, P., da Cunha, S. S., & Ancelotti, A. C. (2018). The use of intelligent computational tools for damage detection and identification with an emphasis on composites – A review. *Composite Structures, 196*, 44–54. https://doi.org/10.1016/j.compstruct.2018.05.002

Harrell, F. E. (2001). *Regression modelling strategies*. Springer-Verlag.

Klapproth, F. (2008). Time and decision making in humans. *Cognitive, Affective, & Behavioral Neuroscience, 8*, 509–524. https://doi.org/10.3758/CABN.8.4.509

Lee, J., Ni, J., Djurdjanovic, D., Qiu, H., & Liao, H. (2006). Intelligent prognostics tools and e-maintenance. *Computers in Industry, 57*, 476–489. https://doi.org/10.1016/j.compind.2006.02.014

Li, Z. N., Tang, J., & Li, Q. S. (2004). Optimal sensor locations for structural vibration measurements. *Applied Acoustics, 65*, 807–818. https://doi.org/10.1016/j.apacoust.2003.12.007

Linares, P. (2002). Multiple criteria decision making and risk analysis as risk management tools for power systems planning. *IEEE Transactions on Power Systems, 17*(3), 895–900. https://doi.org/10.1109/TPWRS.2002.800991

Liu, F., Li, Z., Liang, M., Zhao, B., & Ding, J. (2023). Prediction method of non-stationary random vibration fatigue reliability of turbine runner blade based on transfer learning. *Reliability Engineering & System Safety, 235*, 109215. https://doi.org/10.1016/j.ress.2023.109215

Luo, J., Namburu, M., Pattipati, K., Qiao, L., Kawamoto, M., Chigusa, S. (2003). Model-based prognostic techniques. In *Proceedings of IEEE Systems Readiness Technology Conference* (pp. 330–340). https://doi.org/10.1109/AUTEST.2003.1243596

Mahadevan, M. L., Kumar, S. P., & Vinodh, R. (2010). Preventive maintenance optimization of critical equipments in process plant using heuristic algorithms. In *Proceedings of Conference on Industrial Engineering and Operations Management*, Dhaka.

Mathew, A. D., Zhang, S., Ma, L., Earle, T., & Hargreaves, D. J. (2006). Reducing maintenance cost through effective prediction analysis and process integration. *Advances in Vibration Engineering, 5*(2), 87–96.

Merizalde, Y., Hernández-Callejo, L., Duque-Perez, O., & Alonso-Gómez, V. (2019). Maintenance models applied to wind turbines. A comprehensive overview. *Energies, 12*(2), 225. https://doi.org/10.3390/en12020225

Miettinen, K. (2008). Introduction to multiobjective optimization: Noninteractive approaches. In J. Branke, K. Deb, K. Miettinen, & R. Słowiński (Eds.), *Multiobjective Optimization. Lecture Notes in Computer Science* (Vol. 5252, pp. 1–26). https://doi.org/10.1007/978-3-540-88908-3_1

Mobley, K. R. (2010). *An introduction to predictive maintenance*. Elsevier Science.

Moura, R., Morais, C., Patelli, E., Lewis, J., & Beer, M. (2017). Human factors influencing decision-making: Tendencies from first-line management decisions and implications to reduce major accidents. In *ESREL'2017, Proc. of International Conference on Engineering Sciences and Technologies* (pp. 69–34).

Muller, A., Marquez, A. C., & Iung, B. (2008). On the concept of e-maintenance: Review and current research. *Reliability Engineering and System Safety, 93*, 1165–1187. https://doi.org/10.1016/j.ress.2007.08.006

Mustakerov, I., & Borissova, D. (2013). An intelligent approach to optimal predictive maintenance strategy defining. In *IEEE INISTA* (pp. 1–5). https://doi.org/10.1109/INISTA.2013.6577666

Mustakerov, I., & Borissova, D. (2014). Multi-criteria model for optimal number and placement of sensors for structural health monitoring: Lexicographic method implementation. *AMO – Advanced Modeling and Optimization, 16*(1), 103–112.

Nielsen, U. D., Lajic, Z., & Jensen, J. J. (2012). Towards fault-tolerant decision support systems for ship operator guidance. *Reliability Engineering & System Safety, 104*, 1–14. https://doi.org/10.1016/j.ress.2012.04.009

Pawitan, Y. (2001). *In all likelihood: Statistical modelling and inference using likelihood.* Oxford University Press.

Peneva, V., & Popchev, I. (2006). Models of decision making support by multicriteria problems in fuzzy environment. *Advanced Studies in Contemporary Mathematics, 12*(2), 291–308.

Peneva, V., & Popchev, I. (2008). Multicriteria decision making based on fuzzy relations. *Cybernetics and Information Technologies, 8*(4), 3–12.

Peng, Y., Dong, M., & Zuo, M. J. (2010). Current status of machine prognostics in condition-based maintenance: A review. *International Journal of Advanced Manufacturing Technology, 50*, 297–313. https://doi.org/10.1007/s00170-009-2482-0

Ren, Z., Verma, A. S., Li, Y., Teuwen, J. J. E., & Jiang, Z. (2021). Offshore wind turbine operations and maintenance: A state-of-the-art review. *Renewable and Sustainable Energy Reviews, 144*, 110886. https://doi.org/10.1016/j.rser.2021.110886

Rigopoulos, G., Askounis, D. T. H., & Metaxiotis, K. (2010). NeXCLass: A decision support system for non-ordered multicriteria classification. *International Journal of Information Technology and Decision Making, 9*(53), 53–79. https://doi.org/10.1142/S0219622010003622

Rokicki, E., Gradzki, R., Kulesza, Z., Cecotka, P., & Dec, K. (2023). Frequency and modeshape evaluation of steam turbine blades using the metal magnetic memory method and vibration wave propagation. *Mechanical Systems and Signal Processing, 192*, 110218. https://doi.org/10.1016/j.ymssp.2023.110218

Singh, S., Baglee, D., Michael, K., et al. (2015). Developing RCM strategy for wind turbines utilizing e-condition monitoring. *International Journal of System Assurance Engineering and Management, 6*, 150–156. https://doi.org/10.1007/s13198-014-0259-9

Swanson, D. C. (2001). A general prognostic tracking algorithm for predictive maintenance. In *Proceedings of Aerospace Conference, IEEE* 6 (pp. 2971–2977). https://doi.org/10.1109/AERO.2001.931317

Tavner, P. J., Xiang, J., & Spinato, F. (2007). Reliability analysis for wind turbines. *Wind Energy, 10*(1), 1–18. https://doi.org/10.1002/we.204

Tejada, A. (2009). *A mode-shape-based fault detection methodology for cantilever beams.* NASA/CR-2009-215721. https://ntrs.nasa.gov/citations/20090021633

Velarde-Suarez, S., Ballesteros-Tajadura, R., & Hurtado-Cruz, J. P. (2006). A predictive maintenance procedure using pressure and acceleration signals from a centrifugal fan. *Applied Acoustics, 67*, 49–61. https://doi.org/10.1016/j.apacoust.2005.05.006

Yam, R. C. M., Tse, P. W., Li, L., & Tu, P. (2001). Intelligent predictive decision support system for condition-based maintenance. *The International Journal of Advanced Manufacturing Technology, 17*, 383–391. https://doi.org/10.1007/s001700170173

Zhou, F., Yang, J., Pang, J., & Wang, B. (2023). Research on control methods and technology for reduction of large-scale wind turbine blade vibration. *Energy Reports, 9*, 912–923. https://doi.org/10.1016/j.egyr.2022.12.042

Chapter 6
Economics Aspects and Social Impact of Wind Energy: Determining the Cost of Wind Electricity and the Relative Share of Wind Energy Consumption

Abstract This chapter presents some indicators that can be used to evaluate the feasibility of investments in wind energy projects from different perspectives, such as economic aspects and the social impact. Some economic evaluation indicators of wind power projects are discussed. It is shown how the jobs created and the quality of working conditions as a result of the implementation of wind energy projects can be evaluated. The major groups of risks associated with wind energy are discussed. Finally, the two types of indexes concerning the balance of energy consumption formed by different combinations of conventional and renewable sources and the index for the relative share of wind energy consumption among other renewable sources are proposed.

Keywords Wind energy project indicators · Job creation · Type of wind energy projects risks · Energy consumption balance index

6.1 Wind Energy: Positive and Negative Impacts

Wind electricity generation in 2022 increased by a record 265 TWh (up 14%), reaching more than 2100 TWh.[1] This is the second highest growth among all renewable power technologies, behind solar PV. However, to get on track with the Net Zero Emissions by 2050 Scenario, which envisages approximately 7400 TWh of wind electricity generation in 2030, the average annual generation growth rate needs to increase to about 17%. Achieving this will require increasing annual capacity additions from about 75 GW in 2022 to 350 GW in 2030. Far greater policy and private sector efforts are needed to achieve this level of capacity growth, with the most important areas for improvement facilitating onshore wind and cost reductions for offshore wind.

It should be noted that the development of renewable wind energy (onshore and offshore) is strongly supported by European governments through new policies and

[1] https://www.iea.org/energy-system/renewables/wind

D. Borissova, *Decision-Making in Design, Maintenance, Planning, and Investment of Wind Energy*, International Series in Operations Research & Management Science 355, https://doi.org/10.1007/978-3-031-52219-2_6

various support schemes, notably under the European Green Deal and REPowerEU. The EU Strategy on Offshore Renewable Energy[2] sets targets for an installed capacity of at least 60 GW of offshore wind by 2030 and 300 GW by 2050.

Wind energy is clean, renewable, and eco-friendly which makes it one of the fastest-growing energy sources worldwide. The wind turbines do not cause air and water pollution during operation, hence little to no impact on human health and the economy. Wind energy has immense economic value as its fuel is free, and the operation cost is relatively low.

Low-carbon transition is a multifaceted challenge involving political, technical, and economic elements. Investments in new power generation and energy efficiency are important long-term decisions, locking in both public spending and climate impacts for decades.

The wind-generated electricity is one of the most carbon-efficient forms of clean electricity that has grown significantly over the past few years. The benefits of wind energy can be summarized in the following topics:

- *Better air quality*. Wind power plants reduce air pollution by displacing a more polluting form of electricity generation with wind-generated energy. In practice, they produce no CO_2 emissions. Wind power plants use electricity and generate pollution only when they are being built, maintained, and then demolished at the end of their lifespan.
- *Independent energy solution*. Wind energy a reliable and homegrown source of energy contributes to energy independence by diversifying the energy mix. Wind energy does not use natural resources while generating a lot of electricity.
- *No waste products*. Many of the materials used to build a wind farm can be recycled and reused to produce other goods and products.
- *Climate change mitigation*. Wind energy helps to limit emissions of harmful greenhouse gases and nitrogen oxides.
- *Job availability and employment opportunities*. The renewable energy sector employs more than 11.5 million people globally, and this number is expected to grow steadily in the coming years.
- *Minimal impact on wildlife and environment*. It is well documented that increased carbon in the atmosphere has profound negative effects on the Earth. Ocean acidification is just one example and is a major threat to our complex ecosystems.

Although the list of advantages of wind power is long, it also has some negative impacts.

- *Land use*. Wind farms require a lot of land to maximize their efficiency but, once construction is complete, land use becomes minimal.
- *Impact on some wildlife*. Turbine blades are massive and can pose a threat to flying animals such as birds and bats. If the land has to be cleared, wind farms can

[2] https://energy.ec.europa.eu/topics/renewable-energy/offshore-renewable-energy_en#:~:text=EU%20strategy%20on%20offshore%20renewable%20energy,-To%20ensure%20that&text=The%20strategy%20sets%20targets%20for,GW%2C%20respectively%2C%20by%202050

also disrupt the habitats of native animals. Modern technology has begun to prevent these situations by turning off the turbines during low winds, when bats are most active, and by creating natural migration routes for all winged and four-legged friends.

- *Noise pollution*. Wind farms can be loud and cause distress to people living near them. One method of solving this problem is by placing the farms in remote areas where their sound impact will be reduced.

For the implementation of a project for the building of a wind farm, it is necessary to go through the following stages in succession: research of the necessary conditions and their influence on the design of a wind farm; determining the economics of investing in such a project taking into account economic aspects and wind resources; decision-making for structural health monitoring and predictive maintenance of wind turbines; optimal cutting of the details and planning of their respective treatments for the production of wind turbines; and personnel planning for the construction of such a wind power plant, including determining the price for the construction of such a project. Last but not least, it is needed to make some technical-economic analysis about the project for the wind power plant.

Therefore, this chapter deals with some indicators that can be used to evaluate the feasibility of investments in wind energy projects from different perspectives, such as economic benefits and financial levels. In addition, an approach for assessment of created job and working conditions quality due to wind project realization is described. Some types of risks related to wind energy project performance are presented. The energy consumption balance index formed by various combinations of conventional and renewable sources is formulated and calculated, along with the index of the relative share of wind energy consumption among the other renewable sources.

6.2 Economic Evaluation Indicators of Wind Power Project

The technical-economic feasibility analysis of the wind power plant is to be completed in the second stage of planning and design. The main activities are related to conducting a site survey in order to analyze the feasibility of the construction of the project according to the planning investment costs, energy production throughout the life cycle, and determining the optimal scale of the power plant (Sim, 2023).

In the process of building a wind energy project, an accurate technical and economic assessment is necessary to serve as motivation for the implementation of the project and the achievement of profitability. In the analysis of wind energy projects, different indicators can be used to evaluate the feasibility of investments in such projects from different perspectives, such as economic benefits and financial levels. The used indicators for the economic analysis of wind energy investments include net present value (NPV), present value cost (PVC), PBP (pay-back period),

IRR (internal rate of return), ROI (return on investment), LCOE, etc. (Liu et al., 2023).

6.2.1 Net Present Value

The investment cost and the time horizon are usually known, but the cash flows and the discount rate need to be forecasted and estimated. The investment cost and the time horizon are usually known, but the cash flows and the discount rate need to be forecasted and estimated. NPV refers to the value of future cash flow discounted to the current moment at a discount rate r that can vary. The calculation of NPV depends on the estimation of annual cash flow that considers all variables related to initial capital investment, operation and maintenance cost, and decommissioning cost. The NPV can be calculated as follows (Bourhim et al., 2021):

$$\mathrm{NPV} = -pQ_0 + \sum_{t=1}^{T} \frac{Q_t}{(1+r)^t} \tag{6.1}$$

where Q_0 is the initial cash flow, p expresses the proportion of initial investment, r is the discount rate, and Q_t is the cash flow in year t of the whole life cycle that considers loan repayment, tax, depreciation, and other expenses.

The NPV approach to capital budgeting is widely adopted by practitioners for evaluating investment opportunities because of its relatively simple and intuitive calculations that are understandable to managers and decision-makers. It is not only widely accepted but economically rational as it takes into account the time value of money and provides reasonable decision-making criteria for project selection (Ilhan et al., 2022).

For example, to select the projects that maximize the portfolio value, subject to the budget constraint, the following combinatorial optimization model can be used:

$$\max \sum_{j=1}^{J} x_j \mathrm{NVP}_j \tag{6.2}$$

subject to

$$\sum_{j=1}^{J} x_j I_j \leq B \tag{6.3}$$

$$x_j \in \{0, 1\} \tag{6.4}$$

The investment cost of each project j is expressed by I_j while x_j are the binary decision variables assigned to each project. The objective function (6.2) seeks to select such projects that have better NVP. The constraint (6.3) makes sure the portfolio selected satisfies the restriction of the budget (B).

The simple NPV rule states that one could invest in all projects with a positive NPV and discard those with a negative NPV. In addition, the project should not be undertaken if it prevents the undertaking of other projects with higher NPV.

Along with NPV, IRR and PBP reflect the performance of project construction in terms of cash flow, fund recovery speed, and expected income under the determined market and policy conditions.

6.2.2 *Present Value Cost*

Present value cost (PVC) expresses an estimate of what a future cash flow is worth right now. All future cash flows are to be discounted to the present by an appropriate rate that takes into account the expected rate of return. PVC considers the total cost during the whole life cycle of the wind farm involving the impact of economic factors in the form of interest rates and inflation rates. The calculation of PVC can be done as follows (Ahmed Shata & Hanitsch, 2006; Soulouknga et al., 2020):

$$\mathrm{PVC} = C_I + C_{\mathrm{O\&M}}\left(\frac{1+i}{k-i}\right)\left(1-\left(\frac{1+i}{1+k}\right)^t\right) - S\left(\frac{1+i}{1+k}\right)^T \tag{6.5}$$

where C_I is the investment cost that includes capital cost and loan, $C_{\mathrm{O\ \&\ M}}$ is the operation and maintenance cost, i is the inflation rate, k is the interest rate, T is the lifetime of wind power plant, and S is the residual value of obsolete wind power plant equipment.

6.2.3 *Payback Time Period*

A payback period is widely used when long-term cash flows are difficult to forecast, because no information is required beyond the breakeven point (Coker, 2007). It may be used for preliminary evaluation or as a project screening device for high-risk projects in times of uncertainty. The payback time period is the time required for cash flow to be equal to the investment and can be calculated by the following equation (Khan et al., 2023):

$$\mathrm{PBP} = \frac{C - \mathrm{IC}}{\left(C_{\mathrm{en}} + C_{\mathrm{capa}} + C_{\mathrm{RE}} + C_{\mathrm{GHG}}\right) - \left(C_{\mathrm{O\&M}} + C_{\mathrm{fuel}}\right)} \tag{6.6}$$

where C is the total initial cost of project, IG is the incentives and grant provided for investment, C_{en} is the annual energy saving due to the system, C_{capa} is the annual capacity saving, C_{RE} is the annual renewable energy production credit income, C_{GHG} is the income from greenhouse gas reduction, $C_{O\ \&\ M}$ is the annual operation and maintenance cost, and C_{fuel} is the annual cost of fuel or electricity being used.

6.2.4 Return on Investment

ROI mainly reflects the comprehensive profit of energy investment, which refers to the ratio of pretax profit to total investment during project operation. By comparing the project ROI with the industry average ROI, the profitability of the project investment can be reflected (Kubiszewski et al., 2010). The expression of ROI is as follows:

$$\mathrm{ROI} = \mathrm{EBIT}/\mathrm{TI} \tag{6.7}$$

where the annual profit before tax during the operation period is expressed by EBIT and the total investment cost is denoted by TI.

6.2.5 Cost–Benefit Analysis

Cost–benefit analysis (CBA) is often used as an evaluation index for life cycle analysis, and the lower the CBA, the higher the investment value of the project. CBA involves the comparison of total costs and benefits associated with a project or policy, namely those reflected in market prices (private cost or benefit) and those experienced by the external economic and natural environment without directly influencing the market mechanism (external cost or benefit) (Strantzali & Aravossis, 2016). In this sense, CBA represents an extension of conventional financial analysis that is able to remove market distortions and indicate those actions with the lowest social cost or the highest net social benefit. This CBA is often restricted to only monetized aspects such as capital, operation, and maintenance costs.

The development of wind power plants has many benefits, which can be divided into indirect benefits and direct benefits (Liu et al., 2021). Indirect benefits are related to improvement of national energy security, reducing energy imports from fossil fuels that lead to increasing the degree of energy self-sufficiency. In respect of the economic development, it can increase employment and promote social and economic sustainable development. In response to climate change, the wind power plants can reduce non-renewable energy consumption, reduce carbon dioxide emissions, and help achieve energy conservation and emission reduction goals. Direct benefits from sales of electricity can be calculated using the life cycle cost method using the cost–benefit ratio through the following formula (Kim et al., 2013):

$$C/B_{\text{ratio}} = \left(\sum_{t=1}^{T} \frac{C_{\text{CAPEX}}(t) + C_{\text{OPEX}}(t) + T(t)}{(1+r)^t} \right) / \left(\sum_{t=1}^{T} \frac{E(t)}{(1+r)^t} \right) \tag{6.8}$$

where $C_{\text{CAPEX}}(t)$ and $C_{\text{OPEX}}(t)$ are the total capital expenditure cost and operational expenditure for the year t, r is the annual discount rate, T_t is a corporation tax in the year t, and E_t is the profit in the year t.

When calculating the investment level of the wind power project using the economic evaluation indicator, the detailed information of the annual cash flow and the cost at each stage is required. This is an effective method to establish a life cycle cost model to estimate the cost and cash flow at each stage. The life cycle cost of wind farms can be divided into five parts: (1) predevelopment and consenting cost, (2) production and acquisition cost, (3) installation and commissioning cost, (4) operation and maintenance cost, and (5) decommissioning and disposal cost (Shafiee et al., 2016). The construction and installation cost of offshore wind farms are estimated for more than 30% of the total life cycle cost, which is 1.5 times that of onshore wind farms (Guo et al., 2022). The operation and maintenance cost of offshore wind power plants amounts to more than 20% of the total life cycle cost, while that of onshore wind power plants is only about 5% (Ren et al., 2021). From these studies, it is clear that the construction stage and operation stage are the main research objects of techno-economic analysis, while the predevelopment and consenting part and the decommissioning and disposal part with relatively low cost are mostly processed with simple cost estimation.

Initial capital investment in wind power projects includes the cost during the planning, design, and construction stages. Based on the wind power plant cost model, the initial capital investment can be divided into five separate parts (González et al., 2010):

$$C_{\text{capital}} = C_t + C_f + C_{\text{es}} + C_{\text{ci}} + C_{\text{pc}} \tag{6.9}$$

where C_t expresses the cost of wind turbine, C_f is the cost of foundation, C_{es} is the cost of electrical system, C_{ci} is the construction and installation cost, and C_{pc} is the preliminary engineering preparation cost.

Operation and maintenance cost accounts for a high proportion of the whole life cycle cost of the wind power plant and includes the management of the wind power plant, periodic maintenance plan, and equipment replacement. The operation and maintenance process of wind power plant can be divided into regular maintenance, daily operation and maintenance and component replacement (Ren et al., 2021). The operation and maintenance cost can be expressed as follows (Liu et al., 2023):

$$C_{\text{O\&M}} = C_{\text{rs}} + C_{\text{dom}} + C_r \tag{6.10}$$

where C_{rs} is the regular maintenance cost, C_{dom} is the daily operation and maintenance cost, and C_r expresses the cost related to replacement.

After reaching its operational life, the wind power plant needs to be decommissioned. The decommissioning cost includes cost for site cleaning C_{sc}, cost for equipment disassembly C_{ed}, and recovery cost C_{wr} that is calculated as follows:

$$C_{dc} = C_{sc} + C_{ed} + C_{wr} \tag{6.11}$$

According to traditional economic analysis, the decommissioning cost is usually calculated as 10% of the total initial investment cost. This method only makes a simple estimation of the decommissioning cost and cannot clearly reflect the key factors in the decommissioning cost and the differences between different types of wind power plants.

6.2.6 *Levelized Cost of Electricity from Wind*

In recent years, the renewable energy sector passes through many changes in terms of technological innovation, project financing, and the structuring of public support mechanisms. There are many different parameters to consider when calculating the production costs of the energy sectors, and this is the reason why they differ in different studies and institutions. Despite this, it is possible to divide these costs into three main categories:

- The investment cost "CAPEX" (capital expenditure) is money a company uses to purchase, maintain, or expand fixed assets. These fixed assets are non-current, non-liquid, long-term resources the company intends to use for more than a year. One of the most common types of fixed assets is property, plant, and equipment involved in the production process. In terms of wind energy, the CAPEX represents the cost of acquiring the land and building the energy production unit.
- The operational costs "OPEX" (operating expenses) are the costs that a company incurs for running its day-to-day operations and include the expenses for the upkeep and maintenance of the production unit.
- The cost of inputs: In contrast to the conventional power stations (gas, nuclear, coal, etc.), in the case of wind and sun power plants the costs are equal to zero.

It should be mentioned that wind and solar projects have a very large initial investment (i.e., CAPEX) compared to maintenance, taxes, administration, rents, etc. (i.e., OPEX). Most of the costs come from the infrastructure itself, not from its operation. For renewable energy that comes from wind and sun, the resource is usually free. The profitability of this type of renewable energy project is therefore heavily influenced by the cost of capital that will finance the CAPEX.

The levelized cost of electricity (LCOE) is primarily used by policymakers for long-term planning, as well as for devising incentive mechanisms. The LCOE model is an economic model that is used to calculate the expected cost of power from an unbuilt power plant. The parameter LCOE expresses the measure of the average net present cost of electricity generation for a specific system over its lifetime. It can be defined as the cost of an electricity plant divided by the expected generated electricity sold till the end of its lifetime.

LCOE could serve as a good indicator of cost-effectiveness because it can be calculated without requiring assumptions about the price at which the electricity can be sold to the grid or to an end user, as is the case when calculating the payback period or the net present value. This price affects directly the viability of an investment but varies significantly between different markets (by a factor of 10 or more) or over time. LCOE could be used as an indicator to compare different technologies, without any framework conditions affecting the assessment. With the use of LCOE, the financial viability in specific conditions can be indicated by just comparing directly the LCOE with the price at which electricity could be sold. The simplification formula for the calculation of LCOE is as follows:

$$\mathrm{LCOE} = C_{\mathrm{total}} / E_{\mathrm{total}} \tag{6.12}$$

where C_{total} refers to the wind power cost including investment and construction, operation and maintenance, and financial costs during the entire life cycle of the plant, while E_{total} refers to the power generation amount of the plant during the whole life cycle.

More precisely, the LCOE is defined as the price at which the generated electricity should be sold for the system to break even at the end of its lifetime and can be determined using the following equation (Papapetrou & Kosmadakis, 2022):

$$\mathrm{LCOE} = \left(\sum_{t=0}^{n} \frac{I_t + M_t}{(1+r)^t} \right) \Big/ \left(\sum_{t=0}^{n} \frac{E_t}{(1+r)^t} \right) \tag{6.13}$$

Here relation t expresses the year, and n is the expected plant lifetime, I_t is the investment expenditures in year t, M_t expresses the running costs—operations and maintenance in year t, E_t is the generated electrical energy in year t, and r denotes the discount rate.

One needs to decide whether to use the technical lifetimes of the technologies (which are different) or a common financial lifetime (e.g., the maturity period of the debt finance) of the technologies considered. Usually, the debt of a power plant is fully repaid before the end of its technical life, and hence one needs to assign a scrap value at the end of the financial lifetime. By default, the future fuel and emission costs are also to be discounted over the expected technical lifetime of the technologies.

The discount rate is used to depreciate the cash flows in order to account for the return that the investor would require, and it depends on aspects such as the general economic situation, returns of alternative investment opportunities, and real and perceived risks. This makes the calculations more complex, but it is of high importance considering the anticipated lifetime of 15–20 years or even more. When changing the discount rate, both the annualization of capital costs and the discounting of fuel and CO_2 prices are influenced. The discount rate influences technologies with high investment costs the most, for example, solar PV, wind power, and nuclear power.

The number of annual full load hours is used to express the utilization rate of the power plant. The assumptions regarding full load hours have a high influence on the LCOE due to contributions from fixed costs such as capital cost and fixed operation and maintenance cost. Large onshore wind turbines have a capacity factor of 34%, which is equivalent to 3000 full load hours, while the large offshore wind turbines are stated to have a capacity factor of 46–48%, which is equivalent to 4200 full load hours. In Denmark, large new solar PV systems are expected to have 1130 full load hours per year.[3]

In order to facilitate sound long-term decisions, the *LCoE Calculator*[4] is developed with the possibility to make specific comparisons of the average costs of conventional and new energy solutions. Focusing on not only project-specific costs (investment, O&M, fuels, etc.) but also system and society costs, the developed model gives a holistic assessment of future costs for countries across the globe. Based on an internationally widely used methodology, the LCoE Calculator permits comparison of different electricity generation technologies based on fossil, nuclear, or renewable energy.

The levelized cost of electricity production except the price of the produced electrical energy also considers the economic life of the plant including the costs incurred in the construction, operation, and maintenance and the fuel costs. The cost for fuel is equal to zero in wind power generation, and the wind turbine is directly delivered to the wind farm site, resulting in a short construction time (*t*). This results in the following form for the LCOE for wind power generation (Ragheb, 2017):

$$\mathrm{LCOE_{wind}} = \frac{\sum_{t=1}^{n} \left(I_t + M_t - PTC_t - D_t + T_t + R_t\right)\frac{1}{(1+r)^t}}{IF \sum_{t=0}^{t=n-1} P_t} \tag{6.14}$$

In this relation, t expresses the year, n denotes the duration of the generation period in years, I_t—investment in year t, M_t—operation and maintenance in year t, PTC_t—production tax credit in year t, D_t—depreciation credit in year t, T_t—tax levy in year t, R_t—royalties or land rents in year t, IF—intermittence factor, P_t—electrical generation capacity in year t, and r—discount rate.

The price of electricity from wind turbines does not vary according to their power factor. The price of electricity from wind farms can be extremely competitive, but it is considered less valuable to the grid because wind power is erratic.

[3] https://ens.dk/sites/ens.dk/files/Globalcooperation/vejledning_lcoe_calculator.pdf

[4] https://ens.dk/sites/ens.dk/files/Globalcooperation/lcoe_calculator_with_ee_module.xlsm

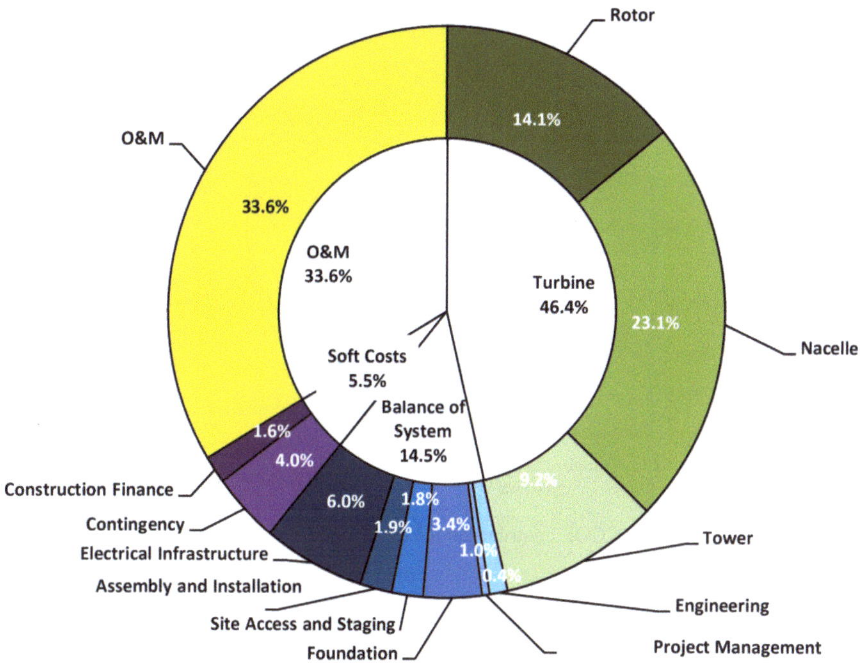

Fig. 6.1 Component-level LCOE contribution for the 2020 land-based wind reference project operating for 25 years

According to the report (Stehly & Duffy, 2022), the standard Annual Technology Baseline, LCOE equation[5] can be simplified for each technology. For wind energy, the following equation could be used to calculate LCOE:

$$\mathrm{LCOE}^{*}_{\mathrm{wind}} = \frac{(\mathrm{CapEx.FCR}) + \mathrm{OpEx}}{(AEP_{\mathrm{net}}/1000)} \tag{6.15}$$

where LCOE is the levelized cost of energy in \$/MWh, FCR is fixed charge rate in %, CapEx is the capital expenditures in \$/kW, $\mathrm{AEP}_{\mathrm{net}}$ is the net average annual energy production in MWh/MW/year, and OpEx is the operational expenditures in \$/kW/yr.

The authors found four major component LCOE categories—wind turbine CapEx, balance-of-system CapEx, finance CapEx, and OpEx—and many subcategories as shown in Figs. 6.1, 6.2, 6.3 and 6.4 (Stehly & Duffy, 2021).

These figures include wind turbine (e.g., wind turbine components), balance-of-system (e.g., development, electrical infrastructure, assembly, and installation), and financial costs (e.g., insurance and construction financing). Most of the land-based reference project LCOE (46.4%) is in the wind turbine itself, whereas the balance of

[5] https://atb.nrel.gov/

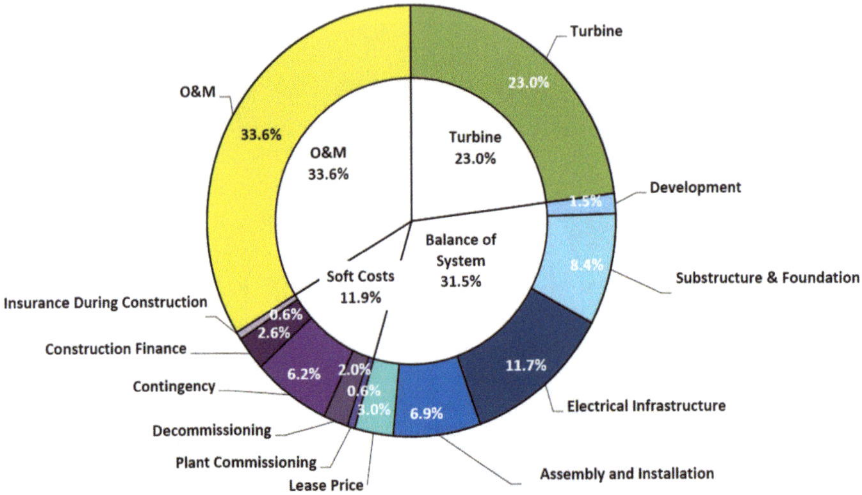

Fig. 6.2 Component-level LCOE contribution for the 2020 fixed-bottom offshore wind reference project operating for 25 years

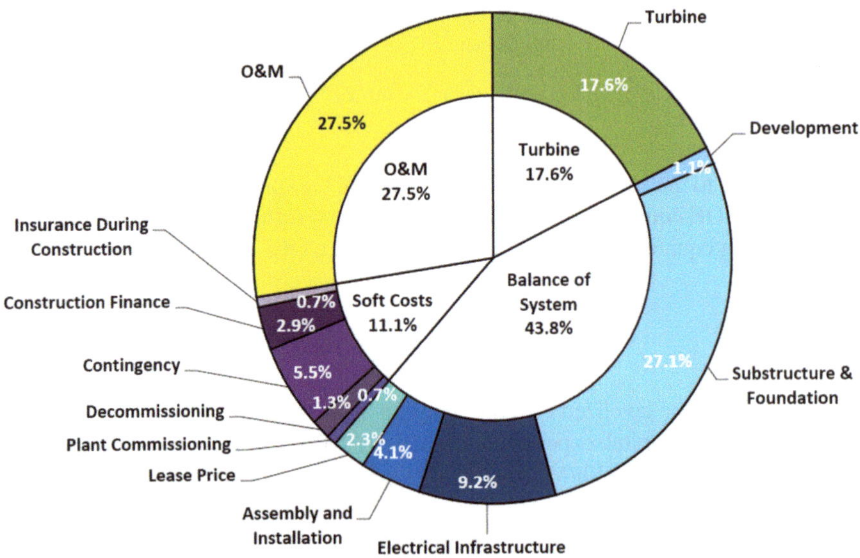

Fig. 6.3 Component-level LCOE contribution for the 2020 floating offshore wind reference project operating for 25 years

the system is the major contributor for the fixed-bottom and floating offshore reference projects, making up 31.5% and 43.8%, respectively. The LCOE categories for the representative distributed wind energy projects only consider the turbine, balance-of-system, and operation and maintenance (O & M) costs because there is

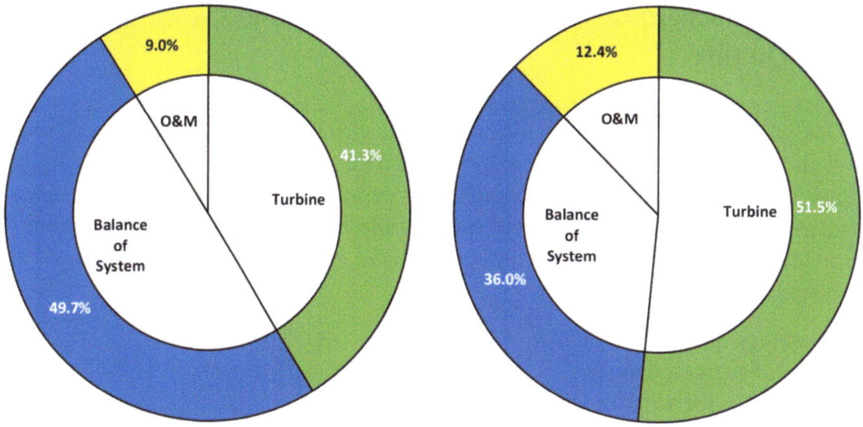

Fig. 6.4 Component-level LCOE contribution for the 2020 residential (left) and commercial (right) distributed wind reference projects operating for 25 years

limited empirical data on more detailed system cost breakdown for distributed wind projects. For the representative residential distributed wind energy project, the balance-of-system cost comprises a majority of the LCOE at 49.8%, whereas the commercially distributed wind project cost is accounted for by the wind turbine itself, contributing 51.6% of the LCOE.

To estimate the cost of wind power in the entire lifetime by calculating the cash flow, the LCOE could be used and based on the method of Xu (Xu et al., 2021), LCOE could be determined as follows (Wang et al., 2022):

$$\mathrm{LCOE}_j = \left(\mathrm{IC}_j - \frac{V_{Rj}}{(1+r)^n} + \sum_{n=1}^{N} \frac{\mathrm{AC}_{nj}}{(1+r)^n}\right)\left(\sum_{n=1}^{N} \frac{\mathrm{GE}_{nj} \times (1 - O_n)}{(1+r)^n}\right)^{-1} \tag{6.16}$$

$$\mathrm{AC}_{nj} = \mathrm{O\&M}_{nj} + \mathrm{IF}_{nj} + \mathrm{IE}_{nj} + \mathrm{Tax}_{nj} \tag{6.17}$$

where the initial construction cost is denoted by IC_j (one-time investment), including the construction and installation cost of turbines, land cost, etc.; V_{Rj} is the residual value of fixed assets; N is the lifetime of wind turbines; n is the n-th year in the life cycle of wind power; AC_{nj} is the annual cost in the n-th year, including O & M_{nj} cost (operation and maintenance cost), IF_{nj} (insurance premium), IE_{nj} (financial expense), and Tax_{nj} (tax); r is the discount rate; GE_{nj} is electricity generation in the n-th year; and O_n is the reduction rate of the wind turbine.

It should be noted that LCOE varies by technology, country, and project, based on the renewable energy resource, capital and operating costs, and technology performance.

6.3 Assessment of Job Creation and Working Conditions Quality

Wind power development provides not only a clean and sustainable energy source but also opportunities for economic growth and job creation. The OECD developed a framework (Organisation for Economic Co-operation and Development) to measure and assess the quality of jobs based on three objective and measurable dimensions to provide a comprehensive assessment. These three objective dimensions are related to:

- Earnings quality
- Labour market security
- Quality of the working environment

Earnings quality refers to earnings contributing to workers' well-being in terms of average earnings and their distribution across the workforce. Labor market security concerns the aspects of economic security related to the risks of job loss and its economic cost for workers. It is defined by the risks of unemployment and the income support received in case of unemployment. Quality of the working environment captures non-economic aspects of job quality and is summarized by the incidence of job strain that can impinge on workers' health and well-being.

To estimate the social impact of wind energy, it is necessary to have a proper measure. Buchmayr et al. (2022) propose an original measure of working conditions quality consisting of the absolute number of jobs created and working condition quality criteria based on the OECD job quality framework (Cazes et al., 2015). This measure is based on the number of created jobs for energy projects and is quantified using employment factors (EF) according to Rutovitz et al. (2015). In the case of the wind sector, the temporary employment $EF_{temp,\ annual}$ can be measured in job-years per MW, where one job-year stands for 1 year of work for one person. The stable employment EF_{st} can be expressed in continuous jobs per MW over the whole lifetime of the plant. In order to facilitate the comparability of the employment numbers, the temporary jobs measure could be averaged over the energy farm lifetime (LT) resulting in EF_{temp} in jobs/MW using the following relation (Buchmayr et al., 2022):

$$EF_{temp} = \frac{EF_{temp,annual}}{LT} \tag{6.18}$$

The factor of total employment represents the sum of both stable jobs and temporary jobs expressed in jobs/MW:

$$EF_{total} = EF_{temp} + EF_{st} \tag{6.19}$$

For more accurate calculations, employment factors for different energy technologies based on the average situation in the OECD member countries are proposed.

To correct the OECD average to the labor market situation in specific countries, the following adjustment factors can be used (Rutovitz et al., 2015):

$$\mathrm{EF}_{\mathrm{total,countyX}} = \frac{\mathrm{GDP}_{\mathrm{OECD}}/E_{\mathrm{OECD}}}{\mathrm{GDP}_{\mathrm{countryX}}/E_{\mathrm{countryX}}} \mathrm{EF}_{\mathrm{total,OECD}} \tag{6.20}$$

where (GDP/E) expresses the average labor productivity calculated by the relation between the gross domestic product (GDP) and the total number of employment (E) in this economy.

The assessment of the quality of working conditions can be made on the basis of the involved job profiles. The evaluation of employment according to the OECD job quality framework is based on criteria for earnings quality, labor market security, and quality of the working environment (Cazes et al., 2015).

The goal is the convergence of these criteria at the level of single job profiles, using the following indicators:

- Salary (S) is an indicator of the quality of income and is expressed in EUR/year of gross annual income.

$$S_w = \sum_i S_i \left(W_{\mathrm{temp},i} \frac{\mathrm{EF}_{\mathrm{temp}}}{\mathrm{EF}_{\mathrm{total}}} + W_{st,i} \frac{\mathrm{EF}_{\mathrm{st}}}{\mathrm{EF}_{\mathrm{total}}} \right) \tag{6.21}$$

- Education level (ED) is assessed using the European Qualifications Framework and is expressed in EQF levels between 1 and 8 (Bogicevic, 2014) as:

$$\mathrm{ED}_w = \sum_i \mathrm{ED}_i \left(W_{\mathrm{temp},i} \frac{\mathrm{EF}_{\mathrm{temp}}}{\mathrm{EF}_{\mathrm{total}}} + W_{st,i} \frac{\mathrm{EF}_{\mathrm{st}}}{\mathrm{EF}_{\mathrm{total}}} \right) \tag{6.22}$$

- Occupational risk indicator (OR) quantifies the quality of the work environment by providing a categorization into three levels: 1—high risk, 2—medium risk, and 3—low risk:

$$\mathrm{OR}_w = \sum_i \mathrm{OR}_i \left(W_{\mathrm{temp},i} \frac{\mathrm{EF}_{\mathrm{temp}}}{\mathrm{EF}_{\mathrm{total}}} + W_{\mathrm{st},i} \frac{\mathrm{EF}_{\mathrm{st}}}{\mathrm{EF}_{\mathrm{total}}} \right) \tag{6.23}$$

The weighting factors W_{temp} and W_{st} are determined by calculating the weighted average over several job profiles (i). The relevance of each job profile for the installation, operation, and maintenance of the energy plant is expressed as a weighting factor, as W_{temp} or W_{st}, in relation to the proportion of the total of temporary and stable jobs created. Temporary jobs are considered in relation to the life cycle of the enterprise and have a lower weight in the total employment factor, as their contribution to the security of the labor market is lower (Buchmayr et al., 2022).

For the wind energy sector, the corresponding wind employment factors for offshore and onshore wind farm are summarized by Rutovitz et al. (2015).

6.4 Types of Risks Associated with Wind Energy Projects

The transition to alternative energy sources continues to be a topic of extreme importance. In this regard, investments in wind energy receive a lot of attention, but the complex combination of risks, both financial and commercial, must also be considered. If the potential investment is related to the construction of wind farms, it is necessary to take into account various factors. The risk management process can be seen as a framework for the actions to be taken related to identifying, assessing, and controlling threats to capital, earnings, and operations. These risks are due to a variety of sources, including financial uncertainties, legal liabilities, technology issues, strategic management errors, accidents, and natural disasters. There are five main steps taken to manage risk that outline the risk management process, namely identification of risks, risk analysis, risk treatment, and monitoring and review (Cooper et al., 2005; Kendrick, 2009). The process of risk analysis includes identifying and quantifying uncertainties, estimating their impact on outcomes that they care about, building a risk analysis model that expresses these elements in quantitative form, exploring the model through simulation and sensitivity analysis, and making risk management decisions that can help to avoid, mitigate, or otherwise deal with risk.

Knowing the severity of the risk and the probability of the risk helps to allocate resources effectively. If businesses understand the risks that affect them, then they will know which risks need the most attention and resources and which the business can ignore. Risk management enables proactive action to mitigate vulnerabilities before major damage is done. Therefore, not only the business processes that are directly related to the project implementation are to be considered but also the external factors that can affect the project implementation in one way or another. The major groups of risks associated with wind energy projects are (1) development and construction risks; (2) economic risks; (3) political risks; (4) social risks; and (5) cyberattack risks as shown in Fig. 6.5.

A business project that can anticipate risk will always have an advantage. Foreseeing the financial risk will limit investments and focus on strengthening finances. If it is possible to assess the impact of the safety risk, a safe way of working can be developed, which is a major competitive advantage. The shortage of skilled workers, strained supply chains, and difficulties with digitalization and cyberattacks are currently among the biggest challenges facing the wind power industry. Cyberattacks on the monitoring and control software of wind farms are a serious threat to the security of supply (Staggs et al., 2017).

Therefore, the purpose of this section is to focus on the types of risks and their indicators that could affect the timely implementation of wind energy projects. The risk in the wind energy industry can be classified according to the wind project life

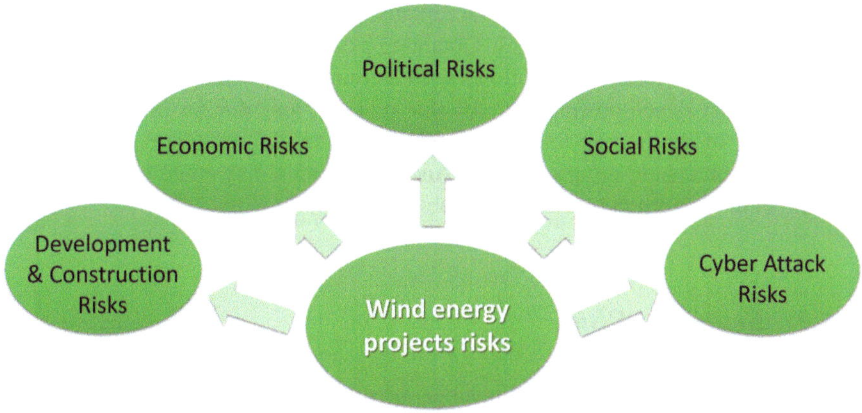

Fig. 6.5 Groups of risks associated with wind energy projects

cycle stages (Enevoldsen, 2016). The risks for distributed wind power investment can be classified into several groups according to different authors, namely political, economic, social, and technical (Zhou & Yang, 2020): economic and business risk, technical risk, political and policy risk, market risk, environmental risk, and social risk according to Kul et al. (2020) or strategic/business risk, transport/construction/completion risk, operation/maintenance risk, liability/legal risk, market/sales risk, counterparty risk, and political, policy, and regulatory risks (Gatzert & Kosub, 2016).

6.4.1 Development and Construction Risks

Development and construction risks encompass all aspects of installing wind turbines. The major technical risk is if the yield and efficiency are lower than the design value, leading to a revenue drop. There is a risk of lower revenues or higher maintenance costs due to the technology's novelty and unpredictability (e.g., faster degradation) (Egli, 2020). Bad weather during the construction and operation period can increase CAPEX, construction costs, and commissioning costs. The problem of transporting the blade can also be a challenge if the roads are narrow or in urban areas. The safety of the transport of the turbines is of particular importance, as any accident will lead to delays in time, and from there to financial losses. In the case of offshore wind projects, the logistics of turbine transportation and needed machinery for the foundation become more complex than onshore. Components of offshore wind power systems have large-scale, non-standard, and vulnerable characteristics, which require meticulous preplanning and complex synchronization operations. Organizational and human factors play an essential role in preventing and managing risks in the handling operations of offshore wind power components in ports (Lin & Lu, 2023).

Getting planning consent and an acceptable grid connection offer is always the biggest project risk for wind energy. Obtaining a connection to the grid is possible if the power distribution network allows such a connection. Therefore, there is a risk for the project development, i.e., risk for revenue loss due to project delay or cancellation of the license for the commercial operation and failure to obtain all required documents to get grid access. Another technical risk could be inconsistent coordination between contractors and sub-contractors.

6.4.2 Economic Risks

Upon starting commercial operations, operators have to analyze the vast variety of possible situations they face. In general, the greatest driver of exposure facing wind farms is equipment breakdown. This involves operating efficiency problems, routine operation and maintenance problems, operation and maintenance problems, force majeure such as catastrophic events, etc. Due to climate change and the impact of pollution, weather conditions sometimes become unpredictable and uncertain. The risk of operational failure can be seen when weather conditions are adverse such as floods, cyclones, and forest fires that can damage components and the system. Another type of economic risk is the possibility of an increase in land rent or interest rate. Scenarios are also possible where a drop in market demand or electricity pricing is observed. The wind energy industry is characterized by large investments and long capital recovery cycles and therefore requires investors to bear risks such as inflation, currency depreciation, and changes in interest rates. Equipment cost risk must be considered when assessing the risk of a wind farm construction project, as it represents a significant investment relative to the total cost of the entire project. It is determined that about 70.4% of the overall cost is for the turbine components for land-based wind farms, 33.6% for fixed-bottom offshore wind systems, and 23.3% for floating offshore wind systems (Stehly & Duffy, 2022). The two main drivers of resource risk are technology characteristics and new developments in a combination of improved assessment tools due to increasing data availability. As the wind is unpredictable, it is possible to reduce power generation and revenue. Price risk is the only risk that increased in relative importance from 2009 to 2013 and from 2013 to 2017 (Egli, 2020).

6.4.3 Political Risks

As renewable technologies expand around the world, so do the risks facing investors, especially as they shift their focus from developed economies to emerging markets. The rapidly evolving regulations and the politicization of wind and solar projects mean that assessing political risks is becoming an increasingly important step in the investment decision-making process. Among the political risks are a volatile regulatory environment, public procurement, and community relations.

One of the key challenges for investors is how to navigate the rapidly changing regulatory environment for wind projects. Understanding the regulatory context is critical to assessing return on investment. It is also one of the main factors in choosing where to invest among the growing number of countries looking to attract foreign direct investment in renewable energy sources. Governments around the world are upgrading their climate change mitigation targets and rolling out new constraints for attracting foreign direct investment in renewable energy projects.

More often country authority relies on competitive auctions for distributing additional renewable energy capacity to foreign investors. In countries with less developed public procurement systems, investors need to navigate complex decision-making processes while trying to preserve their finances and reputation.

Investors need to take into consideration local residents' concerns about the social and environmental impacts of large-scale renewable energy projects. There is also potential for such projects to become politicized at the regional and national levels, posing new risks to investors (Fischhendler et al., 2021). The active involvement of local people, both in terms of ownership and financing of wind energy projects, should be promoted through appropriate regulations (Maleki-Dizaji et al., 2020).

Other types of political risks are the changes in electricity price policy and/or reduction of investment subsidies. The most significant regulatory risk for investors is related to the stability of subsidy regimes. For many years, European governments have subsidized renewable energy projects to ensure their financial viability and attractiveness to private investors. However, the growing financial burdens of these subsidies and technological advances that have made renewable energy production cheaper are forcing governments to abandon subsidy schemes such as feed-in tariffs.

Investors can mitigate regulatory risks by conducting policy-focused due diligence and country-specific scenario planning early in the investment cycle.

6.4.4 Social Risks

The successful implementation of renewable energy depends not only on energy and economic policies and strategies and technical and infrastructure factors but also on social factors such as the dissemination of information and social acceptance. The environmental and social impacts of renewable energy projects on local communities represent some of the most persistent and high risks for investors, not only from a financial perspective but also from a reputational perspective. The involvement of local committees and NGOs in renewable energy project deployment is particularly important for garnering societal acceptance and usage of technologies that are perceived as new and unfamiliar. Therefore, investors need to understand how their projects affect local interests and engage with key opinion leaders, officials, and activists throughout the project's life cycle. The economic benefits and perceived environmental costs influence local acceptance of wind power most significantly. Local acceptance of wind power in some countries can be described as "not in my backyard, but not far away from me" (Guo et al., 2015). This means the acceptance rate is lowest when the wind energy source is located in their village or community, highest when the project is located in their county and city, and decreased for projects that are constructed further away.

Other types of social risks are related to the lack of skilled personnel for construction, operation, and maintenance. Lack of adequate expertise risks refers to the availability of experienced human resources and local knowledge and the maturity of wind energy technology. Operations and maintenance risk refers to the higher operations and maintenance costs due to the novelty and unpredictability of the technology. Therefore, training and human capacity building is also required for large-scale renewable energy production. Setting up a network for specialized capacity in the energy sector is critical, especially for the design, operation, and management of continually advancing technologies (Bamati & Raoofi, 2020).

6.4.5 Cyberattack Risks

With the increasing application of ICT, cyberattacks have become more prevalent against cyber-physical systems such as modern power grids and in particular wind power plants. Cyberattacks against the growing wind industry are rapidly increasing and occurring in parallel with increasing demand. By exploiting the vulnerabilities in cyber components, an attacker could intrude in the wind farm Supervisory Control and Data Acquisition (SCADA) system and Energy Management System (EMS), and maliciously trip one or multiple wind turbines. (Zhang et al., 2017).

To build cyber resilience into digital renewable energy systems, there is a need for understanding the areas of risk that include code vulnerabilities and misconfigurations in embedded software; unsecured APIs; management, control, reporting, and analysis systems; automation; remote access services; physical location; network traffic; Internet connection; legacy infrastructure of electricity grids; lack of regulation and security coordination; etc. (Schachinger, 2023).

End-to-end communication forms the basis for the automation of wind turbines. Thus, through self-optimization, automated wind turbines achieve a greater electricity yield. Measurement data and information about the status of turbines enable the diagnostic of devices in the turbine via wireless LAN (industrial wireless LAN). The simplified commissioning of series-produced Siemens wind turbine equipment relies on the PROFINET protocol running over the Ethernet as shown in Fig. 6.6.

Communication security and operation risk put forward a challenge to the data network of the wind power plant and different decisions are proposed (Lai et al., 2017; Yang et al., 2022). Management and control software, such as SCADA systems, and others that import, analyze, and visualize data from power sources are top targets for cyberattacks as they allow attackers to access the whole system. Systems that integrate data from third-party sources, such as from meteorological towers, offer another route for compromise. All of this means that secure remote

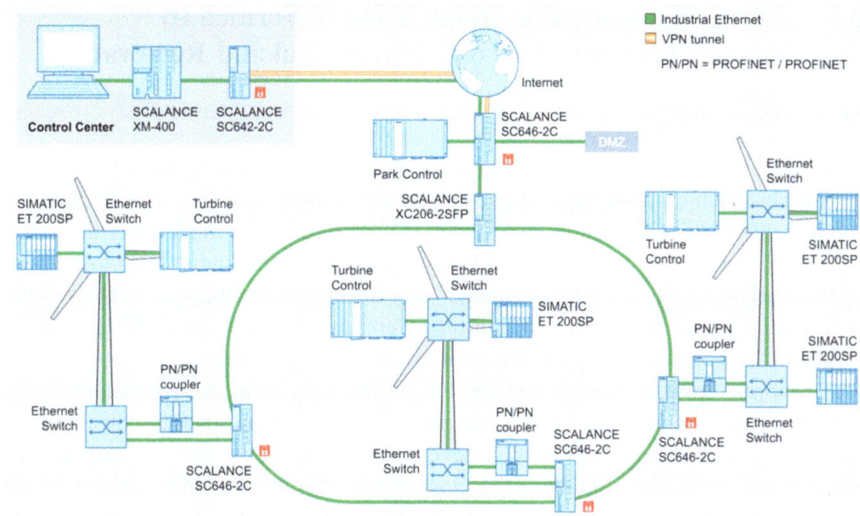

Fig. 6.6 Generalized wind farm communications infrastructure with Siemens equipment (https://www.siemens.com/global/en/industries/wind/equipment/industrial-communication.html)

access is critical for wind power plants, and a key method is the usage of multi-factor authentication. All data that moves across the network should be monitored and encrypted. In response to the growing threats posed by digitization and the rise of cyberattacks, the proposed NIS2 Directive[6] strengthens security requirements aimed at securing supply chains. In this regard, there are already some results about cybersecurity in wind farms. For example, a method using a discrete-time adaptive observer should be relevant for real-world settings to improve the security and reliability of wind farm operations (Ambarita et al., 2023), a reliability assessment technique for cyber-physical power systems that incorporates cybersecurity issues by considering non-normal random variables with nonlinear dependencies (Rostami et al., 2023), power system reliability assessment incorporating cyberattacks against wind farm energy management systems (Zhang et al., 2017), etc.

Security by design makes it possible to implement cybersecurity in the earliest phases of a new project's life cycle. Investors are therefore advised to think, buy, and design cybersecurity as an element of wind power plant functionality. Security requirements need to be established, which can then be used to select software and hardware that have the cybersecurity capabilities necessary to implement defensive, investigative, and remedial security controls aimed at protecting wind assets from cyberattacks.

[6]The NIS2 Directive: A high common level of cybersecurity in the EU, https://www.europarl.europa.eu/thinktank/en/document/EPRS_BRI(2021)689333

6.5 Energy Consumption Balance Index Formed by Various Combinations of Conventional and Renewable Sources

The ongoing climate change provoked the measure expressed by the EU-endorsed Green Deal, which aims for a climate-neutral continent by 2050. The energy mix today consists of a diverse range of sources—coal, oil, gas, nuclear, hydro, solar, wind, and biofuels. The energy transition we need today in moving from fossil fuels to low-carbon energy is a new challenge, very different from the past. In this regard, it is important to know the distribution of energy mix that could be used as an indicator for consumption from different primary energy sources.

Economic growth in recent decades has been driven by increased manufacturing activity, which in turn has been accompanied by significant energy consumption. Technological innovations, financial development, renewable energy consumption, economic growth, and population are closely related and have direct impacts on the ecological footprint (Usman & Hammar, 2021; Shan et al., 2021; Ahmad et al., 2021). Therefore, many countries have started thinking about alternative sources of energy, thereby mitigating environmental pollution without affecting economic growth (Nurul Hossain & Serletis, 2020; Tutak & Brodny, 2022). In this regard, some approaches are proposed to assess renewable energy investment projects (Delapedra-Silva et al., 2022) and some multicriteria decision-making approaches to determine the most reliable supplier for green building (Borissova & Atanassova, 2018). Another direction in scientific research is related to the determination of various indicators for measuring energy efficiency, such as the implementation of the energy trilemma index (Fu et al., 2021) or measuring the energy policy effectiveness (Gatto & Drago, 2021). Other authors investigate the impact of monetary policy uncertainty in the short and long run and found that monetary policy uncertainty has an asymmetric impact on energy consumption (Sohail et al., 2021).

In contrast to renewable sources, non-renewable energy resources emit high-carbon emissions, and on the other hand, they are subject to resource depletion, which affects the safety of the environment and the sustainability of resources. Therefore, it is necessary to adjust the energy mix by reducing the share of non-renewable energy and increasing that of renewable sources (Opeyemi, 2021). The second largest source of energy that has low-carbon emissions is nuclear power, preceded by hydroelectric power. This is why the European Union has agreed that to reduce carbon emissions, natural gas and nuclear power must be a large part of the continent's energy mix. In this regard, the European Commission classifies fossil gas and nuclear as "green" energy.[7] This corresponds to the European Green deal where the roadmap for the EU is to become climate-neutral by 2050.[8] The energy mix

---

[7] EU Taxonomy: The Dirty Politics of Greenwashing Energy, https://www.greeneuropeanjournal.eu/the-dirty-politics-of-greenwashing-energy/

[8] European Green Deal strives for climate neutrality in Europe by 2050, https://stateofgreen.com/en/news/european-green-deal-strives-for-climate-neutrality-in-europe-by-2050/

consists of different sources classified as renewable and conventional or nonrenewable. Among the renewable energy sources, the most commonly used is the solar, wind, and geothermal energy (Li et al., 2015). To achieve better energy effectiveness, different approaches for energy efficiency of buildings are proposed. Some of them rely on formulation of decision-making model for selection of green building project (Borissova, 2020), while others seek energy efficiency optimization of buildings in microgrids by optimizing initial investment costs for the building and seeking environmental pollution as minimum as possible (Guliashki et al., 2019). All this means that it is necessary to make an appropriate choice between renewable and conventional energy sources in order to obtain the desired energy balance. The energy balance itself is composed of all statistically significant energy products used by energy consumers of the entire economy allowing for monitoring the impacts of energy policies.[9] Therefore, the energy balance can be seen as a natural starting point for the study of the energy sector.

The energy mix could also be considered a policy tool composed of cost-efficiency and climate damage to assess the social welfare effects (Ahn et al., 2021). It is found that only a configuration of specific policy instruments targeting both energy efficiency and renewable energy can promote sustained success in the energy transition (Borozan, 2022). To reduce carbon emissions, a complex collection of different strategies, policies, and instruments is needed to support, navigate, and accelerate the energy transition. Integrating different energy transition concepts based on sustainable simulation and exploratory modeling and analysis approaches is also a possible solution to make policy mixes more robust (Castrejon-Campos et al., 2020). In addition, it is shown that the energy mix influences the environmental impacts and varies in each country (Nematchoua et al., 2020). Because of the dynamics of the energy mix, it is important to have a measure that provides information on how much the use of conventional energy sources has decreased and how much the use of renewable energy sources has increased. In this regard, this chapter is focused on the formulation of an objective measure to determine the consumption balance index between conventional and renewable energy. This measure can be used to measure the progress of decarburization.

Energy sources can be roughly divided into two different groups: conventional sources and renewable sources. The first group of energy sources has a strong impact on the global environment and is directly related to an increase in greenhouse emissions. The second group of energy sources emits less to no greenhouse gases. In order to be able to evaluate and determine the energy consumption balance formed by different energy sources, this chapter proposes the formation of an Energy Consumption Balance Index (ECBI). The mathematical representation of this

[9]Energy balance guide. Methodology guide for the construction of energy balances & Operational guide for the energy balance builder tool. 2019, https://ec.europa.eu/eurostat/documents/38154/4956218/ENERGY-BALANCE-GUIDE-DRAFT-31JANUARY2019.pdf/cf121393-919f-4b84-9059-cdf0f69ec045

index is composed of two parts, referring to two groups of energy sources, as follows (Borissova et al., 2022):

$$\mathrm{ECBI} = \frac{\sum_{i=1}^{N} a_i}{\sum_{j=1}^{M} b_j} \tag{6.24}$$

Here, index $i = \{1, 2, \ldots, N\}$ expresses the number of conventional energy sources, while index $j = \{1, 2, \ldots, M\}$ expresses the number of renewable energy sources. The values of parameters a_i represent the measure of consumed energy from different types of conventional sources such as coal, gas, etc., while the parameters b_j represent the measure of consumed energy from different types of renewable sources such as hydropower, wind, solar, etc. The values of both parameters a_i and b_j must have the same dimensions. For example, they could be expressed in kWh or in percentage. The value of the energy consumption balance index determined by (6.24) is expected to decrease as the share of renewable energy sources increases. This means that the desired trend over the years should show a decrease in the ECBI.

6.5.1 Determination of Energy Consumption Balance Index

In order to demonstrate the application of the proposed energy consumption balance index, the following energy sources are considered: coal, oil, gas, nuclear, hydropower, wind, solar, and other renewable sources. Primary energy consumption per capita (by source, 2021) from different sources for a representative sample including Bulgaria is shown in Table 6.1 (Ritchie et al., 2020).

It should be noted that primary energy is calculated by using the "substitution method" considering the inefficiency of the production of fossil fuels by converting energy from non-fossil fuels into the required energy if they had the same conversion losses as fossil fuels (Ritchie et al., 2020).

To evaluate the energy balance index (ECBI) using relation (6.24) there is a need to determine which sources will be considered conventional sources and which of them as renewable sources. In this chapter, two cases are considered as shown in Table 6.2.

Case-1 relies on the following combination of conventional energy sources such as coal and oil and renewable sources such as gas, nuclear, hydropower, wind, solar, and others. In this case, nuclear energy is considered a renewable energy source. In Case-2, gas and nuclear energy is considered conventional energy sources.

The calculated index considering the distribution between conventional and renewable energy sources described by Case-1 and using the data for 4 years 2021, 2020, 2015, and 2010 is shown in Fig. 6.7.

Table 6.1 Primary energy consumption per capita by source, 2021, in KWh

	Coal	Oil	Gas	Nuclear	Hydro-power	Wind	Solar	Other renewables
Bulgaria	8796	7899	4788	6001	1731	544	565	784
EU (27)	4202	13,298	8907	4128	2023	2289	944	1281
Europe	4344	12,134	13,450	3725	2826	1658	647	969
High-income countries	6743	21,509	16,275	3995	2795	1997	1103	989
Lower-middle-income countries	2359	1822	1801	111	281	73	86	70
Upper-middle-income countries	10,788	6218	5116	695	2221	835	401	337
World	5647	6498	5127	893	1420	691	343	301

Table 6.2 Calculated index of energy consumption balance formed by different combinations of energy sources

	ECBI formed by different sources							
	Case-1				Case-2			
	2021	2020	2015	2010	2021	2020	2015	2010
Bulgaria	1.16	1.09	1.38	1.53	7.58	8.26	8.37	12.12
EU (27)	0.89	0.87	1.14	1.17	4.67	4.48	6.11	8.77
Europe	0.71	0.71	0.90	0.91	5.52	5.28	6.98	9.41
High-income countries	1.04	1.00	1.30	1.42	7.05	6.95	9.45	12.19
Lower-middle-income countries	1.73	1.66	1.82	1.66	11.95	11.92	16.33	17.82
Upper-middle-income countries	1.77	1.84	2.29	2.55	6.01	6.11	8.24	9.94
World	1.38	1.38	1.68	1.77	6.59	6.78	9.04	10.96

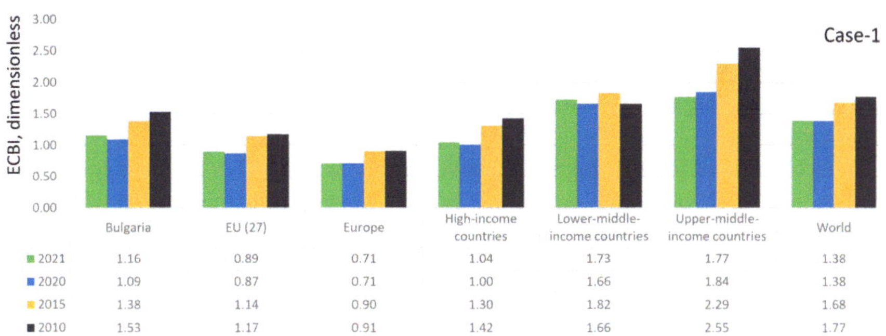

Fig. 6.7 Combination of energy sources under Case-1: (1) Conventional sources: coal and oil; (2) Renewable sources: gas, nuclear, hydropower, wind, solar, and others

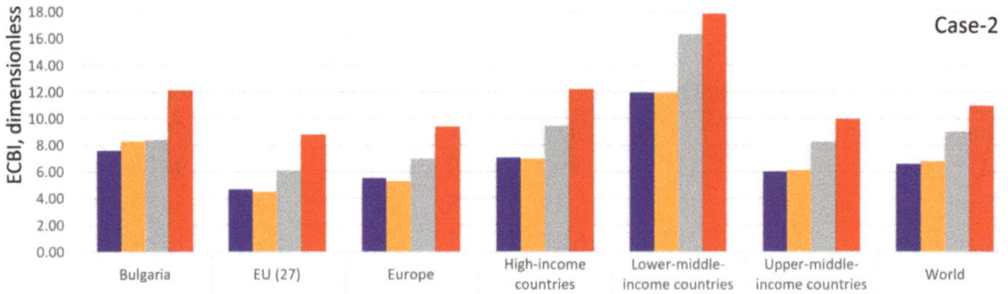

	Bulgaria	EU (27)	Europe	High-income countries	Lower-middle-income countries	Upper-middle-income countries	World
2021	7.58	4.67	5.52	7.05	11.95	6.01	6.59
2020	8.26	4.48	5.28	6.95	11.92	6.11	6.78
2015	8.37	6.11	6.98	9.45	16.33	8.24	9.04
2010	12.12	8.77	9.41	12.19	17.82	9.94	10.96

Fig. 6.8 Combination of energy sources under Case-2: (1) Conventional sources: coal, oil, gas, and nuclear; (2) Renewable sources: hydropower, wind, solar, and others

It should be noted that the combination of energy sources in Case-1 shows the lowest values for Europe in the 4 years studied. Therefore, the values for Europe illustrate that it uses the most renewable energy resources. The reverse of this finding is true for the upper-middle-income countries as they have the bigger ECBI.

The calculated ECBI considering the situation under Case-2 is shown in Fig. 6.8.

The ECBI value obtained, taking into account the distribution between renewable and conventional sources set by Case-2, shows that the highest consumption of renewable energies is observed in the EU (27). In Case-2, where gas and nuclear sources are considered as conventional sources, the highest use of renewable energy is observed for the EU (27). Lower-middle-income countries use the least amount of renewable energy.

The comparison between calculated ECBI determined by different combinations between renewable and conventional sources given by Case-1 and Case-2 is shown in Fig. 6.9.

From the graph in Fig. 6.9, it can be established that regardless of whether gas and nuclear energy is classified as renewable or conventional sources, a clear trend in favor of renewable sources is observed. In both cases, the results for 2020 and 2021 show similar results. To achieve economic sustainability, individual countries agree to reduce the energy extracted from conventional sources at the expense of renewable energy.

This clear trend is encouraging and gives a clear indication of attitudes toward the transition to a low-carbon economy. In this regard, determining the dynamics between the different types of sources could serve as an indicator. The difference in energy consumption between 2021 and 2010 from different renewable energy sources (wind, gas, nuclear, solar, and hydro) is shown in Fig. 6.10.

It is interesting to note that hydropower energy shows a decline in consumption in 2021 compared to 2010, and this is most noticeable in Bulgaria (−5.27%), followed by high-income countries (−0.6%), Upper-middle-income countries (−0.5%) and slightly changes for EU-27 (−0.21%). The same decreasing tendency is observed in gas consumption for lower-middle-income countries (−5.67%). Small changes are

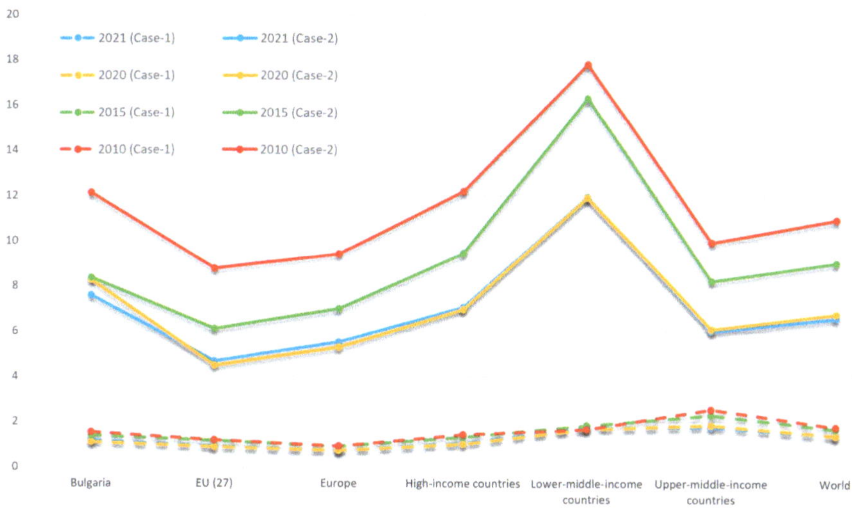

Fig. 6.9 Comparison between calculated ECBI values based on different combinations between renewable and conventional sources

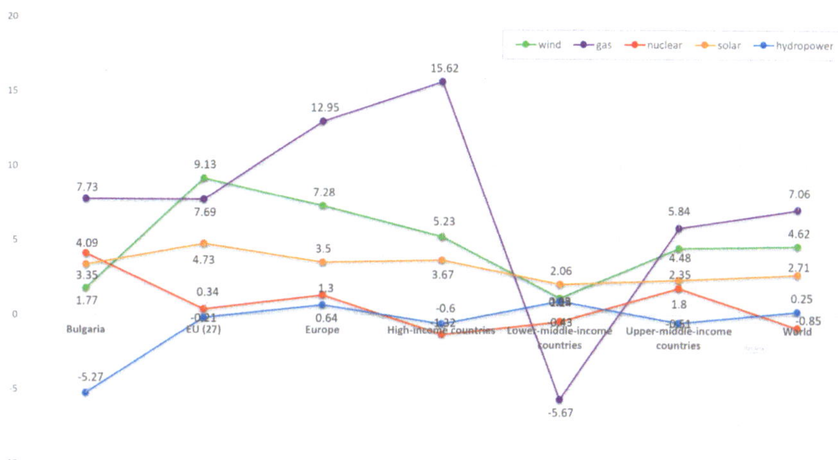

Fig. 6.10 Difference in energy consumption from renewable energy sources for 2021 and 2010

also observed in the consumption of nuclear energy at high-income countries (−1.32%), world (−0.85%), and lower-middle-income countries (−0.43%).

6.6 Index of Relative Share of Wind Energy Consumption Among the Other Renewable Sources

It is also possible to determine the relative share of renewable energy consumption in relation to the energy consumption from conventional sources. For example, the relative share of wind energy consumption can be determined by the following formula:

$$\mathrm{ECBI}^w = \frac{B^w}{\left(\sum_{j=1}^{(M-1)} b_j\right)} .100\% \tag{6.25}$$

where B^w expresses the measure of consumed energy from the wind renewable source, and parameters b_j represent the measure of consumed energy from different types of renewable sources such as hydropower, solar, etc. Index j takes values from 1 to $(M - 1)$, since wind is an energy source from the group of renewable resources.

In contrast to the previously defined index by (6.24), the greater value of the index determined by (6.25), the greater will be the share of the corresponding renewable energy source. These two types of indexes could be used independently of each other to determine the energy consumption balance formed by different energy sources.

6.6.1 Determination of Relative Share of Wind Energy Consumption Among the Other Renewable Sources

Using the relation (6.25) for the renewable sources (wind, hydropower, gas, nuclear, solar), the corresponding renewable energy consumption index for 2021 and 2010 is also determined. The obtained values for the relative share of different renewable sources used in relation to the total consumption from conventional energy sources are shown in Table 6.3.

A significant difference of 9.13 is observed for EU countries (27) in wind energy consumption. For lower-middle-income countries, the main difference is in water energy consumption, equal to 0.95. Another notable change is seen in gas consumption, equal to 12.95 for high-income countries. The most significant change in the consumption of nuclear energy with a value of 4.09 was found for Bulgaria. The largest change in solar energy consumption is observed in EU countries (27) with a value of 4.73. All this is evident of a positive direction, leading to an increase in the consumption of renewable energy.

The relative share of wind energy consumption in 2021, 2020, 2015, and 2010 from the EU (27) and world is shown in Fig. 6.11.

From Fig. 6.11, the sustainable trend of increasing wind energy consumption can be established. It is noteworthy that the relative share of wind energy consumption

Table 6.3 Relative share of renewable energy consumption in relation to the energy consumption from conventional sources for 2021 and 2010

	Wind, %	Hydropower, %	Gas, %	Nuclear, %	Solar, %
	2021				
Bulgaria	3.25	13.64	28.68	35.94	3.38
EU (27)	13.08	9.11	50.90	23.59	5.39
Europe	10.06	11.05	81.62	22.61	3.93
High-income countries	7.07	7.40	57.61	14.14	3.90
Lower-middle-income countries	1.75	7.76	43.08	2.65	2.06
Upper-middle-income countries	4.91	19.60	30.08	4.09	2.36
World	5.69	12.22	42.21	7.35	2.82
	2010				
Bulgaria	1.48	18.91	20.95	31.85	0.03
EU (27)	3.95	9.32	43.21	23.25	0.66
Europe	2.78	10.41	68.67	21.31	0.43
High-income countries	1.84	8.0	41.99	15.46	0.23
Lower-middle-income countries	0.61	6.81	48.75	3.08	0.0
Upper-middle-income countries	0.43	20.11	24.24	2.29	0.01
World	1.07	11.97	35.15	8.20	0.11

from the EU(27) significantly exceeds that of the world, with the percentage ratio for 2010 being 21%–79% in favor of the EU (27) and for 2021 is 30% to 70% also in favor of the EU (27).

The relative share of consumption of different renewable energy sources compared to energy consumption from conventional sources in EU(27) for 2021, 2015, and 2010 is shown in Fig. 6.12.

A trend of sustainable wind energy consumption in the EU (27) among the different renewable energy sources could be noted, which in 2010 was 5%, and in 2020 and 2021 consumption reached 13% (Fig. 6.12).

It is interesting to compare whether the general trend in the proportion of different renewable sources corresponds to those in Bulgaria. Therefore, similar graphs are made of the relative share of consumption of different renewable energy sources compared to energy consumption from conventional sources in Bulgaria for 2021, 2015, and 2010 as shown in Fig. 6.13.

The comparison of Figs. 6.12 and 6.13 shows that the distribution of energy consumption from different sources in Bulgaria follows the EU(27) trend. Significant differences in Bulgaria are observed in solar consumption from 0% in 2010, 4% in 2015, and 8% in 2020. Wind energy consumption shows a sustainable increase from 2% in 2010 to 4% in 2021. Hydropower consumption shows a decreasing trend from 26% to 16% in 2021. The largest consumption of energy is from gas and

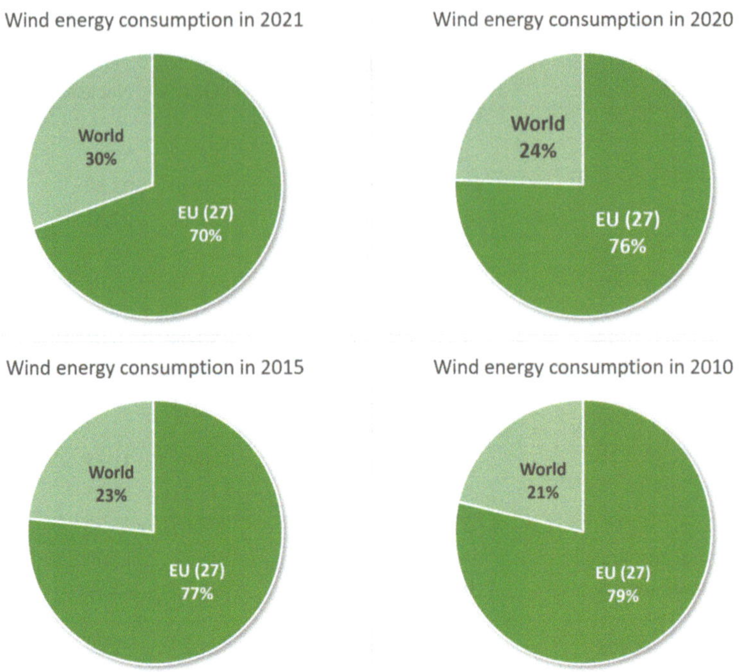

Fig. 6.11 Relative share of wind energy consumption from the EU (27) and world

nuclear energy. A small increase was seen in gas energy consumption from 29% to 34%. The comparison between 2010 and 2021 for nuclear power consumption shows a difference of 1% in favor of 2010.

The energy transition in the global economy must continue in order to protect the environment and more precisely the climate. To realize such an ambitious goal, it is necessary to increase the use of renewable energy sources, which are a key factor in achieving decarbonization. It is therefore necessary to use appropriate indicators to determine appropriate policies to measure the reduction in the use of conventional sources at the expense of renewables. To tackle this task, two types of indicators have been proposed that can be used as a measure to determine the degree of use of renewable energy. The results obtained using these indicators show a clear tendency to reduce the share of conventional sources, regardless of whether gas and nuclear energy is classified as renewable or not. All this means that we are on the right track to not only preserve the Earth but also to achieve better economic sustainability.

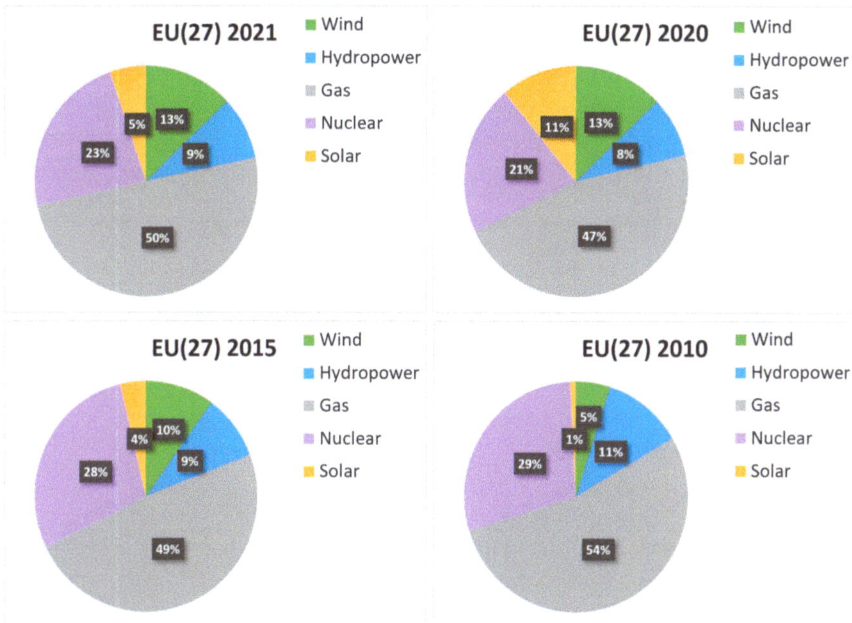

Fig. 6.12 Dynamics of the relative share of consumption from different renewable energy sources in EU (27)

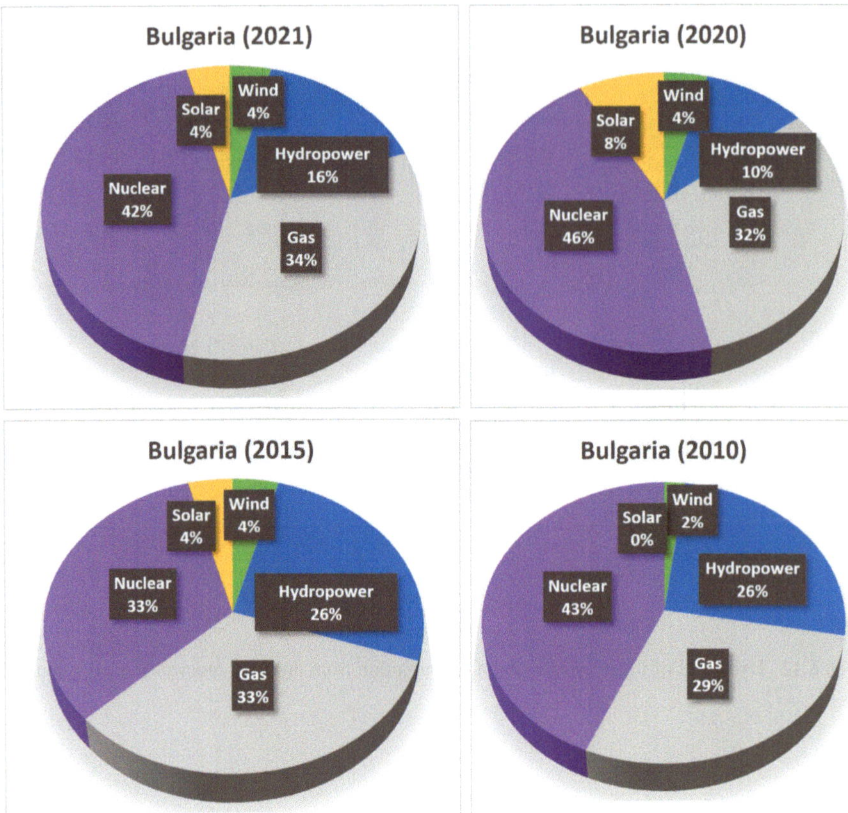

Fig. 6.13 Dynamics of the relative share of consumption from different renewable energy sources in Bulgaria

References

Ahmad, M., Isık, C., Jabeen, G., Ali, T., Ozturk, I., & Atchike, D. W. (2021). Heterogeneous links among urban concentration, non-renewable energy use intensity, economic development, and environmental emissions across regional development levels. *Science of the Total Environment, 765*, 144527. https://doi.org/10.1016/j.scitotenv.2020.144527

Ahmed Shata, A. S., & Hanitsch, R. (2006). Evaluation of wind energy potential and electricity generation on the coast of Mediterranean Sea in Egypt. *Renewable Energy, 31*(8), 1183–1202. https://doi.org/10.1016/j.renene.2005.06.015

Ahn, K., Chu, Z., & Lee, D. (2021). Effects of renewable energy use in the energy mix on social welfare. *Energy Economics, 96*, 105174. https://doi.org/10.1016/j.eneco.2021.105174

Ambarita, E. E., Kuncara, I., Widyotriatmo, A., Karlsen, A., Scibilia, F., & Hasan, A. (2023). On cyber-attacks against wind farms. In *49th Annual Conference of the IEEE Industrial Electronics Society*.

Bamati, N., & Raoofi, A. (2020). Development level and the impact of technological factor on renewable energy production. *Renewable Energy, 151*, 946–955. https://doi.org/10.1016/j.renene.2019.11.098

Bogicevic, Z. (2014). *Referencing the Montenegrin qualifications framework to the European qualifications framework for lifelong learning and the, European Training Foundation*. Italy. Retrieved from https://policycommons.net/artifacts/1943920/referencing-the-montenegrin-qualifications-framework-to-the-european-qualifications-framework-for-lifelong-learning-and-the/2695689/ on 03 Aug 2023. CID: 20.500.12592/b3dnj5

Borissova, D., & Atanassova, Z. (2018). Multi-criteria decision methodology for supplier selection in building industry. *International Journal of 3-D Information Modeling, 7*(4), 49–58. https://doi.org/10.4018/IJ3DIM.2018100103

Borissova, D., Dimitrova, Z., & Dimitrov, V. (2022). Assessing of energy consumption balance index formed by various combinations of conventional and renewable sources. In: 2022 IEEE Sustainable Power and Energy Conference (iSPEC), Perth, Australia, pp. 1–5. https://doi.org/10.1109/iSPEC54162.2022.10033014

Borissova, D. (2020). A multi-criteria group decision making model for selection of green building project. In S. Ofluoglu, O. Ozener, & U. Isikdag (Eds.), *Advances in building information modeling*. EBF 2019. Communications in Computer and Information Science (Vol. 1188, pp. 137–146). https://doi.org/10.1007/978-3-030-42852-5_11

Borozan, D. (2022). Detecting a structure in the European energy transition policy instrument mix: What mix successfully drives the energy transition? *Renewable and Sustainable Energy Reviews, 165*, 112621. https://doi.org/10.1016/j.rser.2022.112621

Bourhim, F.-A., Berrhazi, S., Ouammi, A., & Benchrifa, R. (2021). Decision support model for optimal design of wind technologies based techno–economic approach. *IEEE Access, 9*, 148264–148276. https://doi.org/10.1109/ACCESS.2021.3123561

Buchmayr, A., Verhofstadt, E., Van Ootegem, L., Thomassen, G., Taelman, S. E., & Dewulf, J. (2022). Exploring the global and local social sustainability of wind energy technologies: An application of a social impact assessment framework. *Applied Energy, 312*, 118808. https://doi.org/10.1016/j.apenergy.2022.118808

Castrejon-Campos, O., Aye, L., & Hui, F. K. P. (2020). Making policy mixes more robust: An integrative and interdisciplinary approach for clean energy transitions. *Energy Research & Social Science, 64*, 101425. https://doi.org/10.1016/j.erss.2020.101425

Cazes, S., Hijzen, A., & Saint-Martin, A. (2015). *Measuring and assessing job quality: The OECD job quality framework*. OECD Social, Employment and Migration Working Papers, No. 174. OECD Publishing. https://doi.org/10.1787/5jrp02kjw1mr-en

Coker, A. K. (2007). Cost estimation and economic evaluation. In A. K. Coker (Ed.), *Ludwig's applied process design for chemical and petrochemical plants* (pp. 69–102, 4th ed.). Gulf Professional Publishing. https://doi.org/10.1016/B978-075067766-0/50009-9

Cooper, D., Grey, S., Raymond, G., & Walker, P. (2005). *Project risk management guidelines: Managing risk in large projects and complex procurements*. Wiley.

Delapedra-Silva, V., Ferreira, P., Cunha, J., & Kimura, H. (2022). Methods for financial assessment of renewable energy projects: A review. *PRO, 10*(2), 184. https://doi.org/10.3390/pr10020184

Egli, F. (2020). Renewable energy investment risk: An investigation of changes over time and the underlying drivers. *Energy Policy, 140*, 111428. https://doi.org/10.1016/j.enpol.2020.111428

Enevoldsen, P. (2016). Onshore wind energy in Northern European forests: Reviewing the risks. *Renewable and Sustainable Energy Reviews, 60*, 1251–1262. https://doi.org/10.1016/j.rser.2016.02.027

Fischhendler, I., Herman, L., Barr, A., & Rosen, G. (2021). The impact of community split on the acceptance of wind turbines. *Solar Energy, 220*, 51–62. https://doi.org/10.1016/j.solener.2021.01.055

Fu, F. Y., Alharthi, M., Bhatti, Z., Sun, L., Rasul, F., Hanif, I., & Iqbal, W. (2021). The dynamic role of energy security, energy equity and environmental sustainability in the dilemma of emission reduction and economic growth. *Journal of Environmental Management, 280*, 111828. https://doi.org/10.1016/j.jenvman.2020.111828

Gatto, A., & Drago, C. (2021). When renewable energy, empowerment, and entrepreneurship connect: Measuring energy policy effectiveness in 230 countries. *Energy Research & Social Science, 78*, 101977. https://doi.org/10.1016/j.erss.2021.101977

Gatzert, N., & Kosub, T. (2016). Risks and risk management of renewable energy projects: The case of onshore and offshore wind parks. *Renewable and Sustainable Energy Reviews, 60*, 982–998. https://doi.org/10.1016/j.rser.2016.01.103

González, J. S., Rodriguez, A. G. G., Mora, J. C., Santos, J. R., & Payan, M. R. (2010). Optimization of wind farm turbines layout using an evolutive algorithm. *Renewable Energy, 35*(8), 1671–1681. https://doi.org/10.1016/j.renene.2010.01.010

Guliashki, V., Marinova, G., & Groumpos, P. (2019). Multi-objective optimization approach for energy efficiency in microgrids. *IFAC-PapersOnLine, 52*(25), 477–482. https://doi.org/10.1016/j.ifacol.2019.12.587

Guo, Y., Ru, P., Su, J., & Anadon, L. D. (2015). Not in my backyard, but not far away from me: Local acceptance of wind power in China. *Energy, 82*, 722–733. https://doi.org/10.1016/j.energy.2015.01.082

Guo, Y., Wang, H., & Lian, J. (2022). Review of integrated installation technologies for offshore wind turbines: Current progress and future development trends. *Energy Conversion and Management, 255*, 115319. https://doi.org/10.1016/j.enconman.2022.115319

Ilhan, A., Bilgili, M., Sari, M., & Sahin, B. (2022). Economic analyses of onshore commercial large scale wind power plant installed in Turkey. *NOHU Journal of Engineering Sciences, 11*(1), 170–176.

Kendrick, T. (2009). *Identifying and managing project risk: Essential tools for failure-proofing your project*. Amacom.

Khan, M. Z. M., Rehman, H. M. A., Janjua, A. K., Waqas, A., Shakir, S., & Ali, M. (2023). Techno-economic assessment of wind farm for sustainable power generation in northern coastal region of Arabian Sea. *Energy Reports, 9*, 1278–1290. https://doi.org/10.1016/j.egyr.2022.12.057

Kim, J.-Y., Oh, K.-Y., Kang, K.-S., & Lee, J.-S. (2013). Site selection of offshore wind farms around the Korean Peninsula through economic evaluation. *Renewable Energy, 54*, 189–195. https://doi.org/10.1016/j.renene.2012.08.026

Kubiszewski, I., Cleveland, C. J., & Endres, P. K. (2010). Meta-analysis of net energy return for wind power systems. *Renewable Energy, 35*(1), 218–225. https://doi.org/10.1016/j.renene.2009.01.012

Kul, C., Zhang, L., & Solangi, Y. A. (2020). Assessing the renewable energy investment risk factors for sustainable development in Turkey. *Journal of Cleaner Production, 276*, 124164. https://doi.org/10.1016/j.jclepro.2020.124164

Lai, J., Duan, B., Su, Y., Li, L., & Yin, Q. (2017). An active security defense strategy for wind farm based on automated decision. In *IEEE Power & Energy Society General Meeting*, Chicago, IL (pp. 1–5). https://doi.org/10.1109/PESGM.2017.8274638

Li, K., Bian, H., Liu, C., Zhang, D., & Yang, Y. (2015). Comparison of geothermal with solar and wind power generation systems. *Renewable and Sustainable Energy Reviews, 42*, 1464–1474. https://doi.org/10.1016/j.rser.2014.10.049

Lin, M. S.-M., & Lu, B.-S. (2023). Risk assessment and management in the offshore wind power industry: A focus on component handling operations in ports. *Safety Science, 167*, 106286. https://doi.org/10.1016/j.ssci.2023.106286

Liu, J., Song, D., Li, Q., Yang, J., Hu, Y., Fang, F., & Joo, Y. H. (2023). Life cycle cost modelling and economic analysis of wind power: A state of art review. *Energy Conversion and Management, 277*, 116628. https://doi.org/10.1016/j.enconman.2022.116628

Liu, Q., Sun, Y., & Wu, M. (2021). Decision-making methodologies in offshore wind power investments: A review. *Journal of Cleaner Production, 295*, 126459. https://doi.org/10.1016/j.jclepro.2021.126459

Maleki-Dizaji, P., del Bufalo, N., Di Nucci, M.-R., & Krug, M. (2020). Overcoming barriers to the community acceptance of wind energy: Lessons learnt from a comparative analysis of best practice cases across Europe. *Sustainability, 12*(9), 3562. https://doi.org/10.3390/su12093562

Nematchoua, M. K., Asadi, S., & Reiter, S. (2020). Influence of energy mix on the life cycle of an eco-neighborhood, a case study of 150 countries. *Renewable Energy, 162*, 81–97. https://doi.org/10.1016/j.renene.2020.07.141

Nurul Hossain, A. K. M., & Serletis, A. (2020). Biofuel substitution in the U.S. transportation sector. *The Journal of Economic Asymmetries, 22*, e00161. https://doi.org/10.1016/j.jeca.2020.e00161

Opeyemi, B. M. (2021). Path to sustainable energy consumption: The possibility of substituting renewable energy for non-renewable energy. *Energy, 228*, 120519. https://doi.org/10.1016/j.energy.2021.120519

Papapetrou, M., & Kosmadakis, G. (2022). Resource, environmental, and economic aspects of SGHE. In A. Tamburini, A. Cipollina, & G. Micale (Eds.) *Woodhead Publishing Series in Energy, Salinity Gradient Heat Engines* (pp. 319–353). Woodhead Publishing. https://doi.org/10.1016/B978-0-08-102847-6.00006-1

Ragheb, M. (2017). Economics of wind power generation. In T. M. Letcher (Ed.), *Wind energy engineering* (pp. 537–555). Academic Press. https://doi.org/10.1016/B978-0-12-809451-8.00025-4

Ren, Z., Verma, A. S., Li, Y., Teuwen, J. J. E., & Jiang, Z. (2021). Offshore wind turbine operations and maintenance: A state-of-the-art review. *Renewable and Sustainable Energy Reviews, 144*, 110886. https://doi.org/10.1016/j.rser.2021.110886

Ritchie, H., Roser, M., & Rosado, P. (2020). *"Energy"*, Published online at OurWorldInData.org. Retrieved from: https://ourworldindata.org/energy. [Online Resource] https://ourworldindata.org/energy-mix#energy-mix-what-sources-do-we-get-our-energy-from

Rostami, A., Mohammadi, M., & Karimipour, H. (2023). Reliability assessment of cyber-physical power systems considering the impact of predicted cyber vulnerabilities. *International Journal of Electrical Power & Energy Systems, 147*, 108892. https://doi.org/10.1016/j.ijepes.2022.108892

Rutovitz, J., Dominish, E., & Downes, J. (2015). *Calculating global energy sector jobs: 2015 methodology*. Prepared for Greenpeace International by the Institute for Sustainable Futures, University of Technology Sydney.

Schachinger, S. (2023). *Areas of cyber-risk and resilience in renewable energy*. https://energydigital.com/renewable-energy/areas-of-cyber-risk-and-resilience-in-renewable-energy

Shafiee, M., Brennan, F., & Espinosa, I. A. (2016). A parametric whole life cost model for offshore wind farms. *International Journal of Life Cycle Assessment, 21*, 961–975. https://doi.org/10.1007/s11367-016-1075-z

Shan, S., Genc, S. Y., Kamran, H. W., & Dinca, G. (2021). Role of green technology innovation and renewable energy in carbon neutrality: A sustainable investigation from Turkey. *Journal of Environmental Management, 294*, 113004. https://doi.org/10.1016/j.jenvman.2021.113004

Sim, J. (2023). An economic evaluation of potential offshore wind farm sites in South Korea using a real options approach. *Energy Reports, 10*, 29–37. https://doi.org/10.1016/j.egyr.2023.06.007

Sohail, M. T., Xiuyuan, Y., Usman, A., Majeed, M. T., & Ullah, S. (2021). Renewable energy and non-renewable energy consumption: Assessing the asymmetric role of monetary policy uncertainty in energy consumption. *Environmental Science and Pollution Research, 28*, 31575–31584. https://doi.org/10.1007/s11356-021-12867-0

Soulouknga, M. H., Oyedepo, S. O., Doka, S. Y., et al. (2020). Evaluation of the cost of producing wind-generated electricity in Chad. *International Journal of Energy and Environmental Engineering, 11*, 275–287. https://doi.org/10.1007/s40095-019-00335-y

Staggs, J., Ferlemann, D., & Shenoi, S. (2017). Wind farm security: Attack surface, targets, scenarios and mitigation. *International Journal of Critical Infrastructure Protection, 17*, 3–14. https://doi.org/10.1016/j.ijcip.2017.03.001

Stehly, T., & Duffy, P. (2021). *2020 Cost of wind energy review*. National Renewable Energy Laboratory. NREL/TP-5000-81209. https://www.nrel.gov/docs/fy22osti/81209.pdf

Stehly, T., & Duffy, P. (2022). *2021 Cost of Wind Energy Review*. https://www.nrel.gov/docs/fy23osti/84774.pdf

Strantzali, E., & Aravossis, K. (2016). Decision making in renewable energy investments: A review. *Renewable and Sustainable Energy Reviews, 55*, 885–898. https://doi.org/10.1016/j.rser.2015.11.021

Tutak, M., & Brodny, J. (2022). Renewable energy consumption in economic sectors in the EU-27. The impact on economics, environment and conventional energy sources. A 20-year perspective. *Journal of Cleaner Production, 345*, 131076. https://doi.org/10.1016/j.jclepro.2022.131076

Usman, M., & Hammar, N. (2021). Dynamic relationship between technological innovations, financial development, renewable energy, and ecological footprint: Fresh insights based on the STIRPAT model for Asia Pacific Economic Cooperation countries. *Environmental Science and Pollution Research, 28*, 15519–15536. https://doi.org/10.1007/s11356-020-11640-z

Wang, Y., Qin, Y., Wang, K., Liu, J., Fu, S., Zou, J., & Ding, L. (2022). Where is the most feasible, economical, and green wind energy? Evidence from high-resolution potential mapping in China. *Journal of Cleaner Production, 376*, 134287. https://doi.org/10.1016/j.jclepro.2022.134287

Xu, Y., Yang, K., & Yuan, J. (2021). Levelized cost of offshore wind power in China. *Environmental Science and Pollution Research, 28*, 25614–25627. https://doi.org/10.1007/s11356-021-12382-2

Yang, K., Li, J., Zhang, G., Xing, Y., Bamisile, O., & Huang, Q. (2022). A cooperative control strategy against cyber-attacks for power system with high penetration wind farm. In *4th Asia Energy and Electrical Engineering Symposium (AEEES)*, Chengdu (pp. 321–327). https://doi.org/10.1109/AEEES54426.2022.9759822

Zhang, Y., Xiang, Y., & Wang, L. (2017). Power system reliability assessment incorporating cyber attacks against wind farm energy management systems. *IEEE Transactions on Smart Grid, 8*(5), 2343–2357. https://doi.org/10.1109/TSG.2016.2523515

Zhou, S., & Yang, P. (2020). Risk management in distributed wind energy implementing analytic hierarchy process. *Renewable Energy, 150*, 616–623. https://doi.org/10.1016/j.renene.2019.12.125

MIX
Papier aus verantwortungsvollen Quellen
Paper from responsible sources
FSC® C105338

If you have any concerns about our products,
you can contact us on
ProductSafety@springernature.com

In case Publisher is established outside the EU,
the EU authorized representative is:
Springer Nature Customer Service Center GmbH
Europaplatz 3, 69115 Heidelberg, Germany

Printed by Libri Plureos GmbH
in Hamburg, Germany